T0156069

Monographs in

Volume 8

Topics and Methods
in q-Series

Monographs in Number Theory

ISSN 1793-8341

Monographs in

Number Theory

Volume 8

Topics and Methods in q-Series

James Mc Laughlin

West Chester University, Pennsylvania, USA

 World Scientific

NEW JERSEY · LONDON · SINGAPORE · BEIJING · SHANGHAI · HONG KONG · TAIPEI · CHENNAI

Published by

World Scientific Publishing Co. Pte. Ltd.
5 Toh Tuck Link, Singapore 596224
USA office: 27 Warren Street, Suite 401-402, Hackensack, NJ 07601
UK office: 57 Shelton Street, Covent Garden, London WC2H 9HE

British Library Cataloguing-in-Publication Data
A catalogue record for this book is available from the British Library.

Monographs in Number Theory — Vol. 8
TOPICS AND METHODS IN Q-SERIES

Copyright © 2018 by World Scientific Publishing Co. Pte. Ltd.

ISBN 978-981-3224-17-9
ISBN 978-981-3223-36-3 (pbk)

Printed in Singapore

Foreword

The study of q-series has captured the interest of many mathematicians. Euler (as you will see in the following pages) certainly proved theorems that fall within q-series. Euler was primarily motivated by hoping to find generating functions for various classes of integers. In the mid-nineteenth century, E. Heine first considered q-series as analogs of classical hypergeometric series. Heine's analytic view pervaded the study for at least a century, with somewhat obscure mathematicians like the Rev. F. H. Jackson carrying the torch. During this time a clue to the future flowering of the subject came from G. N. Watson's discovery that a little appreciated theorem of the astronomer F. J. W. Whipple had a q-analog which would yield the celebrated Rogers-Ramanujan identities as a limiting case.

Watson's discovery led W. N. Bailey and his Ph. D. student Lucy Slater to delve deeply into q-series.

In the paper The Well-Poised Thread, I quoted extensively from a lecture given by Bailey, which makes clear his love of this topic, combined with his unstated fear that, in fact, no one was interested.

Bailey's gloom may have been understandable in the 1940s, but things blossomed in the late 1960s with the discovery that the world of partitions had its natural home in q-series. Seemingly esoteric objects like very well-poised q-hypergeometric series turned out to be the central tools for the exploration of partition identities.

With the discoveries of Ramanujan's Lost Notebook in the 1970s and Bailey Chains in the 1980s, q-series have subsequently become a center of intense research.

James (Jimmy) McLaughlin has gathered together the components that fueled this resurgence of q-series and combined them into a natural and coherent text. The central theme revolves around the Bailey Chain, its extensions and implications. The book concludes with a nice development of related continued fractions and an introduction to Ramanujan's mock theta functions. There are sufficient exercises included to allow this book to be used as a text in a graduate course.

This is an intense subject. Don't be put off by the size of some of the formulas. Results like Corollary 5.4 posses a number of symmetries and surprising special cases that, with familiarity, become captivating. The same can be said of many of the more formidable results.

Enjoy!

George E. Andrews

January 17, 2017

Contents

Chapter 1

Introduction

I first encountered the subject of basic hypergeometric series while attending the weekly analytic number theory seminar in the mathematics department at the University of Illinois at Urbana-Champaign (UIUC). I was a graduate student there from 1996 to 2002, and the weekly seminars were something that I looked forward to every week. It was in these seminars that I first heard of such mathematical objects as q-series, mock theta functions, partition congruences and partition identities, infinite q-series-product identities, q-continued fractions, and the fascinating mathematics of Ramanujan. All of these objects were to eventually develop into long-term mathematical interests of my own, and to become important to me in my own research.

At the start of my graduate studies, one of my interests was the regular continued fraction expansions of real numbers (Pell's equation, patterns in the expansion of powers of e, folding patterns in the expansion of various infinite products, etc.), and I had some vague idea of doing some research in the area of continued fractions for my thesis topic. Thus, while not being a regular continued fraction, an immediate object of interest was the Rogers-Ramanujan continued fraction (and later other q-continued fractions of Ramanujan, Gordon, Selberg and others). One of the interesting facts about the Rogers-Ramanujan continued fraction

$$K(q) := 1 + \cfrac{q}{1 + \cfrac{q^2}{1 + \cfrac{q^3}{1 + \cfrac{q^4}{\ddots}}}},$$

1

is that, for $|q| < 1$, it has a representation as a quotient of infinite products:

$$K(q) = \frac{(q^2, q^3; q^5)_\infty}{(q, q^4; q^5)_\infty}.$$

Of course the famous Rogers-Ramanujan identities

$$\sum_{n=0}^{\infty} \frac{q^{n^2}}{(q; q)_n} = \frac{1}{(q, q^4; q^5)_\infty}, \qquad \sum_{n=0}^{\infty} \frac{q^{n(n+1)}}{(q; q)_n} = \frac{1}{(q^2, q^3; q^5)_\infty}$$

mean that $K(q)$ also has a representation as a quotient of basic hyperge-ometric series. The Rogers-Ramanujan identities were just two examples from a list of 130 similar identities stated and proved by Lucy Slater in two papers [236, 237] published in 1951 and 1952. The Rogers-Ramanujan identities, and indeed Slater's entire compendium of such identities, were a mystery to me at the time, and trying to understand the mechanism of the proofs would eventually lead me into the extensive mathematical area involving Bailey pairs, WP-Bailey pairs and Bailey- and WP-Bailey chains and trees.

Another group of mysterious objects that cropped up from time to time in talks that I attended were the mock theta functions. Of course Zwegers [263, 264] has given an explanation of the mock theta functions in terms of Maass forms, but when I first saw examples of the q-series representations of mock theta functions, the only observation I could make was the trivial one that n-th terms of these series were all lacking a term of the form $(q; q)_n$ (or $(q^k; q^k)_n$) in the denominator (unlike all of the basic hypergeometric series in all of the usual q-series identities that I was aware of). This was one small reason, amongst the many deeper reasons, that the identities comprising the mock theta conjectures, for example that if $|q| < 1$, then

$$\sum_{n=0}^{\infty} \frac{q^{n^2}}{(-q; q)_n} = -2q^2 \sum_{n=0}^{\infty} \frac{q^{10n^2+10n}}{(q^2, q^8; q^{10})_{n+1}} + \frac{(q^5; q^5)_\infty (q^5; q^{10})_\infty}{(q, q^4; q^5)_\infty},$$

seemed like very strange beasts indeed! I was still curious enough about the mock theta functions to want to understand more about their basic hypergeometric nature.

However, I did not really start to study q-series in their own right (as op-posed to regarding them as having connections with q-continued fractions) until several years later. I continued to work in the area of continued frac-tions. Doug Bowman, who I knew was interested in continued fractions, became my thesis advisor, and I defended my thesis in May 2002 (the topic related to the convergence behavior of q-continued fractions, such as the Rogers-Ramanujan continued fraction, on the unit circle, at points which were not roots of unity).

I took up a visiting position at Trinity College, in Hartford, Ct, with the aim of continuing my research on continued fractions, collaborating with Nancy Wyshinski at Trinity, whose interests lay in the analytic theory of continued fractions. I continued to work almost entirely in the broad field of continued fractions up until around 2008, when I got interested in the area of basic hypergeometric series again, when trying to prove some identities I had found experimentally, via a computer search. Since then, my primary research interests have generally moved over to the broad field of q-series.

As I read deeper, I discovered many interesting methods for proving basic hypergeometric identities that had appeared during the last 50 or so years. Many new interesting mathematical objects had likewise appeared in the field of q-series, and our understanding of many classical objects has also developed.

Examples of the new methods and techniques (or in some cases, new applications of existing methods and techniques) included the anti-telescoping method of Andrews, Cauchy augmentation, applications of the Cauchy differential operator, q-Lagrange inversion, matrix inverse methods, the WZ-method and the q-Zeilberger algorithm, applications of Abel summation, q-Engel expansion of q-functions, etc.

Apart from the development of new methods to prove q-series identities, other developments in the field include the connection between mock theta functions and Maass forms made by Zwegers, elliptic hypergeometric functions, many developments in the topics of Bailey- and WP-Bailey pairs and chains/lattices, various classes of multi-sum identities such as the Andrews-Gordon identities

$$\sum_{n_{k-1} \geq n_{k-2} \geq \cdots \geq n_1 \geq 0} \frac{q^{n_1^2 + \cdots + n_{k-1}^2 + n_1 + \cdots + n_{k-i}}}{(q;q)_{n_{k-1}-n_{k-2}}(q;q)_{n_{k-2}-n_{k-3}} \cdots (q;q)_{n_2-n_1}(q;q)_{n_1}}$$
$$= \frac{(q^i, q^{2k+1-i}, q^{2k+1}; q^{2k+1})_\infty}{(q;q)_\infty},$$

explicit formulae for radial limits for many of the mock theta functions, congruence relations (and other properties) for the generating functions of new partition functions (the crank, overpartitions, smallest parts function, etc.), various polynomial versions of Slater-type identities, m-versions of Slater-type identities, new modular relations for various q-functions, multi-sectioning of various infinite products, and many others.

Most of these results were scattered throughout the mathematical literature, and as time went on, the thought grew that it would be good to bring many of these developments together and present them in a way that would be accessible to beginners in the field. That idea was the original motivation for this book.[1]

The book starts with the basics, and Chapters 2–8 cover what might be described as classical material in the subject, namely, the q-binomial theorem, Heine's transformation and various summations that derive from it, transformations of Watson ($_8\phi_7$) and Bailey ($_{10}\phi_9$), the Jacobi triple product identity, Ramanujan's $_1\psi_1$ summation and Bailey's $_6\psi_6$ summation. These chapters, with perhaps a selection of topics from Chapter 9 (the Rogers-Fine Identity), Chapter 10 (Bailey Pairs), Chapter 14 (Gaussian Polynomials), Chapter 15 (Bijective Proofs) and Chapter 17 (Lambert Series), could easily provide material for an undergraduate topics course. Every chapter has an extensive selection of exercises.

The remainder of the book contains an extensive treatment of many aspects of Bailey- and WP-Bailey pairs and chains in Chapters 11–13, a quite comprehensive introduction to q-continued fractions in Chapter 16, and an introduction to the mock theta functions in Chapter 18. The book can also serve as a text for a graduate course, the selection of topics beyond the basic ones being up to the instructor.

I have tried to keep the book as self-contained as possible. Apart from some general convergence theorems for continued fractions needed to ensure the convergence of certain q-continued fractions, and which are just quoted, a newcomer to the topic who had the equivalent of an undergraduate course in analysis, and who was prepared to work through the mathematics in the text and the exercises, should be able work their way through most, possibly all, of the book.

I believe the text can also serve as a reference work for researchers in the area. The text includes many results from the literature, both as examples in the text and as exercises at the end of each chapter.

I would like to thank a number of people who were very helpful in bringing this book to completion. At World Scientific, I would like to express deep thanks and appreciation to Rochelle Kronzek for her help, enthusiasm and encouragement all the way through this project. I would also like to thank Bruce Berndt for adopting the book to be part of the Monographs in Number Theory series, of which he is the series editor. A big thank

---

[1]This present volume does not cover all the topics listed above, and I hope to continue this project later, in a second volume.

you is due to George Andrews for agreeing to write a preface for the book, and for the kind remarks he made about the text. Thanks also to Krishna Alladi and Tim Huber who looked at earlier drafts of the manuscript and encouraged me to complete it.

I would also like to take this opportunity to look back in time and thank all my co-authors on papers we wrote together on q-series, and who made the paper-writing process more interesting and enjoyable — (in chronological order) Doug Bowman, Nancy Wyshinski, Andrew Sills, Peter Zimmer, Dennis Eichhorn, Scott Parsell, Jongsil Lee and Jaebum Sohn. I would also like to thank Doug Bowman, who was my thesis advisor when I was a graduate student and also a lot of fun to hang out with and sink a few beers with. Thanks also to some of my former professors at the University of Illinois at Urbana-Champaign, who made my time as a graduate student such an interesting and enjoyable experience — (in alphabetical order) Michael Bennett, Bruce Berndt, Nigel Boston, Harold Diamond, Adolf Hildebrand, Bruce Reznick, Ken Stolarsky and Alexandru Zaharescu. My apologies to anyone that I should have thanked and have forgotten.

A significant part of this book was written while on sabbatical from teaching at West Chester University in Fall 2014. I am very appreciative of the un-interrupted time this sabbatical gave me to concentrate on writing the book. The book may not have happened, or at least would have taken considerably longer, without this sabbatical.

Lastly, I would like to thank my wife Julie for putting up with the various inconveniences caused by the writing of this book, and also for putting up with all the times that I have woken her up talking mathematics in my sleep (including the time she says I was ranting about "... zeros as far as the eye could see ... ").

Exton, PA.
February 2017.

Remark: Despite the best efforts of the author, and others involved in the editing/proofreading process, some typographical errors may have slipped by. These are solely the fault of the author. Any such errors found after the publication of the book will be posted on the "errata page" for the book at http://www.worldscientific.com/worldscibooks/10.1142/10528. If you find such an error, please email me at jmclaughlin2@wcupa.edu.

Chapter 2

Basic Notation

Let a and q be complex numbers. Subscripts i, j, k, m, n, etc. will denote integers, unless specified otherwise. For a non-negative integer n we define the q-*shifted factorial* $(a; q)_n$ by setting $(a, q)_0 := 1$, and for $n \geq 1$,

$$(a; q)_n := (1 - a)(1 - aq) \ldots (1 - aq^{n-1}) = \prod_{j=0}^{n-1} (1 - aq^j). \qquad (2.1)$$

We sometimes write $(a)_n$ for $(a; q)_n$, if the base q is clearly understood. We assume throughout, unless otherwise stated, that the number a (and likewise each of the other parameters stated that appear in a q-shifted factorial) does not make any of the q-products vanish, if they appear in a denominator. Note that $(1; q)_n = 0$, for all $n \geq 1$, more general that $(1/q^k; q)_n = 0$ if $0 \leq k < n$, $(0, q)_n = 1$ for all $n \geq 0$, and

$$\lim_{b \to 0} b^n (a/b; q)_n = (-a)^n q^{n(n-1)/2}, \qquad \forall\, n \geq 0.$$

For ease of notation, set

$$(a_1; q)_n (a_2; q)_n \ldots (a_r; q)_n := (a_1, a_2, \ldots, a_r; q)_n. \qquad (2.2)$$

The corresponding infinite products, for $|q| < 1$, are defined in the obvious way, by letting $n \to \infty$:

$$(a; q)_\infty := \prod_{j=0}^{\infty} (1 - aq^j) \qquad (2.3)$$

$$(a_1; q)_\infty (a_2; q)_\infty \ldots (a_r; q)_\infty := (a_1, a_2, \ldots, a_r; q)_\infty.$$

Upon observing that for $n \geq 0$,

$$(a;q)_n = \frac{(a;q)_\infty}{(aq^n;q)_\infty}, \tag{2.4}$$

we use this formula to extend the definition of $(a;q)_n$ to negative integers. For $n > 0$, define

$$(a;q)_{-n} := \frac{(a;q)_\infty}{(aq^{-n};q)_\infty} = \frac{1}{(aq^{-n};q)_n} = \frac{(-q/a)^n}{(q/a;q)_n} q^{n(n-1)/2}, \tag{2.5}$$

where the last equality follows after some elementary manipulation (see Exercise 1.1). Note that this formula holds for $n < 0$ also (replace a with q/a, set $n = -m$, and rearrange). Note also that this formula implies $1/(q;q)_n = 0$, for all integers $n \leq -1$.

The following properties of q-shifted factorials follow easily from the definitions, and will be used frequently later. In each case, n is an integer and $k \geq 1$ is a positive integer, unless specified otherwise.

For later use, note that

$$\frac{(a^{1/2}q, -a^{1/2}q; q)_n}{(a^{1/2}, -a^{1/2}; q)_n} = \frac{1 - aq^{2n}}{1 - a}, \tag{2.6}$$

$$(a;q)_{2n} = (a, aq; q^2)_n, \tag{2.7}$$

and, more generally,

$$(a;q)_{kn} = (a, aq, \ldots, aq^{k-1}; q^k)_n. \tag{2.8}$$

Factoring gives that

$$(a^2; q^2)_n = (a, -a; q)_n, \tag{2.9}$$

and, more generally,

$$(a^k; q^k)_n = (a, a\omega, a\omega^2, \ldots, a\omega^{k-1}; q)_n, \tag{2.10}$$

where ω is a primitive k-th root of unity. The corresponding identities hold for $n = \infty$.

By using (2.4) as the definition of $(a;q)_n$ for all integers n, it can be shown that

$$(a;q)_{n+k} = (a;q)_n (aq^n; q)_k = (a;q)_k (aq^k; q)_n \tag{2.11}$$

holds for all integers n and k (it is easily seen to hold for all *non-negative* n and k from (2.1)). Then replacing k with $-k$ and using (2.5), it follows that for all integers n and k,

$$(a;q)_{n-k} = \frac{(a;q)_n}{(q^{1-n}/a;q)_k} \left(\frac{-q}{a}\right)^k q^{k(k-1)/2-nk}, \tag{2.12}$$

$$\frac{(a;q)_{n-k}}{(b;q)_{n-k}} = \frac{(a;q)_n}{(b;q)_n} \frac{(q^{1-n}/b;q)_k}{(q^{1-n}/a;q)_k} \left(\frac{b}{a}\right)^k, \tag{2.13}$$

while for $0 \le k \le n$,

$$(q^{-n};q)_k = \frac{(q;q)_n}{(q;q)_{n-k}} (-1)^k q^{k(k-1)/2-nk}. \tag{2.14}$$

In subsequent chapters it will be seen that a simple way to produce a new finite identity from an existing finite identity is to replace q with $1/q$, so the following identity, which holds for all integers n, is recorded for later use:

$$(a;1/q)_n = (1/a;q)_n (-a)^n q^{-n(n-1)/2}. \tag{2.15}$$

For later use we also define the *q-binomial coefficients*, or *Gaussian polynomial*, for non-negative integers n and k (the properties of these will be examined in detail in a later chapter):

$$\begin{bmatrix} n \\ k \end{bmatrix} = \begin{bmatrix} n \\ k \end{bmatrix}_q := \begin{cases} \dfrac{(q;q)_n}{(q;q)_k(q;q)_{n-k}}, & 0 \le k \le n, \\ 0, & \text{otherwise.} \end{cases} \tag{2.16}$$

Note that

$$\lim_{q \to 1} \begin{bmatrix} n \\ k \end{bmatrix} = \binom{n}{k}, \tag{2.17}$$

$$\lim_{n \to \infty} \begin{bmatrix} n \\ k \end{bmatrix} = \frac{1}{(q;q)_k}.$$

The elementary properties stated above, and those stated in the exercises in this section, will be used frequently throughout the book. Question 2.9 should provide a good work-out for a new-comer to the field.

An $_r\phi_s$ basic hypergeometric series is defined by

$$_r\phi_s \begin{bmatrix} a_1, a_2, \ldots, a_r \\ b_1, \ldots, b_s \end{bmatrix}; q, x \end{bmatrix}$$

$$= \sum_{n=0}^{\infty} \frac{(a_1, a_2, \ldots, a_r; q)_n}{(q, b_1, \ldots, b_s; q)_n} \left((-1)^n q^{n(n-1)/2}\right)^{s+1-r} x^n. \quad (2.18)$$

Most commonly, $r = s + 1$ and the definition simplifies to

$$_{s+1}\phi_s \begin{bmatrix} a_1, a_2, \ldots, a_{s+1} \\ b_1, \ldots, b_s \end{bmatrix}; q, x \end{bmatrix} = \sum_{n=0}^{\infty} \frac{(a_1, a_2, \ldots, a_{s+1}; q)_n}{(q, b_1, \ldots, b_s; q)_n} x^n. \quad (2.19)$$

In many identities, one of the a_i has the form $a_i = q^{-N}$ for a positive integer N, so that the series terminates at $n = N$.

The $_{s+1}\phi_s$ series at (2.19) is said to be *well-poised* if

$$b_1 = \frac{qa_1}{a_2}, b_2 = \frac{qa_1}{a_3}, \ldots, b_s = \frac{qa_1}{a_{s+1}}, \quad (2.20)$$

and *very-well-poised* if, in addition, $qa_1^{1/2} = a_2 = -a_3$.

The $_r\psi_s$ bilateral basic hypergeometric series is defined by

$$_r\psi_s \begin{bmatrix} a_1, a_2, \ldots, a_r \\ b_1, b_2, \ldots, b_s \end{bmatrix}; q, z \end{bmatrix} =$$

$$\sum_{n=-\infty}^{\infty} \frac{(a_1, a_2, \ldots, a_r; q)_n}{(b_1, b_2 \ldots, b_s; q)_n} \left((-1)^n q^{n(n-1)/2}\right)^{s-r} z^n. \quad (2.21)$$

Most commonly, $r = s$ and the definition simplifies to

$$_r\psi_r \begin{bmatrix} a_1, a_2, \ldots, a_r \\ b_1, b_2, \ldots, b_r \end{bmatrix}; q, z \end{bmatrix} = \sum_{n=-\infty}^{\infty} \frac{(a_1, a_2, \ldots, a_r; q)_n}{(b_1, b_2 \ldots, b_r; q)_n} z^n. \quad (2.22)$$

Exercises

2.1 Use the definition at (2.1) to prove that if n is a positive integer, then

$$(a; q)_n = (q^{1-n}/a; q)_n (-a)^n q^{n(n-1)/2}.$$

2.2 Use the definition of $(a; q)_{-n}$ on the left side of (2.5), together with the identity in Q1.1 (or otherwise), to prove that if n is a positive integer,

$$(a; q)_{-n} = \frac{(-q/a)^n}{(q/a; q)_n} q^{n(n-1)/2}.$$

2.3 (i) Prove that
$$(-q;q)_\infty(q;q^2)_\infty = 1.$$

(ii) More generally, prove that if $k \geq 2$ is an integer and ω is a primitive k-th root of unity, then

$$\prod_{j=1}^{k-1}(\omega^j q;q)_\infty(q^j;q^k)_\infty = 1.$$

2.4 (i) Prove, for all integers n and k, that

$$(aq^n;q)_k = \frac{(a;q)_k(aq^k;q)_n}{(a;q)_n}.$$

(ii) Hence prove, for all integers n and k, that

$$(aq^{-n};q)_k = \frac{(a;q)_k(q/a;q)_n}{(q^{1-k}/a;q)_n}q^{-nk}.$$

2.5 By starting with (2.11) and employing (2.5), or otherwise, show that

$$(aq^{-kn};q)_n = \frac{(q/a;q)_{kn}(-a)^n}{(q/a;q)_{(k-1)n}}q^{n(n-1)/2-kn^2} \tag{2.23}$$

holds for all integers n and k, and thus that

$$(aq^{-n};q)_n = (q/a;q)_n(-a/q)^n q^{-n(n-1)/2}, \tag{2.24}$$

$$\frac{(aq^{-n};q)_n}{(bq^{-n};q)_n} = \frac{(q/a;q)_n}{(q/b;q)_n}\left(\frac{a}{b}\right)^n. \tag{2.25}$$

2.6 Prove that

$$\left(aq^{-n};q\right)_{n-k} = \frac{(q/a;q)_n}{(q/a;q)_k}\left(-\frac{q}{a}\right)^{k-n}q^{k(k-1)/2-n(n-1)/2}$$

holds for all integers n and k.

2.7 Prove that (2.15) holds for all integers n.

2.8 Prove, for non-negative integers n and k, that

$$(q^{-n};q)_k = \begin{cases} \dfrac{(q;q)_n}{(q;q)_{n-k}}(-1)^k q^{k(k-1)/2-nk}, & 0 \leq k \leq n, \\ 0, & k > n. \end{cases} \tag{2.26}$$

2.9 Prove, for integers $0 \leq j \leq m$, that

$$\frac{1 + aq^{4j}}{1 + a} \frac{\left(-a, q^{-4m}; q^2\right)_{2j} \left(q^{1-4m}/a, -q^2; q^2\right)_{m-j} (-a)^{m-j} q^{4j(m-j)+2j}}{\left(a^2 q^4; q^4\right)_{2j} \left(-q^{1-4m}; q^2\right)_{2j} \left(-q^{4j-4m+1}, aq^{4j+2}; q^2\right)_{m-j}}$$

$$\times \frac{\left(-q; q^2\right)_{2m}}{\left(aq, q^2; q^2\right)_{2m}} = \frac{1}{(q; q)_{2m-2j}(aq; q)_{2j+2m}}.$$

Hint: It may be easier to multiply across so that the right side is 1, and then try to show the left side is 1. It is left in its present form here because of a connection with Q13.14.

2.10 Prove that

$$\sum_{n=0}^{\infty} \frac{q^{n^2+n}}{(x, q/x; q)_{n+1}} = x^{-1}\left(-1 + \sum_{n=0}^{\infty} \frac{q^{n^2}}{(x; q)_{n+1}(q/x; q)_n}\right). \qquad (2.27)$$

Remark: Each side of (2.27) gives a representation of the universal mock theta function defined by Gordon and McIntosh [132] and Hickerson and Mortenson [143], which will be encountered in Chapter 18.

2.11 Prove that

$$\sum_{n=1}^{\infty} \frac{q^{n^2}}{(q; q^2)_n} = q \sum_{n=0}^{\infty} \frac{q^{4n^2+4n}}{(q, q^3; q^4)_{n+1}}. \qquad (2.28)$$

Remark: Each side of (2.28) is equal to the third order mock theta function $\psi(q)$.

Chapter 3

The q-Binomial Theorem

The q-binomial theorem provides one of the most important formulae in q-series. It appears to have been discovered independently by a number of mathematicians, including Cauchy [92], Gauss [124] and Heine [140], while special cases were given by Euler [114] and Rothe [220]. The proof presented here is well known (see for example [34, pp. 489–490], [70, p. 8] or [121, p. 58]). Other proofs are sketched in the exercises for this chapter.

Theorem 3.1. *If* $|z|, |q| < 1$, *then*

$$\sum_{n=0}^{\infty} \frac{(a;q)_n}{(q;q)_n} z^n = \frac{(az;q)_\infty}{(z;q)_\infty}. \tag{3.1}$$

Proof. Let a and q be fixed. The infinite product on the right side of (3.1) converges uniformly and absolutely in any disc $|z| \leq 1 - \epsilon$, so represents and analytic function in $|z| < 1$ and may be represented by its Taylor expansion there. Set

$$F(z) := \frac{(az;q)_\infty}{(z;q)_\infty} = \sum_{n=0}^{\infty} A_n z^n.$$

From the infinite product representation, $F(0) = 1 = A_0$, and

$$F(z) = \frac{1 - az}{1 - z} F(qz),$$

$$(1 - z)F(z) = (1 - az)F(qz),$$

$$A_n - A_{n-1} = q^n A_n - aq^{n-1} A_{n-1} \text{ (equating coefficients of } z^n)$$

$$A_n = \frac{(1 - aq^{n-1})}{1 - q^n} A_{n-1} = \cdots = \frac{(a;q)_n}{(q;q)_n} A_0 = \frac{(a;q)_n}{(q;q)_n},$$

completing the proof. $\qquad\square$

12

Remark: One general topic of interest in q-series is to find "q-versions", "q-analogues" or "q-extensions" of existing identities and functions, in the sense that taking the limit $q \to 1^-$ converts the q-analogue to the regular expression. As an illustration, the expression (for $q \neq 1$)

$$[\alpha]_q := \frac{1 - q^\alpha}{1 - q}$$

is a q-extension of the complex number α. The identity at (3.1) may be regarded as a q-extension of the binomial theorem (hence of course the name), since replacing a with q^α and then letting $q \to 1^-$ produces the binomial series for $(1 - z)^{-\alpha}$ (see, for example, [174], for a rigorous proof):

$$1 + \frac{\alpha}{1}z + \frac{\alpha(\alpha + 1)}{2!}z^2 + \frac{\alpha(\alpha + 1)(\alpha + 2)}{3!}z^3 + \cdots = (1 - z)^{-\alpha}. \quad (3.2)$$

The following special cases of (3.1) are due to Euler [114].

Corollary 3.1.

$$\sum_{n=0}^{\infty} \frac{z^n}{(q; q)_n} = \frac{1}{(z; q)_\infty}, \quad |z| < 1, \ |q| < 1. \quad (3.3)$$

$$\sum_{n=0}^{\infty} \frac{(-z)^n q^{n(n-1)/2}}{(q; q)_n} = (z; q)_\infty, \quad |q| < 1. \quad (3.4)$$

Proof. For (3.3), set $a = 0$ in (3.1), and for (3.4), replace z with z/a in (3.1), and then let $a \to \infty$. $\qquad\square$

The first of the next two special cases is due to Rothe [220] (with a misprint in his statement).

Corollary 3.2. *For each non-negative integer N,*

$$\sum_{n=0}^{N} \begin{bmatrix} N \\ n \end{bmatrix} (-z)^n q^{n(n-1)/2} = (z; q)_N, \quad (3.5)$$

$$\sum_{n=0}^{\infty} \begin{bmatrix} N + n - 1 \\ n \end{bmatrix} z^n = \frac{1}{(z; q)_N}, \quad |z| < 1, \ |q| < 1. \quad (3.6)$$

Proof. For (3.5), set $a = q^{-N}$ and replace z with zq^N in (3.1). For (3.6), replace a with q^N in (3.1). In each case the definition at (2.16) is also used. $\qquad\square$

By expanding each quotient in the identity

$$\frac{(abz;q)_\infty}{(az;q)_\infty}\frac{(az;q)_\infty}{(z;q)_\infty} = \frac{(abz;q)_\infty}{(z;q)_\infty} \tag{3.7}$$

using (3.1), comparing the coefficients of like powers of z on each side, and finally using (2.12), the following corollary results ([121, p. 59]).

Corollary 3.3. *For each non-negative integer n,*

$$\sum_{k=0}^{n}\frac{(b,q^{-n};q)_k}{(q,q^{1-n}/a;q)_k}q^k = \frac{(ab;q)_n}{(a;q)_n}. \tag{3.8}$$

Proof. See Exercise 3.3.   □

Exercises

3.1 (Gasper [121, pp. 56–57]) Set, for $|q|,|z| < 1$,

$$f(a,z) := \sum_{k=0}^{\infty}\frac{(a;q)_k}{(q;q)_k}z^k.$$

(i) Upon noting that $1 - a = (1 - aq^k) - a(1 - q^k)$, deduce that

$$f(a,z) = (1 - az)f(aq,z), \tag{3.9}$$
$$= (az;q)_n f(aq^n,z), \forall n \in \mathbb{N},$$
$$= (az;q)_\infty f(0,z).$$

(ii) By specializing a, show that

$$f(0,z) = \frac{1}{(z,q)_\infty},$$

thus completing a second proof of Theorem 3.1.

3.2 ([34, pp. 488–489], [123, pp. 9–10]). Let $f(a,z)$ be as in Q1.1, noting that $f(a,0) = 1$.
(i) Prove that

$$f(a,z) - f(a,qz) = (1 - a)zf(aq,z). \tag{3.10}$$

(ii) by combining this identity with the first equality in (3.9), show

$$f(a, z) = \frac{1 - az}{1 - z} f(a, qz),$$

$$= \frac{(az; q)_n}{(z; q)_n} f(a, q^n z), \ \forall n \in \mathbb{N},$$

$$= \frac{(az; q)_\infty}{(z; q)_\infty} f(a, 0) = \frac{(az; q)_\infty}{(z; q)_\infty},$$

thus giving another proof of Theorem 3.1.

3.3 Fill in the details in the proof of (3.8).

3.4 Prove that for any integer $n \geq 0$,

$$\sum_{k=0}^{n} \frac{(a, q^{-n}; q)_k}{(q, q^{1-n}/a; q)_k} \left(\frac{-q}{a} \right)^k = \begin{cases} \dfrac{(a^2, q; q^2)_{n/2}}{(a; q)_n}, & n \text{ even,} \\ 0, & n \text{ odd.} \end{cases} \tag{3.11}$$

Hint: Use the infinite version of (2.9), together with (3.1) (with both z and $-z$) and (2.12).

3.5 (Gasper [121, p. 59]) Show that (3.8) is equivalent to the *q-Chu–Vandermonde summation formula*

$$\sum_{k=0}^{n} \frac{(b, q^{-n}; q)_k}{(c, q; q)_k} q^k = \frac{(c/b; q)_n}{(c; q)_n} b^n. \tag{3.12}$$

3.6 (Gasper [121, p. 59]) Show that (3.12) is equivalent to the summation formula

$$\sum_{k=0}^{n} \frac{(b, q^{-n}; q)_k}{(c, q; q)_k} \left(\frac{cq^n}{b} \right)^k = \frac{(c/b; q)_n}{(c; q)_n}. \tag{3.13}$$

Hint: In (3.12), either replace q with $1/q$, or change the order of summation ($\sum_{k=0}^{n} f_k = \sum_{k=0}^{n} f_{n-k}$).

3.7 (i) By factoring $(z; q)_{M+N}$ appropriately, prove for all non-negative integers n, M and N with $0 \leq n \leq M + N$, that

$$\begin{bmatrix} M + N \\ n \end{bmatrix} = \sum_{k=0}^{n} \begin{bmatrix} M \\ k \end{bmatrix} \begin{bmatrix} N \\ n - k \end{bmatrix} q^{(M-k)(n-k)}.$$

(ii) Prove, for all non-negative integers N, that

$$\begin{bmatrix} 2N \\ N \end{bmatrix} = \sum_{k=0}^{N} \begin{bmatrix} N \\ k \end{bmatrix}^2 q^{k^2}.$$

Note that these identities are q-versions of well known identities for ordinary binomial coefficients.

3.8 Define $e_q(z) := ((1-q)z; q)_\infty^{-1}$ and $E_q(z) = (-(1-q)z; q)_\infty$. Show that

$$\lim_{q \to 1} e_q(z) = \lim_{q \to 1} E_q(z) = e^z.$$

(Thus $e_q(z)$ and $E_q(z)$ are both q-extensions of the exponential function e^z.)

3.9 (i) Prove that

$$\sum_{m=0}^{n} \begin{bmatrix} N \\ m \end{bmatrix} \begin{bmatrix} N+n-m-1 \\ n-m \end{bmatrix} (-1)^m q^{m(m-1)/2} = 0$$

for all positive integers n and N with $1 \le n \le N$.

(ii) Prove that

$$\sum_{j=0}^{n} \begin{bmatrix} n \\ j \end{bmatrix} (-1)^j q^{j(j-1)/2} = 0,$$

for all positive integers n.

Hint: For part (i), use both identities in Corollary 3.2.

3.10 The continuous q-ultraspherical polynomials $\{C_n(\cos\theta; \beta|q)\}_{n=0}^\infty$ are defined by

$$C_n(\cos\theta; \beta|q) := \sum_{k=o}^{n} \frac{(\beta; q)_q (\beta; q)_{n-k}}{(q; q)_q (q; q)_{n-k}} e^{i(n-2k)\theta}.$$

Prove that they have the generating function ([218])

$$\sum_{n=0}^{\infty} C_n(\cos\theta; \beta|q) t^n = \frac{(t\beta e^{i\theta}, t\beta e^{-i\theta}; q)_\infty}{(te^{i\theta}, te^{-i\theta}; q)_\infty}.$$

3.11 Prove that if n is a non-negative integer, then

$$\sum_{k=0}^{n} \frac{(a; q)_{n-k}(a; q)_k}{(q; q)_{n-k}(q; q)_k} (-1)^k = \begin{cases} \frac{(a^2; q^2)_{n/2}}{(q^2; q^2)_{n/2}} & n \text{ even,} \\ 0, & n \text{ odd.} \end{cases} \tag{3.14}$$

Chapter 4

Heine's Transformation

In this chapter we derive a number of important transformation- and summation formulae, all of which ultimately derive from, or may be proved by employing, a fundamental identity due to Heine [140], who showed that, for $|b|, |z|, |q| < 1$,

$$\sum_{k=0}^{\infty} \frac{(a, b; q)_k}{(c, q; q)_k} z^k = \frac{(b, az; q)_\infty}{(c, z; q)_\infty} \sum_{k=0}^{\infty} \frac{(c/b, z; q)_k}{(az, q; q)_k} b^k. \tag{4.1}$$

Andrews [9, Fundamental Lemma] proved a very general formula that implies Heine's transformation as a simple special case, and then used this formula to derive [9–11] many transformations similar to Heine's, and also to prove many of the mock-theta identities. Here we are content to prove a special case of this general formula, which is a slight extension of Heine's transformation ((4.1) is the case $h = 1$ of (4.2)).

Theorem 4.1 ([10], Lemma 1). *For $|b|, |z|, |q| < 1$, and h a positive integer,*

$$\sum_{k=0}^{\infty} \frac{(a; q^h)_k (b; q)_{hk}}{(q^h; q^h)_k (c; q)_{hk}} z^k = \frac{(b; q)_\infty (az; q^h)_\infty}{(c; q)_\infty (z; q^h)_\infty} \sum_{j=0}^{\infty} \frac{(c/b; q)_j (z; q^h)_j}{(q; q)_j (az; q^h)_j} b^j. \tag{4.2}$$

Proof. By employing the q-binomial theorem (3.1) twice below, it follows that

$$\sum_{k=0}^{\infty} \frac{(a; q^h)_k (b; q)_{hk}}{(q^h; q^h)_k (c; q)_{hk}} z^k = \frac{(b; q)_\infty}{(c; q)_\infty} \sum_{k=0}^{\infty} \frac{(a; q^h)_k (cq^{hk}; q)_\infty}{(q^h; q^h)_k (bq^{hk}; q)_\infty} z^k$$

$$= \frac{(b; q)_\infty}{(c; q)_\infty} \sum_{k=0}^{\infty} \frac{(a; q^h)_k}{(q^h; q^h)_k} z^k \sum_{j=0}^{\infty} \frac{(c/b; q)_j}{(q; q)_j} \left(bq^{hk} \right)^j$$

17

$$= \frac{(b;q)_\infty}{(c;q)_\infty} \sum_{j=0}^{\infty} \frac{(c/b;q)_j}{(q;q)_j} b^j \sum_{k=0}^{\infty} \frac{(a;q^h)_k}{(q^h;q^h)_k} \left(zq^{hj}\right)^k$$

$$= \frac{(b;q)_\infty}{(c;q)_\infty} \sum_{j=0}^{\infty} \frac{(c/b;q)_j}{(q;q)_j} b^j \frac{(azq^{hj};q^h)_\infty}{(zq^{hj};q^h)_\infty}$$

$$= \frac{(b;q)_\infty (az;q^h)_\infty}{(c;q)_\infty (z;q^h)_\infty} \sum_{j=0}^{\infty} \frac{(c/b;q)_j (z;q^h)_j}{(q;q)_j (az;q^h)_j} b^j.$$

$\square$

Remark: The authors in [40, Chapter 1] use (4.2), together with similar results from [9–11], to prove a large number of identities stated by Ramanujan in his lost notebook.

Iteration of Heine's transformation (4.1) produces two other expressions for the series on the right side of (4.1).

Corollary 4.1. *(i) For $|c/b|, |z|, |q| < 1$,*

$$\sum_{k=0}^{\infty} \frac{(a,b;q)_k}{(c,q;q)_k} z^k = \frac{(c/b, bz;q)_\infty}{(c, z;q)_\infty} \sum_{k=0}^{\infty} \frac{(abz/c, b;q)_k}{(bz, q;q)_k} \left(\frac{c}{b}\right)^k. \tag{4.3}$$

(i) For $|abz/c|, |z|, |q| < 1$,

$$\sum_{k=0}^{\infty} \frac{(a,b;q)_k}{(c,q;q)_k} z^k = \frac{(abz/c;q)_\infty}{(z;q)_\infty} \sum_{k=0}^{\infty} \frac{(c/a, c/b;q)_k}{(c,q;q)_k} \left(\frac{abz}{c}\right)^k. \tag{4.4}$$

Proof. For (4.3), apply (4.1) to the series on the right side of (4.1), with a, b, c and z replaced with, respectively, z, c/b, az and b. For (4.4), apply (4.1) to the series on the right side of (4.3), with a, b, c and z replaced with, respectively, b, abz/c, bz and c/b. Initially, it is necessary to have $|b| < 1$ to apply (4.1), but this restriction can be dropped by analytic continuation. $\square$

4.1   The q-Pfaff–Saalschütz sum

The q-Pfaff–Saalschütz sum (Corollary 4.2) may be derived by equating coefficients of like powers of z on each side of (4.4).

Corollary 4.2. *For each non-negative integer n,*

$$\sum_{k=0}^{n} \frac{(a,b,q^{-n};q)_k}{(c, abq^{1-n}/c, q;q)_k} q^k = \frac{(c/a, c/b;q)_n}{(c, c/ab;q)_n}. \tag{4.5}$$

Proof. After using the q-binomial theorem (3.1) to expand the infinite product on the right side of (4.4), and comparing coefficients of z^n on each side of the resulting expression, the following identity results:

$$\frac{(a,b;q)_n}{(c,q;q)_n} = \sum_{m=0}^{n} \frac{(c/a,c/b;q)_m}{(c,q;q)_m} \left(\frac{ab}{c}\right)^m \frac{(ab/c:q)_{n-m}}{(q;q)_{n-m}}.$$

The result now follows after using (2.12), and then replacing a with c/a and b with c/b. $\qquad\square$

Remark: Note that letting $n \to \infty$ in q-Pfaff–Saalschütz sum (4.5) recovers the q-Gauss summation formula (4.8) below.

The next corollary, due to Sears [236], is an extension of the q-Pfaff–Saalschütz sum, in the sense that specializing the parameters in the identity of Sears in a particular way recovers the q-Pfaff–Saalschütz sum. We follow the proof in [34, pp. 524–525].

Corollary 4.3. *For each positive integer n,*

$$\sum_{k=0}^{n} \frac{(a,b,c,q^{-n};q)_k}{(d,e,f,q;q)_k} q^k = a^n \frac{(e/a,f/a;q)_n}{(e,f;q)_n} \sum_{k=0}^{n} \frac{(a,d/b,d/c,q^{-n};q)_k}{(d,aq^{1-n}/e,aq^{1-n}/f,q;q)_k} q^k,$$
(4.6)

$$\sum_{k=0}^{n} \frac{(a,b,c,q^{-n};q)_k}{(d,e,f,q;q)_k} q^k$$

$$= \frac{(a,ef/ab,ef/ac;q)_n}{(e,f,ef/abc;q)_n} \sum_{k=0}^{n} \frac{(e/a,f/a,ef/abc,q^{-n};q)_k}{(ef/ab,ef/ac,q^{1-n}/a,q;q)_k} q^k, \quad (4.7)$$

if $def = abcq^{1-n}$.

Proof. In (4.4), replace a with d, b with e and c with f, with the restriction $ab/c = de/f$. Multiply the left side of the original identity by the right side of the new identity, and vice-versa, then cancel the infinite products to get

$$\sum_{k=0}^{\infty} \frac{(a,b;q)_k}{(c,q;q)_k} z^k \sum_{k=0}^{\infty} \frac{(f/d,f/e;q)_k}{(f,q;q)_k} \left(\frac{dez}{f}\right)^k$$

$$= \sum_{k=0}^{\infty} \frac{(d,e;q)_k}{(f,q;q)_k} z^k \sum_{k=0}^{\infty} \frac{(c/a,c/b;q)_k}{(c,q;q)_k} \left(\frac{abz}{c}\right)^k.$$

Upon equating the coefficient of z^n, and replacing each of the letters $a,\ldots,$ f by its upper-case equivalent (to make subsequent transformations easier to track), it follows that

$$\sum_{k=0}^{m} \frac{(A,B;q)_{n-k}}{(C,q;q)_{n-k}} \frac{(F/D,F/E;q)_k}{(F,q;q)_k} \left(\frac{DE}{F}\right)^k$$

$$= \sum_{k=0}^{n} \frac{(D,E;q)_k}{(F,q;q)_k} \frac{(C/A,C/B;q)_{n-k}}{(C,q;q)_{n-k}} \left(\frac{AB}{C}\right)^{n-k}.$$

Now apply (2.12), and make the following transformations: $A \to q^{1-n}/e$, $B \to q^{1-n}/f$, $C \to q^{1-n}/a$, $D \to d/b$, $E \to d/c$ and $F \to d$. Finally employ the $k = 1$ case of (2.23), and (4.6) follows.

The proof of (4.7) is similar, and is left as an exercise. $\qquad\square$

4.2 Heine's q-Gauss sum and the q-Kummer sum

Another important implication is Heine's q-Gauss summation formula.

Corollary 4.4. *For* $|c/ab|, |q| < 1$,

$$\sum_{k=0}^{\infty} \frac{(a,b;q)_k}{(c,q;q)_k} \left(\frac{c}{ab}\right)^k = \frac{(c/a,c/b;q)_\infty}{(c,c/ab;q)_\infty}. \tag{4.8}$$

Proof. Set $z = c/ab$ in (4.4), so that $(abz/c,b;q)_k = (1;q)_k = 0$ for $k > 0$, and the series on the right side collapses to have the value 1. $\qquad\square$

Note that setting $c = q$ and letting $a,b \to \infty$ leads to Euler's identity:

$$\sum_{k=0}^{\infty} \frac{q^{k^2}}{(q;q)_k^2} = \frac{1}{(q;q)_\infty}. \tag{4.9}$$

The q-analogue of Kummer's summation formula, found independently by Bailey [57] and Daum [109], may also proved using Heine's transformation formula.

Corollary 4.5. *For* $|q/b|, |q| < 1$

$$\sum_{k=0}^{\infty} \frac{(a,b;q)_k}{(aq/b,q;q)_k} \left(\frac{-q}{b}\right)^k = \frac{(-q;q)_\infty (aq,aq^2/b^2;q^2)_\infty}{(aq/b,-q/b;q)_\infty}. \tag{4.10}$$

Proof. Upon noting that the left side of (4.1) is symmetric in a and b, so that the right side is also, first switch a and b on the right side. Then replace c with aq/b and z with $-q/b$ to get

$$\sum_{k=0}^{\infty} \frac{(a,b;q)_k}{(aq/b,q;q)_k} \left(\frac{-q}{b}\right)^k = \frac{(a,-q;q)_\infty}{(aq/b,-q/b;q)_\infty} \sum_{k=0}^{\infty} \frac{(q/b,-q/b;q)_k}{(-q,q;q)_k} a^k$$

$$= \frac{(a,-q;q)_\infty}{(aq/b,-q/b;q)_\infty} \sum_{k=0}^{\infty} \frac{(q^2/b^2;q^2)_k}{(q^2;q^2)_k} a^k$$

$$= \frac{(a,-q;q)_\infty}{(aq/b,-q/b;q)_\infty} \frac{(aq^2/b^2;q^2)_\infty}{(a;q^2)_\infty},$$

where the last equality is by the q-binomial theorem (3.1). The result now follows upon writing $(a;q)_\infty = (a,aq;q^2)_\infty$ and simplifying. $\square$

Before proving Jackson's [150] transformation formula, we recall the q-analogue of the q-Chu–Vandermonde summation formula ((3.13), with b replaced with c/b):

$$\frac{(b;q)_n}{(c;q)_n} = \sum_{k=0}^{n} \frac{(c/b,q^{-n};q)_k}{(c,q;q)_k} (bq^n)^k = \sum_{k=0}^{n} \frac{(q;q)_n(c/b;q)_k}{(q;q)_{n-k}(c,q;q)_k} (-b)^k q^{k(k-1)/2},$$

$$(4.11)$$

with the final equality following from (2.26).

Corollary 4.6. (*Jackson, [150]*) *For* $|z|, |q| < 1$,

$$\sum_{n=0}^{\infty} \frac{(a,b;q)_n}{(c,q;q)_n} z^n = \frac{(az;q)_\infty}{(z;q)_\infty} \sum_{k=0}^{\infty} \frac{(a,c/b;q)_k}{(c,az,q;q)_k} (-bz)^k q^{k(k-1)/2}. \qquad (4.12)$$

Proof.

$$\sum_{n=0}^{\infty} \frac{(a,b;q)_n}{(c,q;q)_n} z^n = \sum_{n=0}^{\infty} (a;q)_n z^n \sum_{k=0}^{n} \frac{(c/b;q)_k}{(q;q)_{n-k}(c,q;q)_k} (-b)^k q^{k(k-1)/2}$$

$$= \sum_{k=0}^{\infty} \sum_{n=k}^{\infty} \frac{(a;q)_n(c/b;q)_k}{(q;q)_{n-k}(c,q;q)_k} z^n (-b)^k q^{k(k-1)/2}$$

$$= \sum_{k=0}^{\infty} \sum_{n=0}^{\infty} \frac{(a;q)_{n+k}(c/b;q)_k}{(q;q)_n(c,q;q)_k} z^{n+k} (-b)^k q^{k(k-1)/2}$$

$$= \sum_{k=0}^{\infty} \frac{(a,c/b;q)_k}{(c,q;q)_k} (-bz)^k q^{k(k-1)/2} \sum_{n=0}^{\infty} \frac{(aq^k;q)_n}{(q;q)_n} z^n$$

$$= \sum_{k=0}^{\infty} \frac{(a, c/b; q)_k}{(c, q; q)_k} (-bz)^k q^{k(k-1)/2} \frac{(aq^k z; q)_\infty}{(z; q)_\infty}$$

$$= \frac{(az; q)_\infty}{(z; q)_\infty} \sum_{k=0}^{\infty} \frac{(a, c/b; q)_k}{(c, az, q; q)_k} (-bz)^k q^{k(k-1)/2}.$$

□

Exercises

4.1 Prove that the Sears transformation at (4.6) implies the q-Pfaff–Saalschütz sum (4.5).

4.2 Prove the transformation formula at (4.7).

4.3 (i) (Andrews [9], Theorem A_1) By modifying the proof of (4.2), prove that

$$\sum_{k=0}^{\infty} \frac{(a; q^2)_k (b; q)_k}{(q^2; q^2)_k (c; q)_k} z^k = \frac{(b; q)_\infty (az; q^2)_\infty}{(c; q)_\infty (z; q^2)_\infty} \sum_{j=0}^{\infty} \frac{(c/b; q)_{2j} (z; q^2)_j}{(q; q)_{2j} (az; q^2)_j} b^{2j}$$

$$+ \frac{(b; q)_\infty (azq; q^2)_\infty}{(c; q)_\infty (zq; q^2)_\infty} \sum_{j=0}^{\infty} \frac{(c/b; q)_{2j+1} (zq; q^2)_j}{(q; q)_{2j+1} (azq; q^2)_j} b^{2j+1}. \quad (4.13)$$

(ii) Derive the special case of (4.13) (to be used later):

$$\sum_{k=0}^{\infty} \frac{(a; q^2)_k}{(q^2; q^2)_k (c; q)_k} z^k = \frac{(az; q^2)_\infty}{(c; q)_\infty (z; q^2)_\infty} \sum_{j=0}^{\infty} \frac{(z; q^2)_j c^{2j} q^{2j^2 - j}}{(q; q)_{2j} (az; q^2)_j}$$

$$- \frac{(azq; q^2)_\infty}{(c; q)_\infty (zq; q^2)_\infty} \sum_{j=0}^{\infty} \frac{(zq; q^2)_j c^{2j+1} q^{2j^2 + j}}{(q; q)_{2j+1} (azq; q^2)_j}. \quad (4.14)$$

(iii) More generally, show that for each integer $h \geq 2$,

$$\sum_{k=0}^{\infty} \frac{(a; q^h)_k (b; q)_k}{(q^h; q^h)_k (c; q)_k} z^k$$

$$= \frac{(b; q)_\infty}{(c; q)_\infty} \sum_{r=0}^{h-1} \frac{(azq^r; q^h)_\infty}{(zq^r; q^h)_\infty} \sum_{j=0}^{\infty} \frac{(c/b; q)_{hj+r} (zq^r; q^h)_j}{(q; q)_{hj+r} (azq^r; q^h)_j} b^{hj+r}.$$

4.4 Prove each of the following statements by specializing the parameters in the Sears transformation (4.6) (see [34, p. 525]):

$$\sum_{k=0}^{n} \frac{(a,b,q^{-n};q)_k}{(d,e,q;q)_k} q^k = \frac{(e/a;q)_n a^n}{(e;q)_n} \sum_{k=0}^{n} \frac{(a,d/b,q^{-n};q)_k}{(d,aq^{1-n}/e,q;q)_k} \left(\frac{bq}{e}\right)^k.$$

$$\text{(4.15)}$$

$$\sum_{k=0}^{n} \frac{(a,b,q^{-n};q)_k}{(e,f,q;q)_k} q^k = \frac{(e/a,f/a;q)_n a^n}{(e,f;q)_n} \sum_{k=0}^{n} \frac{(a,abq^{1-n}/ef,q^{-n};q)_k}{(aq^{1-n}/e,aq^{1-n}/f,q;q)_k} q^k.$$

$$\text{(4.16)}$$

$$\sum_{k=0}^{\infty} \frac{(a,b,c;q)_k}{(d,e,q;q)_k} \left(\frac{de}{abc}\right)^k = \frac{(e/a,de/bc;q)_\infty}{(e,de/abc;q)_\infty} \sum_{k=0}^{\infty} \frac{(a,d/b,d/c;q)_k}{(d,de/bc,q;q)_k} \left(\frac{e}{a}\right)^k.$$

$$\text{(4.17)}$$

$$\sum_{k=0}^{\infty} \frac{(a,b,c;q)_k}{(d,e,q;q)_k} \left(\frac{de}{abc}\right)^k = \frac{(a,de/ab,de/ac;q)_\infty}{(d,e,de/abc;q)_\infty} \sum_{k=0}^{\infty} \frac{(d/a,e/a,de/abc;q)_k}{(de/ab,de/ac,q;q)_k} a^k.$$

$$\text{(4.18)}$$

4.5 (Alladi [6]) By employing (4.3) or otherwise, prove that

$$\sum_{n=0}^{\infty} \frac{(a;q)_n q^{mn}}{(bq;q)_n} = b^{-m}(1-b)\frac{(q;q)_{m-1}}{(a/b;q)_m} \left(\frac{(a;q)_\infty}{(b;q)_\infty} - \sum_{j=0}^{m-1} \frac{(a/b;q)_j b^j}{(q;q)_j}\right)$$

holds for each positive integer m, and, in particular, that

$$\sum_{n=0}^{\infty} \frac{(a;q)_n q^n}{(bq;q)_n} = \frac{(1-b)}{(b-a)} \left(\frac{(a;q)_\infty}{(b;q)_\infty} - 1\right).$$

4.6 Prove that

$$\sum_{n=0}^{\infty} \frac{(b;q)_n t^n q^{n(n+1)/2}}{(c,q;q)_n} = \frac{(b,-tq;q)_\infty}{(c;q)_\infty} \sum_{n=0}^{\infty} \frac{(c/b;q)_n b^n}{(-tq,q;q)_n}.$$

Hence, or otherwise, prove Lebesgue's identity ([176])

$$\sum_{n=0}^{\infty} \frac{(-aq;q)_n q^{n(n+1)/2}}{(q;q)_n} = (-q;q)_\infty (-aq^2;q^2)_\infty.$$

4.7 (Alladi [5]) Prove that

$$\sum_{n=0}^{\infty} q^{2n}(aq^{2n+1},q^{2n+2};q^2)_\infty = \sum_{n=0}^{\infty} (-a)^n q^{n^2}.$$

4.8 By using Heine's transformation or otherwise, prove Ramanujan's identity ([39, p. 10], [213, Entry 1.4.9])

$$\sum_{n=0}^{\infty} \frac{q^{n(n+1)/2}}{(q;q)_n^2} = \frac{(-q;q)_\infty}{(q;q)_\infty} \sum_{n=0}^{\infty} \frac{(-1)^n q^{n(n+1)/2}}{(q^2;q^2)_n}.$$

4.9 (Andrews [16]) By using Jackson's identity (4.12), or otherwise, prove the following q-analogue of a sum due to Bailey:

$$\sum_{k=0}^{\infty} \frac{(b, q/b; q)_k c^k q^{k(k-1)/2}}{(c;q)_k (q^2;q^2)_k} = \frac{(bc, qc/b; q^2)_\infty}{(c;q)_\infty}.$$

4.10 Prove the following identities for Ramanujan's order-3 mock theta function, $\phi(q)$.

$$\phi(q) := \sum_{n=0}^{\infty} \frac{q^{n^2}}{(-q^2;q^2)_n} = \frac{(-q;q^2)_\infty}{(-q^2;q^2)_\infty} \sum_{n=0}^{\infty} \frac{(q;q^2)_n q^{n^2+n}}{(-q, q^2;q^2)_n}$$

$$= (q;q^2)_\infty \sum_{n=0}^{\infty} \frac{(-1;q^2)_n q^n}{(q^4;q^4)_n}.$$

4.11 Prove, for all $n \geq 1$, that

$$\sum_{j=0}^{n} \frac{(-1;q)_{n-j}(-1,q)_j}{(q;q)_{n-j}(q,q)_j}(-1)^j = 0.$$

4.12 By first specializing a in (4.13) and them applying (4.2), prove the identities

(i)
$$\sum_{k=0}^{\infty} \frac{(cq/bz;q^2)_k (b;q)_k}{(q^2;q^2)_k (c;q)_k} z^k = \frac{(cb, cq/b;q^2)_\infty}{(c, -b;q)_\infty (q;q^2)_\infty} \sum_{j=0}^{\infty} \frac{(q/z, b^2;q^2)_j}{(cb, q^2;q^2)_j} z^j$$

$$+ \frac{b(cqb, c/b;q^2)_\infty}{(c, -b;q)_\infty (q;q^2)_\infty} \sum_{j=0}^{\infty} \frac{(q^2/z, b^2;q^2)_j}{(cbq, q^2;q^2)_j}(zq)^j;$$

$$\tag{4.19}$$

(ii)
$$\sum_{k=0}^{\infty} \frac{(b;q)_k}{(q^2;q^2)_k (c;q)_k}\left(\frac{-c}{b}\right)^k q^{k^2} = \frac{(cb, cq/b;q^2)_\infty}{(c, -b;q)_\infty (q;q^2)_\infty}$$

$$\times \sum_{j=0}^{\infty} \frac{(b^2;q^2)_j (-1)^j q^{j^2}}{(cb, q^2;q^2)_j} + \frac{b(cqb, c/b;q^2)_\infty}{(c, -b;q)_\infty (q;q^2)_\infty} \sum_{j=0}^{\infty} \frac{(b^2;q^2)_j (-1)^j q^{j^2+2j}}{(cbq, q^2;q^2)_j}.$$

$$\tag{4.20}$$

4.13 Derive the following special case of Heine's transformation (4.1),

$$\sum_{k=0}^{\infty} \frac{(b;q)_k q^{k(k+1)/2}}{(c,q;q)_k} = \frac{(b,-q;q)_\infty}{(c;q)_\infty} \sum_{k=0}^{\infty} \frac{(c^2/b^2;q^2)_k}{(q^2;q^2)_k(-c/b;q)_k} b^k, \qquad (4.21)$$

and use this identity together with (4.14) to prove that

$$\sum_{k=0}^{\infty} \frac{(b;q)_k q^{k(k+1)/2}}{(c,q;q)_k} = \frac{(-q;q)_\infty (bq, c^2/b; q^2)_\infty}{(c, -c/b; q)_\infty} \sum_{k=0}^{\infty} \frac{(b;q^2)_k (c/b)^{2k} q^{2k^2-k}}{(c^2/b; q^2)_k (q;q)_{2k}}$$

$$+ \frac{(-q;q)_\infty (b, c^2 q/b; q^2)_\infty}{(c, -c/b; q)_\infty} \sum_{k=0}^{\infty} \frac{(bq;q^2)_k (c/b)^{2k+1} q^{2k^2+k}}{(c^2 q/b; q^2)_k (q;q)_{2k+1}}. \qquad (4.22)$$

4.14 (i) By moving the infinite product in the $h = 2$ case of (4.2) to the left side and then making a change of parameters, prove that if $s \in \{0, 1\}$, then

$$\sum_{k=0}^{\infty} \frac{(z;q^2)_k q^{k(k+1)/2+sk}}{(a;q^2)_k (q;q)_k} = \frac{(z;q^2)_\infty (-q^{1+s};q)_\infty}{(a;q^2)_\infty} \sum_{k=0}^{\infty} \frac{(a/z;q^2)_k z^k}{(q^4;q^4)_k (-q^{1+2s};q^2)_k}. \qquad (4.23)$$

(ii) By using (4.23) in conjunction with (4.13) (with q replaced with q^2), prove that if $s \in \{0, 1\}$, then

$$\sum_{k=0}^{\infty} \frac{(z;q^2)_k q^{k(k+1)/2+sk}}{(a;q^2)_k (q;q)_k}$$

$$= \frac{(zq^2;q^4)_\infty (-q^2, a/z; q^2)_\infty}{(a;q^2)_\infty} \sum_{j=0}^{\infty} \frac{(-zq^{1+2s}/a; q^2)_{2j} (z;q^4)_j (a/z)^{2j}}{(q^2;q^2)_{2j}}$$

$$+ \frac{(z;q^4)_\infty (-q^2, a/z; q^2)_\infty}{(a;q^2)_\infty} \sum_{j=0}^{\infty} \frac{(-zq^{1+2s}/a; q^2)_{2j+1} (zq^2;q^4)_j (a/z)^{2j+1}}{(q^2;q^2)_{2j+1}}. \qquad (4.24)$$

4.15 (i) Derive from (4.13) that if $s \in \{0, 1\}$, then

$$\sum_{k=0}^{\infty} \frac{(b;q)_k q^{k^2+sk}}{(q^2;q^2)_k} = (b;q)_\infty (-q;q^2)_\infty \sum_{j=0}^{\infty} \frac{b^{2j+s}}{(q^2;q^2)_j (q^2;q^4)_{j+s}}$$

$$+ (b;q)_\infty (-q^2;q^2)_\infty \sum_{j=0}^{\infty} \frac{b^{2j+1-s}}{(q^4;q^4)_j (q;q^2)_{j+1-s}}. \qquad (4.25)$$

(ii) By applying the $h = 2$ case of (4.2) (with the infinite product moved the left side and q replaced with q^2) to the first sum in (4.25), and applying (4.13) (with q replaced with q^2) to the second sum, prove that if $s \in \{0, 1\}$, then

$$
\sum_{k=0}^{\infty} \frac{(b; q)_k q^{k^2 + sk}}{(q^2; q^2)_k} = \frac{b^s}{(-b; q)_\infty (q; q^2)_\infty} \sum_{k=0}^{\infty} \frac{(b^2; q^2)_{2k} (-1)^k q^{2k^2 + 4sk}}{(q^4; q^4)_k}
$$

$$
+ \frac{b^{1-s}(b; q)_\infty (-q^2; q^2)_\infty}{(q; q^2)_\infty (b^2; q^4)_\infty} \sum_{k=0}^{\infty} \frac{(b^2, q^4)_k q^{4k^2 + 4k(1-s)}}{(q^2; q^2)_{2k}}
$$

$$
- \frac{b^{1-s}(b; q)_\infty (-q^2; q^2)_\infty}{(q; q^2)_\infty (b^2 q^2; q^4)_\infty} \sum_{k=0}^{\infty} \frac{(b^2 q^2, q^4)_k q^{(2k+1)(2k+3-2s)}}{(q^2; q^2)_{2k+1}}. \quad (4.26)
$$

Remarks: The identities at (4.20), (4.22), (4.24) and (4.26) were proved by Berndt ([71, Theorem 1.2.3], [71, Theorem 1.2.4], [71, Theorem 2.2.2] and [71, Theorem 2.2.3], respectively). These identities will be used to prove some relations between third order mock theta functions in Chapter 18.

4.16 (i) (Lovejoy [186, Eq. (1.11)]) Derive from the $h = 2$ case of (4.2), (with q replaced with q^2) the identity

$$
\sum_{k=0}^{\infty} \frac{(a^2 q^2; q^2)_{2k}(-1)^k q^{2k^2}}{(q^4; q^4)_k (-a^2 q/x; q^2)_{2k+1}} = \frac{(a^2 q^2; q^2)_\infty (q^2; q^4)_\infty}{(-a^2 q/x; q^2)_\infty} \sum_{k=0}^{\infty} \frac{(-q/x; q^2)_k a^{2k} q^{2k}}{(q^2; q^2)_k (q^2; q^4)_k}. \quad (4.27)
$$

(ii) (Lovejoy [186, Eq. (1.12)]) Similarly derive from the $h = 2$ case of (4.2), (this time with q eventually replaced with $-q$) the identity

$$
\sum_{k=0}^{\infty} \frac{(-q^2/x; q^2)_k (aq)^{2k+1}}{(-q^2; q^2)_k (q; q)_{2k+1}}
$$

$$
= \frac{-a(-a^2 q^2/x, -q; q^2)_\infty}{(a^2 q^2; q^2)_\infty} \sum_{k=1}^{\infty} \frac{(a^2 q^2; q^2)_{k-1}(-q)^{k(k+1)/2}}{(-q; -q)_{k-1}(-a^2 q^2/x; q^2)_k}. \quad (4.28)
$$

(iii) (Lovejoy [186, Eq. (1.13)]) Derive from (4.13) the identity

$$
\frac{(-aq; q)_\infty}{(-q; q)_\infty} \sum_{k=0}^{\infty} \frac{(x; q^2)_k (aq; q)_k (-q/x)^k}{(q^2; q^2)_k}
$$

$$
= \frac{(a^2 q^2; q^2)_\infty}{(-q^2, -q/x; q^2)_\infty} \sum_{k=0}^{\infty} \frac{(-q/x; q^2)_k (aq)^{2k}}{(q; q)_{2k}(-q; q^2)_k}
$$

$$+ \frac{(a^2q^2;q^2)_\infty}{(-q,-q^2/x;q^2)_\infty} \sum_{k=0}^{\infty} \frac{(-q^2/x;q^2)_k(aq)^{2k+1}}{(q;q)_{2k+1}(-q^2;q^2)_k}. \quad (4.29)$$

(iv) (Lovejoy [186, Eq. (1.5)]) By combining the identities in parts (i), (ii) and (iii), prove the identity

$$\frac{(-aq;q)_\infty}{(-q;q)_\infty} \sum_{k=0}^{\infty} \frac{(x;q^2)_k(aq;q)_k(-q/x)^k}{(q^2;q^2)_k}$$

$$= \frac{(-a^2q/x;q^2)_\infty}{(-q/x;q^2)_\infty} \sum_{k=0}^{\infty} \frac{(a^2q^2;q^2)_{2k}(-1)^kq^{2k^2}}{(-a^2q/x;q^2)_{2k+1}(q^4;q^4)_k}$$

$$- \frac{a(-a^2q^2/x;q^2)_\infty}{(-q^2/x;q^2)_\infty} \sum_{k=1}^{\infty} \frac{(a^2q^2;q^2)_{k-1}(-q)^{k(k+1)/2}}{(-q;-q)_{k-1}(-a^2q^2/x;q^2)_k}. \quad (4.30)$$

4.17 By paralleling the methods of Q4.16, prove the identities (sketched in Lovejoy [186], although with some minor typographic errors).

$$a\sum_{k=0}^{\infty} \frac{(a^2q^2;q^2)_{2k}(-1)^kq^{2k^2+4k+1}}{(q^4;q^4)_k(-a^2q^3/x;q^2)_{2k+1}}$$

$$= \frac{(a^2q^2;q^2)_\infty(q^2;q^4)_\infty}{(-a^2q^3/x;q^2)_\infty} \sum_{k=0}^{\infty} \frac{(-q^3/x;q^2)_ka^{2k+1}q^{2k+1}}{(q^2;q^2)_k(q^2;q^4)_{k+1}}. \quad (4.31)$$

$$\sum_{k=0}^{\infty} \frac{(-q^2/x;q^2)_k(aq)^{2k}}{(-q^2;q^2)_k(q;q)_{2k}}$$

$$= \frac{(-a^2q^2/x,-q;q^2)_\infty}{(a^2q^2;q^2)_\infty} \sum_{k=0}^{\infty} \frac{(a^2q^2;q^2)_k(-q)^{k(k+1)/2}}{(-q;-q)_k(-a^2q^2/x;q^2)_{k+1}}. \quad (4.32)$$

$$\frac{(-aq;q)_\infty}{(-q;q)_\infty} \sum_{k=0}^{\infty} \frac{(x;q^2)_k(aq;q)_k(-q^2/x)^k}{(q^2;q^2)_k}$$

$$= \frac{(a^2q^2;q^2)_\infty}{(-q,-q^2/x;q^2)_\infty} \sum_{k=0}^{\infty} \frac{(-q^2/x;q^2)_k(aq)^{2k}}{(q;q)_{2k}(-q^2;q^2)_k}$$

$$+ \frac{(a^2q^2;q^2)_\infty}{(-q^2,-q^3/x;q^2)_\infty} \sum_{k=0}^{\infty} \frac{(-q^3/x;q^2)_k(aq)^{2k+1}}{(q;q)_{2k+1}(-q;q^2)_{k+1}}. \quad (4.33)$$

$$\frac{(-aq;q)_\infty}{(-q;q)_\infty} \sum_{k=0}^{\infty} \frac{(x;q^2)_k(aq;q)_k(-q^2/x)^k}{(q^2;q^2)_k}$$

$$= \frac{a(-a^2q^3/x;q^2)_\infty}{(-q^3/x;q^2)_\infty} \sum_{k=0}^{\infty} \frac{(a^2q^2;q^2)_{2k}(-1)^k q^{2k^2+4k+1}}{(-a^2q^3/x;q^2)_{2k+1}(q^4;q^4)_k}$$

$$+ \frac{(-a^2q^2/x;q^2)_\infty}{(-q^2/x;q^2)_\infty} \sum_{k=0}^{\infty} \frac{(a^2q^2;q^2)_k(-q)^{k(k+1)/2}}{(-q;-q)_k(-a^2q^2/x;q^2)_{k+1}}. \qquad (4.34)$$

Remarks : 1) Some of the identities in Q4.16 and Q4.17 will be used in Chapter 18 to prove some identities of Ramanujan for sixth order mock theta functions.

*2) The identities resulting from letting $x \to \infty$ in (4.34) and (4.30) were stated by Ramanujan in the Lost Notebook [213], and proved by Andrews and Berndt [39, Entries **1.4.6, 1.4.7**].*

Chapter 5

Other Important Basic Hypergeometric Transformations

In this chapter we derive several more basic hypergeometric identities that may be regarded as fundamental, or of historical significance. In contrast to the previous chapter, where all of the identities ultimately derived from manipulating Heine's transformation at (4.1) or its extension (4.2), in this chapter we use several different methods. This will also serve to give an indication of the variety of methods to be encountered in later chapters.

5.1 Gasper's bibasic telescoping sum

In [120], Gasper proved the following bibasic summation formula.

Theorem 5.1. *For each positive integer n,*

$$\sum_{k=0}^{n} \frac{(1 - ap^k q^k)(1 - bp^k q^{-k})}{(1 - a)(1 - b)} \frac{(a, b; p)_k (c, a/bc; q)_k}{(ap/c, bcp; p)_k (aq/b, q; q)_k} q^k$$

$$= \frac{(ap, bp; p)_n (cq, aq/bc; q)_n}{(ap/c, bcp; p)_n (aq/b, q; q)_n}. \quad (5.1)$$

Proof. If $\{s_k\}_{k=0}^{n}$ is a sequence, $t_0 = s_0$ and $t_k = s_k - s_{k-1}$ for $k \geq 1$, then

$$\sum_{k=0}^{k} t_k = s_0 + (s_1 - s_0) + \cdots + (s_n - s_{n-1}) = s_n.$$

29

If
$$s_k = \frac{(ap, bp; p)_k (cq, aq/bc; q)_n}{(ap/c, bcp; p)_n (aq/b, q; q)_k},$$

then
$$t_k = s_k - s_{k-1} = s_{k-1} \left(\frac{(1 - ap^k)(1 - bp^k)(1 - cq^k)(1 - aq^k/bc)}{(1 - ap^k/c)(1 - bcp^k)(1 - aq^k/b)(1 - q^k)} - 1 \right)$$

$$= s_{k-1} \frac{(1 - c)(1 - a/bc) \left(1 - ap^k q^k\right) \left(1 - bp^k q^{-k}\right)}{(1 - q^k) \left(1 - aq^k/b\right) \left(1 - ap^k/c\right) \left(1 - bcp^k\right)} q^k$$

$$= \frac{\left(1 - ap^k q^k\right) \left(1 - bp^k q^{-k}\right)}{(1 - a)(1 - b)} \frac{(a, b; p)_k (c, a/bc; q)_k}{(ap/c, bcp; p)_k (aq/b, q; q)_k} q^k.$$

With this formula, $t_0 = 1 = s_0$, and (5.1) follows (the third equality follows after some simple algebra). ☐

Corollary 5.1.

$$\sum_{k=0}^{n} \frac{(1 - aq^{2k})}{(1 - a)} \frac{(a, b, q^{-n}, aq^n/b; q)_k}{(aq^{n+1}, bq^{1-n}, aq/b, q; q)_k} q^k = \delta_{0,n}, \qquad (5.2)$$

where

$$\delta_{m,n} := \begin{cases} 1, & m = n, \\ 0, & otherwise, \end{cases}$$

is the Kronecker delta function.

Proof. Set $p = q$ and $c = q^{-n}$ in (5.1), making the right side vanish if $n > 0$, since $(q^{1-n}; q)_n = 0$. ☐

Note for use below the special case (let $b \to 0$ or $b \to \infty$ in (5.2))

$$\sum_{k=0}^{n} \frac{(1 - aq^{2k})}{(1 - a)} \frac{(a, q^{-n}; q)_k}{(aq^{n+1}, q; q)_k} q^{nk} = \delta_{0,n}. \qquad (5.3)$$

5.2   A summation formula for a very-well-poised $_6\phi_5$ series

To prove the summation in the next theorem, we need a particular case of the q-Pfaff–Saalschütz sum (replace a with aq/bc, b with bq^n, c with aq/b in (4.5), and finally use (2.25)):

$$\sum_{k=0}^{n} \frac{(aq/bc, aq^n, q^{-n}; q)_k}{(aq/b, aq/c, q; q)_k} q^k = \frac{(c, q^{1-n}/b; q)_n}{(aq/b, cq^{-n}/a; q)_n} = \frac{(c, b; q)_n}{(aq/b, aq/c; q)_n} \left(\frac{aq}{bc} \right)^n.$$

$$(5.4)$$

Theorem 5.2.

$$\sum_{k=0}^{n} \frac{1-aq^{2k}}{1-a} \frac{(a,b,c,q^{-n};q)_k}{(aq/b,aq/c,aq^{n+1},q;q)_k} \left(\frac{aq^{n+1}}{bc}\right)^k = \frac{(aq,aq/bc;q)_n}{(aq/b,aq/c;q)_n}. \quad (5.5)$$

Proof.

$$\sum_{k=0}^{n} \frac{1-aq^{2k}}{1-a} \frac{(a,b,c,q^{-n};q)_k}{(aq/b,aq/c,aq^{n+1},q;q)_k} \left(\frac{aq^{n+1}}{bc}\right)^k$$

$$= \sum_{k=0}^{n} \frac{1-aq^{2k}}{1-a} \frac{(a,q^{-n};q)_k}{(aq^{n+1},q;q)_k} q^{nk} \sum_{j=0}^{k} \frac{(aq/bc,aq^k,q^{-k};q)_j}{(aq/b,aq/c,q;q)_j} q^j \quad \text{(by (5.4))}$$

$$= \sum_{j=0}^{n} \frac{(aq/bc,a;q)_j (-1)^j q^{j(j+1)/2}}{(aq/b,aq/c,q;q)_j} \sum_{k=j}^{n} \frac{1-aq^{2k}}{1-a} \frac{(aq^j,q^{-n};q)_k q^{k(n-j)}}{(aq^{n+1};q)_k (q;q)_{k-j}}$$

$$= \sum_{j=0}^{n} \frac{(aq/bc,a;q)_j (-1)^j q^{j(j+1)/2}}{(aq/b,aq/c,q;q)_j}$$

$$\times \sum_{k=0}^{n-j} \frac{1-aq^{2k+2j}}{1-a} \frac{(aq^j,q^{-n};q)_{k+j} q^{(k+j)(n-j)}}{(aq^{n+1};q)_{k+j}(q;q)_k}$$

$$= \sum_{j=0}^{n} \frac{(aq/bc,q^{-n};q)_j (a;q)_{2j} (-1)^j q^{j(j+1)/2+j(n-j)}}{(aq/b,aq/c,aq^{n+1},q;q)_j} \frac{1-aq^{2j}}{1-a}$$

$$\times \sum_{k=0}^{n-j} \frac{1-aq^{2j}q^{2k}}{1-aq^{2j}} \frac{(aq^{2j},q^{-(n-j)};q)_k q^{(n-j)k}}{(aq^{2j}q^{n-j+1},q;q)_k}$$

$$= \frac{(aq/bc,q^{-n};q)_n (a;q)_{2n} (-1)^n q^{n(n+1)/2}}{(aq/b,aq/c,aq^{n+1},q;q)_n} \frac{1-aq^{2n}}{1-a}.$$

The result now follows after using the second formula at (2.12) (with $k = n$), and simplifying. The first equality follows by (5.4), the second uses the second formula at (2.12), and last equality follows by (5.2). $\square$

As Gasper points out in [121, p. 64], a non-terminating extension of this sum may be derived by an appeal to the identity theorem.

Corollary 5.2. *If* $|aq/bcd| < 1$, *then*

$$\sum_{k=0}^{\infty} \frac{1-aq^{2k}}{1-a} \frac{(a,b,c,d;q)_k}{(aq/b,aq/c,aq/d,q;q)_k} \left(\frac{aq}{bcd}\right)^k$$

$$= \frac{(aq,aq/bc,aq/bd,aq/cd;q)_\infty}{(aq/b,aq/c,aq/d,aq/bcd;q)_\infty}. \quad (5.6)$$

Proof. We will prove the equivalent form

$$\sum_{k=0}^{\infty} \frac{1-aq^{2k}}{1-a} \frac{(a,b,c,1/D;q)_k}{(aq/b,aq/c,aqD,q;q)_k} \left(\frac{aqD}{bc}\right)^k$$

$$= \frac{(aq,aq/bc,aqD/b,aqD/c;q)_\infty}{(aq/b,aq/c,aqD,aqD/bc;q)_\infty}.$$

By (5.5), the identity above holds for $D = q^n$, $n = 0,1,2,\ldots$, and by an appeal to Tannery's theorem [247], it holds for $\lim_{n\to\infty} q^n = 0$, and thus (5.6) holds by the identity theorem. □

The q-Dixon sum is a special case of this $_6\phi_5$ sum.

Corollary 5.3. *If* $|\sqrt{a}q/bc| < 1$, *then*

$$\sum_{k=0}^{\infty} \frac{(a,-\sqrt{a}q,b,c;q)_k}{(-\sqrt{a},aq/b,aq/c,q;q)_k} \left(\frac{\sqrt{a}q}{bc}\right)^k = \frac{(aq,aq/bc,\sqrt{a}q/b,\sqrt{a}q/c;q)_\infty}{(aq/b,aq/c,\sqrt{a}q,\sqrt{a}q/bc;q)_\infty}.$$

$$(5.7)$$

$$\sum_{k=0}^{n} \frac{(a,-\sqrt{a}q,b,q^{-n};q)_k}{(-\sqrt{a},aq/b,aq^{n+1},q;q)_k} \left(\frac{\sqrt{a}q^{n+1}}{b}\right)^k = \frac{(aq,\sqrt{a}q/b;q)_n}{(aq/b,\sqrt{a}q;q)_n}. \qquad (5.8)$$

Proof. To get (5.7), let $d = \sqrt{a}$ in (5.6), and (5.8) then follows from the substitution $c = q^{-n}$. □

5.3 Watson's transformation for an $_8\phi_7$ series

The method used to prove (5.5) is more generally applicable, as the authors describe in [123, pp. 40-41]. We modify their method slightly.

Theorem 5.3. *Let* $\{A_k\}_{k=0}^{\infty}$ *be an arbitrary sequence, then*

$$\sum_{k=0}^{n} \frac{(a,b,c,q^{-n};q)_k}{(aq/b,aq/c,aq^{n+1},q;q)_k} A_k$$

$$= \sum_{j=0}^{n} \frac{(aq/bc,q^{-n};q)_j (a;q)_{2j}}{(aq/b,aq/c,aq^{n+1},q;q)_j} \left(\frac{-bc}{a}\right)^j q^{-j(j+1)/2}$$

$$\times \sum_{k=0}^{n-j} \frac{(aq^{2j},q^{-(n-j)};q)_k}{(aq^{n+j+1},q;q)_k} \left(\frac{bc}{aq}\right)^k q^{-kj} A_{k+j}. \qquad (5.9)$$

Proof. The proof is similar to the proof of (5.5), and the details are left as an exercise (see question 5.1).   □

Watson's [257] transformation for a terminating $_8\phi_7$ series follows by specializing the A_k.

Corollary 5.4. *For each non-negative integer n,*

$$\sum_{k=0}^{n} \frac{1-aq^{2k}}{1-a} \frac{(a,b,c,d,e,q^{-n};q)_k}{(aq/b,aq/c,aq/d,aq/e,aq^{n+1},q;q)_k} \left(\frac{a^2q^{2+n}}{bcde}\right)^k$$

$$= \frac{(aq,aq/de;q)_n}{(aq/d,aq/e;q)_n} \sum_{j=0}^{n} \frac{(aq/bc,d,e,q^{-n};q)_j}{(aq/b,aq/c,deq^{-n}/a,q;q)_j} q^j. \quad (5.10)$$

Proof. In (5.9), set

$$A_k = \frac{1-aq^{2k}}{1-a} \frac{(d,e;q)_k}{(aq/d,aq/e;q)_k} \left(\frac{a^2q^{2+n}}{bcde}\right)^k,$$

so that the left side becomes the left side of (5.10), and the inner sum on the right side of (5.9) becomes

$$\frac{1-aq^{2j}}{1-a} \frac{(d,e;q)_j}{(aq/d,aq/e;q)_j} \left(\frac{a^2q^{2+n}}{bcde}\right)^j$$

$$\times \sum_{k=0}^{n-j} \frac{1-aq^{2j+2k}}{1-aq^{2j}} \frac{(aq^{2j},dq^j,eq^j,q^{-(n-j)};q)_k}{(aq^{n+j+1},aq^{j+1}/d,aq^{j+1}/e,q;q)_k} \left(\frac{aq^{n-j+1}}{de}\right)^k$$

$$= \frac{1-aq^{2j}}{1-a} \frac{(d,e;q)_j}{(aq/d,aq/e;q)_j} \left(\frac{a^2q^{2+n}}{bcde}\right)^j \frac{(aq^{2j+1},aq/de;q)_{n-j}}{(aq^{j+1}/d,aq^{j+1}/e;q)_{n-j}},$$

where the equality follows from (5.5) (after replacing a, b, c and n with, respectively, aq^{2j}, dq^j, eq^j and $n-j$). After some elementary q-product manipulations and simple algebra, the right side of (5.9) becomes the right side of (5.10).   □

Jackson's $_8\phi_7$ summation formula is now a simple consequence of (5.10).

Corollary 5.5. *If n is a non-negative integer and $a^2q^{n+1} = bcde$, then*

$$\sum_{k=0}^{n} \frac{1-aq^{2k}}{1-a} \frac{(a,b,c,d,e,q^{-n};q)_k}{(aq/b,aq/c,aq/d,aq/e,aq^{n+1},q;q)_k} q^k$$

$$= \frac{(aq,aq/bc,aq/bd,aq/cd;q)_n}{(aq/b,aq/c,aq/d,aq/bcd;q)_n}. \quad (5.11)$$

Proof. If $a^2q^{n+1} = bcde$, then the left side of (5.10) becomes the left side of (5.11), and the $_4\phi_3$ on the right side of (5.10) simplifies to

$$\sum_{j=0}^{n} \frac{(d,e,q^{-n};q)_j}{(aq/b,aq/c,q;q)_j} q^j = \frac{(aq/bd,aq/be;q)_n}{(aq,aq/bde;q)_n},$$

where the equality follows from the q-Pfaff–Saalschütz sum (4.2). Since the left side of (5.11) is invariant under permutations of $\{b,c,d,e\}$, the result follows.   □

A transformation formula relating two $_8\phi_7$ series is now also easy to prove.

Corollary 5.6. *If* $|a^2q^2/(bcdef)|, |aq/ef| < 1$ *and* $\lambda = qa^2/bcd$, *then*

$$\sum_{k=0}^{\infty} \frac{1-aq^{2k}}{1-a} \frac{(a,b,c,d,e,f;q)_k}{(aq/b,aq/c,aq/d,aq/e,aq/f,q;q)_k} \left(\frac{a^2q^2}{bcdef}\right)^k$$

$$= \frac{(aq,aq/ef,\lambda q/e,\lambda q/f;q)_\infty}{(\lambda q,\lambda q/ef,aq/e,aq/f;q)_\infty}$$

$$\times \sum_{k=0}^{\infty} \frac{1-\lambda q^{2k}}{1-\lambda} \frac{(\lambda,\lambda b/a,\lambda c/a,\lambda d/a,e,f;q)_k}{(aq/b,aq/c,aq/d,\lambda q/e,\lambda q/f,q;q)_k} \left(\frac{aq}{ef}\right)^k. \quad (5.12)$$

Proof. We first prove a terminating version of (5.12) (which follows from setting $f = q^{-n}$ in that identity). For $n = 0, 1, 2, \ldots$,

$$\sum_{k=0}^{n} \frac{1-aq^{2k}}{1-a} \frac{(a,b,c,d,e,q^{-n};q)_k}{(aq/b,aq/c,aq/d,aq/e,aq^{n+1},q;q)_k} \left(\frac{a^2q^{2+n}}{bcde}\right)^k$$

$$= \frac{(aq,\lambda q/e;q)_n}{(\lambda q,aq/e;q)_n}$$

$$\times \sum_{k=0}^{n} \frac{1-\lambda q^{2k}}{1-\lambda} \frac{(\lambda,\lambda b/a,\lambda c/a,\lambda d/a,e,q^{-n};q)_k}{(aq/b,aq/c,aq/d,\lambda q/e,\lambda q^{n+1},q;q)_k} \left(\frac{aq^{n+1}}{e}\right)^k. \quad (5.13)$$

To prove (5.13), observe that by (5.10), the series on the left side of (5.13) equals

$$\frac{(aq,aq/de;q)_n}{(aq/d,aq/e;q)_n} \sum_{j=0}^{n} \frac{(aq/bc,d,e,q^{-n};q)_j}{(aq/b,aq/c,deq^{-n}/a,q;q)_j} q^j,$$

and, after a little simplification, the series on the right side equals

$$\frac{(\lambda q,aq/de;q)_n}{(aq/d,\lambda q/e;q)_n} \sum_{j=0}^{n} \frac{(d,aq/bc,e,q^{-n};q)_j}{(aq/b,aq/c,deq^{-n}/a,q;q)_j} q^j,$$

and the identity at (5.13) is now immediate. The full proof now follows once again by an appeal to the identity theorem, as in the proof of (5.6) (the details are omitted).   □

5.4 Bailey's transformation for an $_{10}\phi_9$ series

Bailey's [54] transformation relating two $_{10}\phi_9$ series is one of the most important in the field of basic hypergeometric series, in the sense that several of the identities already encountered may be derived as special cases.

Theorem 5.4. *For each non-negative integer n,*

$$\sum_{k=0}^{n} \frac{1-aq^{2k}}{1-a} \frac{(a,b,c,d,e,f,\lambda aq^{n+1}/ef,q^{-n};q)_k}{(aq/b,aq/c,aq/d,aq/e,aq/f,efq^{-n}/\lambda,aq^{n+1},q;q)_k} q^k$$

$$= \frac{(aq,aq/ef,\lambda q/e,\lambda q/f;q)_n}{(\lambda q,\lambda q/ef,aq/e,aq/f;q)_n}$$

$$\times \sum_{k=0}^{n} \frac{1-\lambda q^{2k}}{1-\lambda} \frac{(\lambda,\lambda b/a,\lambda c/a,\lambda d/a,e,f,\lambda aq^{n+1}/ef,q^{-n};q)_k}{(aq/b,aq/c,aq/d,\lambda q/e,\lambda q/f,efq^{-n}/a,\lambda q^{n+1},q;q)_k} q^k,$$

$$(5.14)$$

where $\lambda = qa^2/bcd$.

Proof. We give the proof in [123, pp. 47–48], with slightly more details. First note that by (5.11),

$$\sum_{j=0}^{k} \frac{1-\lambda q^{2j}}{1-\lambda} \frac{(\lambda,\lambda b/a,\lambda c/a,\lambda d/a,aq^k,q^{-k};q)_j}{(aq/b,aq/c,aq/d,\lambda q^{1-k}/a,\lambda q^{k+1},q;q)_j} q^j$$

$$= \frac{(b,c,d,\lambda q;q)_k}{(aq/b,aq/c,aq/d,a/\lambda;q)_k}, \quad (5.15)$$

and so the left side of (5.14) may be written as

$$\sum_{k=0}^{n} \frac{1-aq^{2k}}{1-a} \frac{(a,e,f,\lambda aq^{n+1}/ef,q^{-n},a/\lambda;q)_k}{(aq/e,aq/f,efq^{-n}/\lambda,aq^{n+1},q,\lambda q;q)_k} q^k$$

$$\times \sum_{j=0}^{k} \frac{1-\lambda q^{2j}}{1-\lambda} \frac{(\lambda,\lambda b/a,\lambda c/a,\lambda d/a,aq^k,q^{-k};q)_j}{(aq/b,aq/c,aq/d,\lambda q^{1-k}/a,\lambda q^{k+1},q;q)_j} q^j$$

$$= \sum_{j=0}^{n} \frac{1-\lambda q^{2j}}{1-\lambda} \frac{(\lambda,\lambda b/a,\lambda c/a,\lambda d/a,a;q)_j}{(aq/b,aq/c,aq/d,\lambda q,q;q)_j} \left(\frac{a}{\lambda}\right)^j$$

$$\times \sum_{k=j}^{n} \frac{1 - aq^{2k}}{1 - a} \frac{(aq^j, e, f, \lambda aq^{n+1}/ef, q^{-n}; q)_k (a/\lambda; q)_{k-j}}{(aq/e, aq/f, efq^{-n}/\lambda, aq^{n+1}, \lambda q^{j+1}; q)_k (q; q)_{k-j}} q^k$$

$$= \sum_{j=0}^{n} \frac{1 - \lambda q^{2j}}{1 - \lambda} \frac{(\lambda, \lambda b/a, \lambda c/a, \lambda d/a, a; q)_j}{(aq/b, aq/c, aq/d, \lambda q, q; q)_j} \left(\frac{a}{\lambda}\right)^j$$

$$\times \sum_{k=0}^{n-j} \frac{1 - aq^{2k+2j}}{1 - a} \frac{(aq^j, e, f, \lambda aq^{n+1}/ef, q^{-n}; q)_{j+k} (a/\lambda; q)_k}{(aq/e, aq/f, efq^{-n}/\lambda, aq^{n+1}, \lambda q^{j+1}; q)_{j+k} (q; q)_k} q^{j+k}$$

$$= \sum_{j=0}^{n} \frac{1 - \lambda q^{2j}}{1 - \lambda} \frac{(\lambda, \lambda b/a, \lambda c/a, \lambda d/a, e, f, \lambda aq^{n+1}/ef, q^{-n}; q)_j}{(aq/b, aq/c, aq/d, aq/e, aq/f, efq^{-n}/\lambda, aq^{n+1}, q; q)_j} \left(\frac{aq}{\lambda}\right)^j$$

$$\frac{(aq; q)_{2j}}{(\lambda q; q)_{2j}} \sum_{k=0}^{n-j} \frac{(1 - aq^{2k+2j})(aq^{2j}, eq^j, fq^j, \lambda aq^{n+j+1}/ef, a/\lambda, q^{-(n-j)}; q)_k q^k}{(1 - aq^{2j})(aq^{j+1}/e, aq^{j+1}/f, efq^{-n+j}/\lambda, aq^{n+j+1}, \lambda q^{2j+1}, q; q)_k}.$$

The result follows after once again using (5.11) to sum the inner series, and then simplifying. □

Remark: Observe that letting $n \to \infty$ in (5.14) recovers (5.12), while letting b, c or $d \to \infty$ recovers (5.10).

Bailey [55] iterated (5.14) and then used the resulting identity to produce a new transformation for the $_8\phi_7$ series on the left side of (5.12).

Corollary 5.7. *For each non-negative integer n,*

$$\sum_{k=0}^{n} \frac{1 - aq^{2k}}{1 - a} \frac{(a, b, c, d, e, f, \lambda aq^{n+1}/ef, q^{-n}; q)_k}{(aq/b, aq/c, aq/d, aq/e, aq/f, efq^{-n}/\lambda, aq^{n+1}, q; q)_k} q^k$$

$$= \frac{(aq, aq/de, aq/df, aq/ef; q)_n}{(aq/d, aq/e, aq/f, aq/def; q)_n} \times$$

$$\sum_{k=0}^{n} \frac{1 - \mu q^{2k}}{1 - \mu} \frac{(\mu, \mu b/a, \mu c/a, d, e, f, aq/bc, q^{-n}; q)_k q^k}{(aq/b, aq/c, efq^{-n}/a, dfq^{-n}/a, deq^{-n}/a, bc\mu/a, def/a, q; q)_k},$$

$$\tag{5.16}$$

where $\lambda = qa^2/bcd$ and $\mu = defq^{-n-1}/a$.

Proof. Apply (5.14) again, with the replacements $a \to \lambda$, $b \to \lambda b/a$, $c \to \lambda c/a$, $d \to \lambda aq^{n+1}/ef$, $e \to e$ and $f \to f$, so that $\lambda \to defq^{-n-1}/a =: \mu$, and simplify. □

Bailey wished to let $n \to \infty$ to derive a new transformation for the series on the left side of (5.12), but as will be seen in the proof of the next corollary, it was necessary to treat the right side of (5.16) more carefully. We state this result not just for its own sake, but also to illustrate the method Bailey used to deal with the difficulties involved.

Corollary 5.8. *If* $|a^2q^2/(bcdef)| < 1$, *then*

$$\sum_{k=0}^{\infty} \frac{1 - aq^{2k}}{1 - a} \frac{(a, b, c, d, e, f; q)_k}{(aq/b, aq/c, aq/d, aq/e, aq/f, q; q)_k} \left(\frac{a^2q^2}{bcdef} \right)^k$$

$$= \frac{(aq, aq/de, aq/df, aq/ef; q)_\infty}{(aq/d, aq/e, aq/f, aq/def; q)_\infty} \sum_{k=0}^{\infty} \frac{(aq/bc, d, e, f; q)_k}{(aq/b, aq/c, def/a, q; q)_k} q^k$$

$$+ \frac{(aq, aq/bc, d, e, f, a^2q^2/bdef, a^2q^2/cdef; q)_\infty}{(aq/b, aq/c, aq/d, aq/e, aq/f, a^2q^2/bcdef, def/aq; q)_\infty}$$

$$\times \sum_{k=0}^{\infty} \frac{(aq/de, aq/df, aq/ef, a^2q^2/bcdef; q)_k}{(a^2q^2/bdef, a^2q^2/cdef, aq^2/def, q; q)_k} q^k. \quad (5.17)$$

Sketch of Proof. As previously noted, letting $n \to \infty$ on the left side of (5.16) gives the left side of (5.17). The difficulty on the right side is that, as Bailey notes, for large n terms near both ends of the series on the right are large compared with the terms in the middle. To deal with the right side, Bailey assumed that n was odd ($n = 2m + 1$), and divided the series on the right side of (5.16) into two series:

$$\sum_{k=0}^{2m+1} f_k = \sum_{k=0}^{m} f_k + \sum_{k=m+1}^{2m+1} f_k = \sum_{k=0}^{m} f_k + \sum_{k=0}^{m} f_{2m+1-k}, \quad (5.18)$$

where f_k denotes the k-th term in the series on the right side of (5.16). The limit as n (or m) $\to \infty$ can now be taken term-wise, and one can check that since $\mu = defq^{-2m-2}/a$ when $n = 2m + 1$,

$$\lim_{m \to \infty} \frac{1 - \mu q^{2k}}{1 - \mu} \frac{(\mu, \mu b/a, \mu c/a, q^{-2m-1}; q)_k}{(efq^{-2m-1}/a, dfq^{-2m-1}/a, deq^{-2m-1}/a, bc\mu/a; q)_k} = 1$$

so that the first sum in (5.18) tends to the first product-series combination on the right side of (5.17).

The limit can also be taken termwise in the second series, and the only complication in finding a limit of the form $\lim_{m \to \infty} (x; q)_{2m+1-k}$ arises if $x = yq^{-2m-1}$ for some constant y. In this case use (2.12) followed by (2.23), simplify before letting $m \to \infty$, and the second product-series combination on the right side of (5.17) results.   □

Bailey [56, 58] elsewhere developed a transformation formula relating three $_8\phi_7$ series, and another relating four $_{10}\phi_9$ series (see also [123, pp. 53–58]).

Exercises

5.1 Prove the identity at (5.9).

5.2 By Bailey's transformation (5.14), the identity

$$\sum_{k=0}^{n} \frac{1-aq^{2k}}{1-a} \frac{(a,b,c,d,e,f,\lambda agq/ef,1/g;q)_k}{(aq/b,aq/c,aq/d,aq/e,aq/f,ef/\lambda g,agq,q;q)_k} q^k$$

$$= \frac{(aq,aq/ef,agq/e,agq/f,\lambda q/e,\lambda q/f,\lambda gq,\lambda gq/ef;q)_\infty}{(\lambda q,\lambda q/ef,\lambda gq/e,\lambda gq/f,aq/e,aq/f,agq,agq/ef;q)_\infty}$$

$$\times \sum_{k=0}^{n} \frac{1-\lambda q^{2k}}{1-\lambda} \frac{(\lambda,\lambda b/a,\lambda c/a,\lambda d/a,e,f,\lambda agq/ef,1/g;q)_k}{(aq/b,aq/c,aq/d,\lambda q/e,\lambda q/f,ef/ag,\lambda gq,q;q)_k} q^k$$

holds for $g = q^n, n = 0, 1, 2, \ldots$ (check!), and it holds for $g = 0 = \lim_{n\to\infty} q^n$ (the value $g = 0$ giving the identity at (5.12)). Explain why it is not possible to conclude by the identity theorem that the identity holds for arbitrary values of g in some open neighborhood of zero.

5.3 Use Watson's [257] transformation (5.10) in conjunction with the Jacobi triple product identity (6.1) to reproduce his proof of the famous Rogers–Ramanujan identities:

$$\sum_{n=0}^{\infty} \frac{q^{n^2}}{(q;q)_n} = \frac{(q^2,q^3,q^5;q)_\infty}{(q;q)_\infty},$$

$$\sum_{n=0}^{\infty} \frac{q^{n^2+n}}{(q;q)_n} = \frac{(q,q^4,q^5;q)_\infty}{(q;q)_\infty}.$$

5.4 Prove that if n is a non-negative integer, then

$$\sum_{k=0}^{n} \frac{1-aq^{2k}}{1-a} \frac{(a,q^{-n};q)_k}{(aq^{n+1},q;q)_k} q^{nk} = \delta_{n,0}.$$

5.5 (i) By specializing the parameters in (5.13) and employing (5.8), or otherwise, prove that

$$
\sum_{k=0}^{n} \frac{1 - \dfrac{a^2 q^{2k+n+1}}{d^2}}{1 - \dfrac{a^2 q^{n+1}}{d^2}} \frac{\left(\dfrac{a^2 q^{n+1}}{d^2}, \dfrac{-aq}{d^2}, \dfrac{-aq^{n+1}}{d}, \dfrac{aq^{n+1}}{d}, -a, q^{-n}; q\right)_k \left(-q^{n+1}\right)^k}{\left(-aq^{n+1}, \dfrac{-aq}{d}, \dfrac{aq}{d}, \dfrac{-aq^{2+n}}{d^2}, \dfrac{a^2 q^{2n+2}}{d^2}, q; q\right)_k}
$$

$$
= \frac{(-aq, -aq^2/d^2, -q; q)_n (a^2 q^3/d^2; q^2)_n}{(-aq^3/d^2, -aq^2; q^2)_n (a^2 q^2/d^2; q)_n}.
$$

(ii) Hence prove

$$
\sum_{k=0}^{\infty} \frac{(-a, -aq/d^2; q)_k q^{k(k+1)/2}}{(a^2 q^2/d^2; q^2)_k (q; q)_k} = \frac{(-aq, -aq^2/d^2, -q; q)_\infty (a^2 q^3/d^2; q^2)_\infty}{(-aq^3/d^2, -aq^2; q^2)_\infty (a^2 q^2/d^2; q)_\infty}.
$$

(iii) Finally, use the identity theorem to show

$$
\sum_{k=0}^{\infty} \frac{1 - \dfrac{a^2 fq^{2k+1}}{d^2}}{1 - \dfrac{a^2 fq}{d^2}} \frac{\left(\dfrac{a^2 fq}{d^2}, \dfrac{-aq}{d^2}, \dfrac{-afq}{d}, \dfrac{afq}{d}, -a, \dfrac{1}{f}; q\right)_k \left(-fq\right)^k}{\left(-afq, \dfrac{-aq}{d}, \dfrac{aq}{d}, \dfrac{-afq^2}{d^2}, \dfrac{a^2 f^2 q^2}{d^2}, q; q\right)_k}
$$

$$
= \frac{(-q, a^2 q^2 f/d^2; q)_\infty (-aq, -aq^2/d^2, -aq^3 f^2/d^2, -aq^2 f^2; q^2)_\infty}{(-aqf, -aq^2 f/d^2, -qf; q)_\infty (a^2 q^3 f^2/d^2, a^2 q^2/d^2; q^2)_\infty}.
$$

Remark: This identity is equivalent to the q-analogue of Watson's $_3F_2$ sum (5.19) (see, for example, [123, p. 355, (II.16)]).

5.6 Instead of the method used in Q5.5, specialize the parameters in (5.12) and use (5.7) to prove the q-analogue of Watson's $_3F_2$ sum directly (see [122]):

$$
\sum_{k=0}^{\infty} \frac{1 - \lambda q^{2k}}{1 - \lambda} \frac{(\lambda, a, b, c, -c, \lambda q/c^2; q)_k}{(\lambda q/a, \lambda q/b, \lambda q/c, -\lambda q/c, c^2, q; q)_k} \left(-\frac{\lambda q}{ab}\right)^k
$$

$$
= \frac{(\lambda q, c^2/\lambda; q)_\infty (aq, bq, c^2 q/a, c^2 q/b; q^2)_\infty}{(\lambda q/a, \lambda q/b; q)_\infty (q, abq, c^2 q, c^2 q/ab; q^2)_\infty}, \quad (5.19)
$$

where $\lambda = -c(ab/q)^{1/2}$, and $|\lambda q/ab| < 1$.

5.7 By specializing the parameters in (5.12) in a different way, and again using (5.7), prove that if $|\lambda| < 1$, then

$$\sum_{k=0}^{\infty} \frac{1-\lambda q^{2k}}{1-\lambda} \frac{(\lambda, -\lambda, q/e, q/f, e, f; q)_k}{(-aq/e, -aq/f, -q, \lambda q/e, \lambda q/f, q; q)_k}(-\lambda)^k$$
$$= \frac{(\lambda, \lambda q; q)_\infty(-aq, -\lambda^2 q/a, -aq^2/e^2, -aq^2/f^2; q^2)_\infty}{(-aq/e, -aq/f, \lambda q/e, \lambda q/f; q)_\infty},$$

where $\lambda = -aq/fe$.

Remark: This identity is equivalent to the q-analogue of Whipple's $_3F_2$ sum (see, for example, [123, p. 355, (II.18)] or [122]):

$$\sum_{k=0}^{\infty} \frac{1-(-c)q^{2k}}{1-(-c)} \frac{(-c, a, q/a, c, -d, -q/d; q)_k}{(-cq/a, -ac, -q, cq/d, cd, q; q)_k}c^k$$
$$= \frac{(-c, -cq; q)_\infty(acd, acq/d, cdq/a, cq^2/ad; q^2)_\infty}{(cd, cq/d, -ac, -cq/a; q)_\infty}. \quad (5.20)$$

5.8 (i) By using the identity in Q5.7, or otherwise, prove that if n is a non-negative integer and $|q/ge| < 1$, then

$$\sum_{k=0}^{\infty} \frac{1-q^{2k+1}/ge}{1-q/ge} \frac{(q/ge, -q/ge, e, q/e, q^{1+n}/g, gq^{-n}; q)_k}{(q^{1-n}/e, q/g, -q, q^2/ge^2, q^{2+n}/eg^2, q; q)_k}\left(\frac{-q}{ge}\right)^k$$

$$= \begin{cases} 0, & n \text{ odd}, \\ \dfrac{(q/eg, q^2/eg; q)_\infty(q, q^2/e^2, q^2/g^2, q^3/e^2g^2; q^2)_\infty}{(q/e, q/g, q^2/e^2g, q^2/eg^2; q)_\infty} & \\ \quad \times \dfrac{(q, e^2; q^2)_{n/2}(q^2/eg^2; q)_n}{(q^2/g^2, q^3/e^2g^2; q^2)_{n/2}(e; q)_n}, & n \text{ even.} \end{cases}$$

(ii) Hence show that

$$\sum_{k=0}^{n} \frac{1-q^{2k+1}/e}{1-q/e} \frac{(q/e, -q/e, e, q/e, q^{1+n}, q^{-n}; q)_k}{(q^{1-n}/e, q, -q, q^2/e^2, q^{2+n}/e, q; q)_k}\left(\frac{-q}{e}\right)^k$$

$$= \begin{cases} 0, & n \text{ odd}, \\ \dfrac{(q, e^2; q^2)_{n/2}(q^2/e; q)_n}{(q^2, q^3/e^2; q^2)_{n/2}(e; q)_n}, & n \text{ even.} \end{cases}$$

5.9 Prove that (5.10) follows from (5.17).

5.10 Prove, for each integer $n \geq 0$, that

$$\sum_{k=0}^{n} \frac{1-aq^{2k}}{1-a} \frac{(a, b, c, a/bc; q)_k}{(aq/c, bcq, aq/b, q; q)_k}q^k = \frac{(aq, bq, cq, aq/bc; q)_n}{(aq/c, bcq, aq/b, q; q)_n}. \quad (5.21)$$

5.11 ([123, p. 51]) (i) By starting with (5.17) or otherwise, prove that

$$\frac{(aq, aq/be, aq/bf, aq/ef; q)_\infty}{(aq/b, aq/e, aq/f, aq/bef; q)_\infty}$$

$$= \frac{(aq, aq/de, aq/df, aq/ef; q)_\infty}{(aq/d, aq/e, aq/f, aq/def; q)_\infty} \sum_{k=0}^{\infty} \frac{(d/b, e, f; q)_k}{(aq/b, def/a, q; q)_k} q^k$$

$$+ \frac{(aq, d/b, e, f, aq/ef, a^2q^2/bdef; q)_\infty}{(aq/b, aq/d, aq/e, aq/f, aq/bef, def/aq; q)_\infty}$$

$$\times \sum_{k=0}^{\infty} \frac{(aq/de, aq/df, aq/bef; q)_k}{(a^2q^2/bdef, aq^2/def, q; q)_k} q^k.$$

(ii) Hence prove the follow non-terminating extension of the q-Pfaff–Saalschütz sum (see also Sears [236]):

$$\sum_{k=0}^{\infty} \frac{(a, b, c; q)_k}{(e, abcq/e, q; q)_k} q^k = \frac{(e/a, e/b, e/c, e/abc; q)_\infty}{(e, e/ab, e/ac, e/bc; q)_\infty}$$

$$- \frac{(a, b, c, e/abc, e^2/abc; q)_\infty}{(e, e/ab, e/ac, e/bc, abc/e; q)_\infty} \times \sum_{k=0}^{\infty} \frac{(e/ab, e/ac, e/bc; q)_k}{(e^2/abc, eq/abc, q; q)_k} q^k.$$

5.12 Prove that

$$\sum_{k=0}^{\infty} \frac{(c, -c, d, q/d; q)_k}{(e, c^2q/e, -q, q; q)_k} q^k = 2 \frac{(-q, -q; q)_\infty (de, eq/d, c^2dq/e, c^2q^2/de; q^2)_\infty}{(-d, -q/d, e, c^2q/e; q)_\infty}$$

$$- \frac{(d, q/d, -e, -c^2q/e; q)_\infty}{(-d, -q/d, e, c^2q/e; q)_\infty} \sum_{k=0}^{\infty} \frac{(c, -c, -d, -q/d; q)_k}{(-e, -c^2q/e, -q, q; q)_k} q^k.$$

5.13 By starting with (5.17) or otherwise, prove that

$$\frac{(a^2q^2, -aq^2/d^2, -aq^2/f^2; q^2)_\infty}{(-aq^2, -aq^2/d^2f^2, a^2q^2/d^2; q^2)_\infty}$$

$$= \frac{(aq, -aq/f^2; q)_\infty}{(aq/d, -aq/df^2; q)_\infty} \sum_{k=0}^{\infty} \frac{(d, q/d, f, -f; q)_k}{(-aq/d, -df^2/a, -q, q; q)_k} q^k$$

$$+ \frac{(aq, d, q/d, aq^2/df^2; q)_\infty (f^2, a^2q^3/d^2f^2; q^2)_\infty}{(-q, -aq^2/d^2f^2, -df^2/aq; q)_\infty (a^2q^2/d^2; q^2)_\infty}$$

$$\times \sum_{k=0}^{\infty} \frac{(aq/df, -aq/df, -aq/f^2, -aq^2/d^2f^2; q)_k}{(aq^2/df^2, -aq^2/df^2, a^2q^2/d^2f^2, q; q)_k} q^k. \quad (5.22)$$

5.14 By specializing the parameters in (5.22), or otherwise, prove
(i) the terminating q-analogue of Watson's $_3F_2$ sum (Andrews [19]):

$$\sum_{k=0}^{n} \frac{(c, -c, a^2 q^{n+1}, q^{-n}; q)_k}{(aq, -aq, c^2, q; q)_k} q^k = \begin{cases} 0, & n \text{ odd}, \\ \dfrac{(q, a^2 q^2/c^2; q^2)_{n/2} c^n}{(c^2 q, a^2 q^2; q^2)_{n/2}}, & n \text{ even}. \end{cases}$$

(ii) the terminating q-analogue of Whipple's $_3F_2$ sum (Andrews [19]):

$$\sum_{k=0}^{n} \frac{(f, -f, q^{n+1}, q^{-n}; q)_k}{(e, f^2 q/e, -q, q; q)_k} q^k = \frac{(q/e; q)_n (eq^{1-n}/f^2; q^2)_n f^{2n}}{(f^2 q/e; q)_n (eq^{1-n}; q^2)_n}.$$

5.15 Prove the Sears $_4\phi_3$ transformation:

$$\sum_{k=0}^{n} \frac{(a, b, c, q^{-n}; q)_k q^k}{(d, e, f, q; q)_k} = \frac{(e/c, de/ab; q)_n}{(e, de/abc; q)_n} \sum_{k=0}^{n} \frac{(c, d/a, d/b, q^{-n}; q)_k q^k}{(d, de/ab, df/ab, q; q)_k}, \quad (5.23)$$

for $n \geq 0$ and $def = abcq^{1-n}$.

The identities in the next two questions will be used in the proofs of two of Warnaar's WP-Bailey chains in Chapter 12.

5.16 (Warnaar [256, Lemma 2.2]) By setting $a = dq$ and $c = d^2$ in the identity in Q15, so that $d/a = 1/q$ (causing the series on the right to terminate after $k = 1$), prove that for each non-negative integer n,

$$\sum_{k=0}^{n} \frac{(dq, b, d^2, q^{-n}; q)_k q^k}{(d, e, bd^2 q^{2-n}/e, q; q)_k} = \frac{1 + dq^n/b}{1 + d/b} \frac{(e/d^2 q, e/bq; q)_n}{(e, e/bd^2 q; q)_n}, \quad (5.24)$$

when $e = -bdq$ or $e = d^2 q/b$.

5.17 (Warnaar [256, Lemma 2.3]) By combining the transformation at (5.10) with (5.24), prove that for each non-negative integer n,

$$\sum_{k=0}^{n} \frac{1 - aq^{2k}}{1 - a} \frac{(a, b, aq^n/b^{1/2}, -aq^n/b^{1/2}, -q^{-n}, q^{-n}; q)_k}{(aq/b, b^{1/2} q^{1-n}, -b^{1/2} q^{1-n}, -aq^{n+1}, aq^{n+1}, q; q)_k} q^{2k}$$

$$= \frac{(-a/b; q)_{2n}}{(-aq; q)_{2n}} \frac{(a^2 q^2, b; q^2)_n}{(1/b, a^2 q^2/b^2; q^2)_n} \left(\frac{q}{b}\right)^n. \quad (5.25)$$

5.18 (i) By letting $n \to \infty$ in (5.10) derive the transformation

$$\sum_{k=0}^{\infty} \frac{1 - aq^{2k}}{1 - a} \frac{(a, b, c, d, e; q)_k}{(aq/b, aq/c, aq/d, aq/e, q; q)_k} \left(\frac{-a^2}{bcde}\right)^k q^{k(k+3)/2}$$

$$= \frac{(aq, aq/de; q)_\infty}{(aq/d, aq/e; q)_\infty} \sum_{k=0}^{\infty} \frac{(aq/bc, d, e; q)_k}{(aq/b, aq/c, q; q)_k} \left(\frac{aq}{de}\right)^k. \quad (5.26)$$

(ii) By redefining the parameters a, b and c in the identity above, prove that if $|fg/ade|, |q| < 1$, then

$$\sum_{k=0}^{\infty} \frac{(a, d, e; q)_k}{(f, g, q; q)_k} \left(\frac{fg}{ade}\right)^k = \frac{(fg/ad, fg/ae; q)_\infty}{(fg/a, fg/ade; q)_\infty}$$

$$\times \sum_{k=0}^{\infty} \frac{1 - fgq^{2k}/aq}{1 - fg/aq} \frac{(fg/aq, f/a, g/a, d, e; q)_k}{(f, g, fg/ad, fg/ae, q; q)_k} \left(\frac{-fg}{de}\right)^k q^{k(k-1)/2}. \quad (5.27)$$

Remark: This result will be used in the proof of Theorem 13.10.

Chapter 6

The Jacobi Triple Product Identity

The triple product identity, stated by Jacobi [153], but first proved by Gauss [125], is another fundamental identity in the field of q-series. As will be seen later, it has applications to proving identities of the Rogers–Ramanujan type, representing theta functions as infinite products and proving identities about integer partitions. We will give two proofs, and outline another in the exercises.

Theorem 6.1. *For $|q| < 1$ and $z \neq 0$,*

$$\sum_{n=-\infty}^{\infty} (-z)^n q^{n^2} = (zq, q/z, q^2; q^2)_\infty. \tag{6.1}$$

Remark: We will call a quantity such as $(zq, q/z, q^2; q^2)_\infty$ a "triple product" in what follows.

Proof 1. The first proof presented here was found independently by Andrews [8] and Menon [203]. By the special case of the q-binomial theorem at (3.4) (with z replaced with zq and q replaced with q^2),

$$(zq; q^2)_\infty = \sum_{n=0}^{\infty} \frac{(-zq)^n q^{n^2-n}}{(q^2; q^2)_n} = \frac{1}{(q^2; q^2)_\infty} \sum_{n=0}^{\infty} (-z)^n q^{n^2} (q^{2+2n}; q^2)_\infty$$

$$= \frac{1}{(q^2; q^2)_\infty} \sum_{n=-\infty}^{\infty} (-z)^n q^{n^2} (q^{2+2n}; q^2)_\infty$$

$$= \frac{1}{(q^2; q^2)_\infty} \sum_{n=-\infty}^{\infty} (-z)^n q^{n^2} \sum_{m=0}^{\infty} \frac{(-q^{2n+2})^m q^{m^2-m}}{(q^2; q^2)_m}$$

$$= \frac{1}{(q^2;q^2)_\infty} \sum_{m=0}^{\infty} \frac{(q/z)^m}{(q^2;q^2)_m} \sum_{n=-\infty}^{\infty} (-z)^{n+m} q^{(n+m)^2}$$

$$= \frac{1}{(q^2;q^2)_\infty} \sum_{m=0}^{\infty} \frac{(q/z)^m}{(q^2;q^2)_m} \sum_{n=-\infty}^{\infty} (-z)^n q^{n^2}$$

$$= \frac{1}{(q^2, q/z; q^2)_\infty} \sum_{n=-\infty}^{\infty} (-z)^n q^{n^2}.$$

The second equality follows since $(q^{2+2n};q^2)_\infty = 0$ for $n < 0$, the third holds by (3.4) again (this time with z replaced with q^{2+2n} and q replaced with q^2), and the last equality holds by (3.3) (with z replaced with q/z and q replaced with q^2, provided $|q/z| < 1$). Thus (6.1) holds for $|q/z| < 1$, and by analytic continuation for $z \neq 0$. □

Proof 2. As the authors in [34] point out, the following proof was known to both Gauss [124] and Cauchy [93]. By (3.5),

$$\sum_{k=0}^{2n} \begin{bmatrix} 2n \\ k \end{bmatrix}_{q^2} (-z/q^{2n-1})^k q^{k^2-k} = (z/q^{2n-1};q^2)_{2n} = (z/q^{2n-1};q^2)_n (zq;q^2)_n$$

$$\overset{(2.24)}{=} (q/z;q^2)_n (-z/q)^n q^{-n^2+n} (zq;q^2)_n.$$

After some rearrangement, this gives

$$\sum_{k=0}^{2n} \begin{bmatrix} 2n \\ k \end{bmatrix}_{q^2} (-z)^{-n+k} q^{(n-k)^2} = (q/z;q^2)_n (zq;q^2)_n$$

$$\implies \sum_{k=-n}^{n} \begin{bmatrix} 2n \\ n+k \end{bmatrix}_{q^2} (-z)^k q^{k^2} = (q/z;q^2)_n (zq;q^2)_n. \qquad (6.2)$$

The result now follows after an application of Tannery's theorem [247] and letting $n \to \infty$, since

$$\begin{bmatrix} 2n \\ n+k \end{bmatrix}_{q^2} = \frac{(q^2;q^2)_{2n}}{(q^2;q^2)_{n+k}(q^2;q^2)_{n-k}} \longrightarrow \frac{1}{(q^2;q^2)_\infty} \text{ as } n \to \infty.$$

□

An equivalent expression for the Jacobi triple product identity is the following (in (6.1), replace q with $q^{1/2}$ and z with $zq^{1/2}$):

$$\sum_{n=-\infty}^{\infty} (-z)^n q^{n(n+1)/2} = (zq, 1/z, q; q)_\infty. \qquad (6.3)$$

The following identity, commonly referred to as "Jacobi's identity", is also important.

Corollary 6.1. *If $|q| < 1$, then*

$$\sum_{n=0}^{\infty}(-1)^n(2n+1)q^{n(n+1)/2} = (q;q)_\infty^3. \tag{6.4}$$

Proof. From (6.3), it follows that

$$\sum_{n=0}^{\infty}(-z)^n q^{n(n+1)/2} + \sum_{n=-\infty}^{-1}(-z)^n q^{n(n+1)/2} = (1-1/z)(zq, q/z, q; q)_\infty.$$

Next, in the second sum on the left, replace n with $-n$ (to get $1 \le n < \infty$) and then n with $n+1$ (to get $0 \le n < \infty$), divide both sides by $1 - 1/z$ so that

$$\sum_{n=0}^{\infty}\frac{(-z)^{n+1}+(-z)^{-n}}{1-z}q^{n(n+1)/2} = (zq, q/z, q; q)_\infty.$$

Now take the limit as $z \to 1$, using L'Hospital's rule on the left side, and the result follows. $\square$

The following elementary "sectioning" identity is sometimes useful.

Corollary 6.2. *If $|q| < 1$, $z \neq 0$ and m is a positive integer, then*

$$(zq, q/z, q^2; q^2)_\infty$$
$$= \sum_{r=0}^{m-1}(-z)^r q^{r^2}\left((-(-z)^m q^{m^2+2mr}, -\frac{q^{m^2-2mr}}{(-z)^m}, q^{2m^2}; q^{2m^2}\right)_\infty. \tag{6.5}$$

Proof. By (6.1),

$$(zq, q/z, q^2; q^2)_\infty = \sum_{n=-\infty}^{\infty}(-z)^n q^{n^2} = \sum_{r=0}^{m-1}\sum_{k=-\infty}^{\infty}(-z)^{mk+r}q^{(mk+r)^2}$$

$$= \sum_{r=0}^{m-1}(-z)^r q^{r^2}\sum_{k=-\infty}^{\infty}((-z)^m q^{2mr})^k (q^{m^2})^{k^2},$$

and (6.5) follows upon once again using (6.1) on the inner sums. $\square$

6.1 Ramanujan's theta functions

Ramanujan's general theta function is defined by

$$f(a,b) := \sum_{n=-\infty}^{\infty} a^{n(n+1)/2} b^{n(n-1)/2}. \tag{6.6}$$

The Jacobi triple product identity may be stated in terms of $f(a,b)$ as

$$f(a,b) = (-a, -b, ab; ab)_\infty. \tag{6.7}$$

Ramanujan defined three special cases of this general theta function as follows:

$$\phi(q) := f(q,q) = \sum_{n=-\infty}^{\infty} q^{n^2}, \tag{6.8}$$

$$\psi(q) := f(q,q^3) = \sum_{n=-\infty}^{\infty} q^{2n^2-n} = \sum_{n=0}^{\infty} q^{n(n+1)/2}, \tag{6.9}$$

$$f(-q) := f(-q, -q^2) = \sum_{n=-\infty}^{\infty} (-1)^n q^{n(3n-1)/2}. \tag{6.10}$$

These theta functions have the following infinite product representations:

Corollary 6.3.

$$\phi(q) = (-q, -q, q^2; q^2)_\infty = \frac{(-q, q^2; q^2)_\infty}{(q, -q^2; q^2)_\infty}, \tag{6.11}$$

$$\psi(q) = \frac{(q^2; q^2)_\infty}{(q; q^2)_\infty}, \tag{6.12}$$

$$f(-q) = (q; q)_\infty. \tag{6.13}$$

Remarks: (1) The identity

$$\sum_{n=-\infty}^{\infty} (-1)^n q^{n(3n-1)/2} = (q; q)_\infty \tag{6.14}$$

is known as *Euler's pentagonal number theorem*, and is an important identity in the theory of integer partitions.

(2) If $r_k(n)$ counts the number of representations of the integer n as a sum of k squares, where order and sign matter, then

$$\sum_{n=0}^{\infty} r_k(n)q^n = \phi(q)^k. \tag{6.15}$$

(3) If $t_k(n)$ counts the number of representations of the integer n as a sum of k triangular numbers (numbers of the form $m(m+1)/2$ for $m \geq 0$), where order matters, then

$$\sum_{n=0}^{\infty} t_k(n)q^n = \psi(q)^k. \tag{6.16}$$

Ramanujan stated many elegant identities for his theta functions — see [69, Chapter 16], and we include some examples to illustrate the methods of proof.

Corollary 6.4. *For $|q| < 1$, the following hold;*

$$\phi(q) + \phi(-q) = 2\phi(q^4),$$
$$\phi(q)\psi(q^2) = \psi^2(q),$$
$$\phi^2(q) + \phi^2(-q) = 2\phi^2(q^2).$$

Proof. (i) From (6.8),

$$\phi(q) + \phi(-q) = \sum_{n=-\infty}^{\infty} q^{n^2} + \sum_{n=-\infty}^{\infty} (-1)^n q^{n^2} = 2 \sum_{\substack{n=-\infty \\ n \text{ even}}}^{\infty} q^{n^2}$$

$$= 2 \sum_{m=-\infty}^{\infty} q^{(2m)^2} = 2 \sum_{m=-\infty}^{\infty} (q^4)^{m^2} = 2\phi(q^4).$$

(ii) By (6.11) and (6.12),

$$\phi(q)\psi(q^2) = (-q, -q, q^2; q^2)_\infty \frac{(q^4; q^4)_\infty}{(q^2; q^4)_\infty} = (-q, -q, q^2; q^2)_\infty \frac{(-q^2, q^2; q^2)_\infty}{(-q, q; q^2)_\infty}$$

$$= \frac{(q^2, q^2; q^2)_\infty (-q; q)_\infty}{(q; q^2)_\infty} = \frac{(q^2, q^2; q^2)_\infty}{(q, q; q^2)_\infty} = \psi^2(q).$$

(iii) From (6.8) again,

$$\phi^2(q) + \phi^2(-q) = \left(\sum_{m=-\infty}^{\infty} q^{m^2} \right) \left(\sum_{n=-\infty}^{\infty} q^{n^2} \right)$$

$$+ \left(\sum_{m=-\infty}^{\infty} (-1)^m q^{m^2} \right) \left(\sum_{n=-\infty}^{\infty} (-1)^n q^{n^2} \right)$$

$$= \sum_{m,n=-\infty}^{\infty} q^{m^2+n^2} + \sum_{m,n=-\infty}^{\infty} (-1)^{m+n} q^{m^2+n^2} = 2 \sum_{\substack{m,n=-\infty \\ m+n \text{ even}}}^{\infty} q^{m^2+n^2}$$

$$= 2 \sum_{j,k=-\infty}^{\infty} q^{(j+k)^2+(j-k)^2} = 2 \sum_{j,k=-\infty}^{\infty} q^{2j^2+2k^2}$$

$$= 2 \left(\sum_{j=-\infty}^{\infty} q^{2j^2} \right) \left(\sum_{k=-\infty}^{\infty} q^{2k^2} \right) = 2\phi^2(q^2).$$

For the fourth equality we substituted $m = j+k$ and $n = j-k$ (so $m+n = 2j$ and $m - n = 2k$). $\qquad\qquad\square$

6.2 Schröter's formula

Schröter's identity (see [82, page 111]), which first appeared in Schröter's 1854 dissertation, gives a way of representing a product of two triple products in terms of other triple products. We will use the following identity (see exercise 6.2) in the proof:

$$(zq^{1+2m}, q^{1-2m}/z, q^2; q^2)_\infty = (-z)^{-m} q^{-m^2} (zq, q/z, q^2; q^2)_\infty \qquad (6.17)$$

holds for each $m \in \mathbb{Z}$.

Remark; Extensions of Schröter's formula to arbitrarily many triple products may be found in Cao [89] and Mc Laughlin [194].

Theorem 6.2. *Let a and b be non-zero complex numbers, $|q| < 1$, and u, v be positive integers. Then*

$$\left(-q^u a, \frac{-q^u}{a}, q^{2u}; q^{2u} \right)_\infty \left(-q^v b, \frac{-q^v}{b}, q^{2v}; q^{2v} \right)_\infty$$

$$= \sum_{j=0}^{u+v-1} q^{uj^2} a^j \left(\frac{-q^{u+v+2uj} a}{b}, -\frac{q^{u+v-2uj} b}{a}, q^{2(u+v)}; q^{2(u+v)} \right)_\infty$$

$$\times \left(-q^{(u+v+2j)uv} a^v b^u, -\frac{q^{(u+v-2j)uv}}{a^v b^u}, q^{2(u+v)uv}; q^{2(u+v)uv} \right)_\infty. \qquad (6.18)$$

Proof. After two applications of the Jacobi triple product identity,

$$
\begin{aligned}
F &:= \left(-q^u a, \frac{-q^u}{a}, q^{2u}; q^{2u} \right)_\infty \left(-q^v b, \frac{-q^v}{b}, q^{2v}; q^{2v} \right)_\infty \\
&= \sum_{m,p\in\mathbb{Z}} q^{um^2} a^m \frac{q^{vp^2}}{b^p} \\
&= \sum_{t,p\in\mathbb{Z}} a^t q^{u(t^2+2tp+p^2)} \frac{q^{vp^2} a^p}{b^p} \quad (t=m-p) \\
&= \sum_{t,p\in\mathbb{Z}} a^t q^{ut^2} q^{(u+v)p^2} \left(\frac{aq^{2ut}}{b} \right)^p \\
&= \sum_{t\in\mathbb{Z}} a^t q^{ut^2} \left(\frac{-q^{u+v+2ut}a}{b}, \frac{-q^{u+v-2ut}b}{a}, q^{2(u+v)}; q^{2(u+v)} \right)_\infty .
\end{aligned}
$$

Now set $t = (u+v)r + j$, $r \in \mathbb{Z}$, $0 \le j \le (u+v)-1$, and apply (6.17) (with q^{u+v}, ur and $-aq^{2uj}/b$ instead of q, m and z, respectively) to the triple products to get

$$
\begin{aligned}
F &= \sum_{j=0}^{u+v-1} \sum_{r\in\mathbb{Z}} q^{u((u+v)r+j)^2} a^{(u+v)r+j} \left(\frac{q^{2uj}a}{b} \right)^{-ur} (q^{u+v})^{-(ur)^2} \\
&\qquad \times \left(\frac{-q^{u+v+2uj}a}{b}, -\frac{q^{u+v-2uj}b}{a}, q^{2(u+v)}; q^{2(u+v)} \right)_\infty \\
&= \sum_{j=0}^{u+v-1} q^{uj^2} a^j \left(\frac{-q^{u+v+2uj}a}{b}, -\frac{q^{u+v-2uj}b}{a}, q^{2(u+v)}; q^{2(u+v)} \right)_\infty \\
&\qquad\qquad \times \sum_{r\in\mathbb{Z}} q^{(u+v)uvr^2} (a^v b^u q^{2uvj})^r .
\end{aligned}
$$

The result follows after one further application of the Jacobi triple product identity to the inner sum. $\qquad\square$

6.2.1   The quintuple product identity

One useful application of Schröter's identity is to give a quick proof of the well-known quintuple product identity, stated by Watson [258], and redis-covered independently by Bailey [64] and Gordon [127]. This identity may be stated as

$$(-z, -q/z, q; q)_\infty (qz^2, q/z^2; q^2)_\infty = \sum_{n=-\infty}^{\infty} (-1)^n q^{n(3n-1)/2} z^{3n} (1 + zq^n).$$
(6.19)

An equivalent representation is given by

$$(zq, q/z, q^2; q^2)_\infty (q^4 z^2, 1/z^2; q^4)_\infty = \sum_{n=-\infty}^{\infty} q^{n(3n+2)} z^{3n} (1 - zq^{2n+1}). \quad (6.20)$$

For a survey that includes a history and twenty nine proofs of this identity, the reader should consult [108]. We will prove the quintuple product identity in the following form (after replacing q with q^2, z with $-z$ in (6.19), and using the Jacobi triple product identity to sum the resulting series on the right side).

Corollary 6.5. (*Quintuple Product Identity*)

$$(z, q^2/z, q^2; q^2)_\infty (q^2 z^2, q^2/z^2; q^4)_\infty$$
$$= (-q^2 z^3, -q^4/z^3, q^6; q^6)_\infty - z(-q^4 z^3, -q^2/z^3, q^6; q^6)_\infty. \quad (6.21)$$

Proof. In (6.18), set $u = 1$, $v = 2$, $a = -z/q^3$, $b = -1/z^2$ (so that $a/b = z^3/q^3$ and $a^v b^u = (-z/q^3)^2 (-1/z^2) = -1/q^6$) to get

$$\left(z/q^2, q^4/z, q^2; q^2\right)_\infty \left(q^2/z^2, q^2 z^2, q^4; q^4\right)_\infty$$
$$= \sum_{j=0}^{2} q^{j^2} (-z/q^3)^j \left(\frac{-q^{3+2j} z^3}{q^3}, -\frac{q^{3-2j} q^3}{z^3}, q^6; q^6 \right)_\infty$$
$$\times \left(-q^{(3+2j)2} \left(\frac{-1}{q^6} \right), -q^{(3-2j)2} (-q^6), q^{12}; q^{12} \right)_\infty$$
$$= \sum_{j=0}^{2} q^{j^2 - 3j} (-z)^j \left(-q^{2j} z^3, -\frac{q^{6-2j}}{z^3}, q^6; q^6 \right)_\infty \left(q^{4j}, q^{12-4j}, q^{12}; q^{12} \right)_\infty.$$

The $j = 0$ term vanishes since $(1; q^{12})_\infty = 0$ and (6.21) follows, upon noting that $(q^4, q^8, q^{12}; q^{12})_\infty = (q^4; q^4)_\infty$ (so a factor of $(q^4; q^4)_\infty$ may be cancelled on each side) and that

$$\left(z/q^2, q^4/z, q^2; q^2 \right)_\infty = -z/q^2 \left(z, q^2/z, q^2; q^2 \right)_\infty$$

(so a factor of $-z/q^2$ may be cancelled each side also). $\qquad \square$

In Schröter's identity (6.2), the upper limit on the sum (namely, $u+v-1$) depends on the exponents on the bases in the triple products (namely, u and v in q^u and q^v). It is also possible to state an identity in which two triple products may be expanded in a sum with arbitrarily many triple product (somewhat analogous to the expansion at (6.5) for a single triple product).

Corollary 6.6. *Let a and b be non-zero complex numbers, $|q| < 1$, and u, m be positive integers. Then*

$$\left(-qa, \frac{-q}{a}, q^2; q^2\right)_\infty \left(-q^m b, \frac{-q^m}{b}, q^{2m}; q^{2m}\right)_\infty$$

$$= \sum_{j=0}^{u(1+m)-1} q^{j^2} a^j \left(\frac{-q^{1+m+2j}a}{b}, -\frac{q^{1+m-2j}b}{a}, q^{2(1+m)}; q^{2(1+m)}\right)_\infty$$

$$\times \left(-q^{(u+mu+2j)mu}(a^m b)^u, -\frac{q^{(u+mu-2j)mu}}{(a^m b)^u}, q^{2(1+m)mu^2}; q^{2(1+m)mu^2}\right)_\infty .$$

$$(6.22)$$

Proof. In Theorem (6.2), replace v with mu, and then replace q^u with q. $\square$

Remark: Note that the left side of (6.22) is independent of u. Note also the special case $m = 1$:

$$\left(-qa, \frac{-q}{a}, q^2; q^2\right)_\infty \left(-qb, \frac{-q}{b}, q^2; q^2\right)_\infty$$

$$= \sum_{j=0}^{2u-1} q^{j^2} a^j \left(\frac{-q^{2+2j}a}{b}, -\frac{q^{2-2j}b}{a}, q^4; q^4\right)_\infty$$

$$\times \left(-q^{(2u+2j)u}(ab)^u, -\frac{q^{(2u-2j)u}}{(ab)^u}, q^{4u^2}; q^{4u^2}\right)_\infty , \quad (6.23)$$

for any positive integer u.

Remark: We conclude this chapter by noting that a generalization of the Jacobi triple product has been found by Warnaar [255, Theorem 2.5] (see Q13.20 for a statement of Warnaar's result). However, we defer its proof until Chapter 13, where a variation of the method of proof used by Warnaar is used to prove another result in that chapter — see Lemma 13.8.

Exercises

6.1 The following proof of the Jacobi triple product identity is a variation of the proof given in Hardy and Wright [139, pp. 282–283].
Let $f(z) = (zq, q/z; q^2)_\infty$. Since the infinite products are absolutely convergent for $|q| < 1$ and $z \neq 0$, $f(z)$ may be expanded as a Laurent series

$$f(z) = \sum_{k=-\infty}^{\infty} A_k z^k,$$

where A_k is independent of z.
(i) Deduce that $A_{-k}(q) = A_k(q)$ for $k \geq 1$.
(ii) Show that $f(z) = -zqf(zq^2)$, and hence that $A_k = -q^{2k-1}A_{k-1}$.
Conclude that $A_k = (-1)^k q^{k^2} A_0$.
(iii) By using the special case of the q-binomial theorem at (3.4) to expand the infinite products $(q/z; q^2)_\infty$ and $(zq; q^2)_\infty$, conclude that

$$A_0 = \sum_{n=0}^{\infty} \frac{q^{2n^2}}{(q^2, q^2; q^2)_n}.$$

(iv) By considering a special case of Heine's summation formula (4.8), show that

$$\sum_{n=0}^{\infty} \frac{q^{2n^2}}{(q^2, q^2; q^2)_n} = \frac{1}{(q^2; q^2)_\infty},$$

thus concluding another proof (6.1).

6.2 For each integer m, prove that

$$(zq, q/z, q^2; q^2)_\infty = (-z)^m q^{m^2}(zq^{1+2m}, q^{1-2m}/z, q^2; q^2)_\infty.$$

6.3 Prove the identities at (6.9), (6.11), (6.12) and (6.13).

6.4 (i) Prove

$$\phi(q) - \phi(-q) = 4q\psi(q^8).$$

(ii) By employing the identities in Corollary 6.4 or otherwise, prove

$$\phi^2(q) - \phi^2(-q) = 8q\psi^2(q^4).$$

(iii) Hence prove

$$\phi^4(q) = \phi^4(-q) + 16q\psi^4(q^2),$$

and thus that
$$(-q; q^2)_\infty^8 = (q; q^2)_\infty^8 + 16q(-q^2; q^2)_\infty^8. \tag{6.24}$$

6.5 By mirroring the proof of Theorem 6.2, prove the following extension of Schröter's formula (here k is any positive integer).

$$\left(-q^u a, \frac{-q^u}{a}, q^{2u}; q^{2u}\right)_\infty \left(-q^v b, \frac{-q^v}{b}, q^{2v}; q^{2v}\right)_\infty$$

$$= \sum_{j=0}^{k(u+v)-1} q^{uj^2} a^j \left(\frac{-q^{u+v+2uj}a}{b}, -\frac{q^{u+v-2uj}b}{a}, q^{2(u+v)}; q^{2(u+v)}\right)_\infty$$

$$\times \left(-q^{(k(u+v)+2j)kuv} a^{kv} b^{ku}, -\frac{q^{(k(u+v)-2j)kuv}}{a^{kv} b^{ku}}, q^{2k^2(u+v)uv}; q^{2k^2(u+v)uv}\right)_\infty.$$

6.6 (Fine [115, p. 83]) By starting with (6.19) and proceeding similarly to the proof of (6.4), or otherwise, prove that

$$\sum_{n=-\infty}^{\infty} (6n + 1)q^{n(3n+1)/2} = (q; q)_\infty^3 (q; q^2)_\infty^2. \tag{6.25}$$

6.7 In a similar fashion, prove that

$$\sum_{n=-\infty}^{\infty} (3n + 1)q^{n(3n+2)} = (q; q)_\infty^2 \frac{(q^4; q^4)_\infty}{(q^2; q^4)_\infty}. \tag{6.26}$$

6.8 (Cooper [108, Equation (9.3)]) By replacing q with q^3 in (6.19), then specializing z, and finally manipulating the resulting infinite series on the right side, show

$$\psi(q) - 3q\psi(q^9) = (q; q)_\infty (q, q^5; q^6)_\infty.$$

6.9 (Bailey's [64] proof of the quintuple product identity). Define

$$f(x) := (1/x, qx; q)_\infty (qx^2, q/x^2; q^2)_\infty = \sum_{n=-\infty}^{\infty} c_n z^n,$$

where the infinite series is the expansion of $f(x)$ as a Laurent series.
(i) Show that $f(x) = q^2 x^3 f(xq)$, and thus that $c_n = q^{n-1} c_{n-3}$.
(ii) Iterate this relation and show for all $n \in \mathbb{Z}$ that

$$c_{3n} = q^{n(3n+1)/2} c_0,$$
$$c_{3n+1} = q^{n(3n+3)/2} c_1,$$
$$c_{3n+2} = q^{n(3n-1)/2} c_{-1}.$$

(iii) Show $f(x) = -x^{-1}f(x^{-1})$, which implies $c_0 = -c_{-1}$, $c_1 = -c_{-2}$, and thus that $c_1 = 0$.

(iv) Use the Jacobi triple product to evaluate $f(-1)$, and show that $c_0 = 1/(q;q)_\infty$, thus proving the quintuple product identity in the form

$$(1/x, qx, q; q)_\infty (qx^2, q/x^2; q^2)_\infty = \sum_{n=-\infty}^{\infty} q^{n(3n+1)/2}(x^{3n} - x^{-3n-1}).$$

6.10 By using Schröter's identity or otherwise, prove

(i) $(-zq, -q/z, q^2; q^2)_\infty (-yq, -q/y, q^2; q^2)_\infty$
        $\quad + (zq, q/z, q^2; q^2)_\infty (yq, q/y, q^2; q^2)_\infty$
        $= 2(-yzq^2, -q^2/yz, q^4; q^4)_\infty (-q^2z/y, -q^2y/z, q^4; q^4)_\infty;$ (6.27)

(ii) $(-zq, -q/z, q^2; q^2)_\infty (-yq, -q/y, q^2; q^2)_\infty$
        $\quad - (zq, q/z, q^2; q^2)_\infty (yq, q/y, q^2; q^2)_\infty$
        $= 2qz(-yzq^4, -1/yz, q^4; q^4)_\infty (-q^4z/y, -y/z, q^4; q^4)_\infty.$ (6.28)

The identities in question 6.10 are equivalent to two identities of Ramanujan in **Entry 29** *in chapter 16 of Ramanujan's second notebook (see [69, p. 45]).*

Chapter 7

Ramanujan's $_1\psi_1$ Summation Formula

Ramanujan's $_1\psi_1$ summation formula may be stated as

$$_1\psi_1(a;b;q,z) := \sum_{n=-\infty}^{\infty} \frac{(a;q)_n}{(b;q)_n} z^n = \frac{(q,b/a,az,q/az;q)_\infty}{(b,q/a,z,b/az;q)_\infty}. \qquad (7.1)$$

Note that this formula is a common extension of both the q-binomial theorem (3.1) and the Jacobi triple product identity (6.1) — see exercise 7.1. Another special case that will be used below is the following (valid for $|b/z| < 1$):

$$\sum_{n=-\infty}^{\infty} \frac{(-1)^n q^{n(n-1)/2} z^n}{(b;q)_n} = \frac{(q,z,q/z;q)_\infty}{(b,b/z;q)_\infty}. \qquad (7.2)$$

By (2.5), the bilateral series may be written as

$$\sum_{n=-\infty}^{\infty} \frac{(a;q)_n}{(b;q)_n} z^n = \sum_{n=0}^{\infty} \frac{(a;q)_n}{(b;q)_n} z^n + \sum_{n=1}^{\infty} \frac{(q/b;q)_n}{(q/a;q)_n} \left(\frac{b}{az}\right)^n, \qquad (7.3)$$

so the formula is valid provided $|z|, |b/az| < 1$.

This identity was recorded by Ramanujan in his second notebook [212, Chapter 16, Entry 17], but first appeared in print in [138, Eq. (12.12.2)], where Hardy described it as a "remarkable formula with many parameters". The first proofs were given by Hahn [137] and Jackson [152], and there are presently many proofs — see [105] for a recent survey. The following proof is essentially that of Jackson [152]. Some of the other proofs are sketched in the exercises.

We shall prove (7.1) in the following form.

Theorem 7.1. *For $|b/a| < |z| < 1$,*

$$\frac{(az, q/az; q)_\infty}{(z, b/az; q)_\infty} = \frac{(b, q/a; q)_\infty}{(q, b/a,; q)_\infty}\left(\sum_{n=0}^\infty \frac{(a; q)_n}{(b; q)_n} z^n + \sum_{n=1}^\infty \frac{(q/b; q)_n}{(q/a; q)_n}\left(\frac{b}{az}\right)^n\right).$$

$$(7.4)$$

Proof. By two applications of (3.1),

$$\frac{(az, q/az; q)_\infty}{(z, b/az; q)_\infty} = \sum_{m,n\geq 0} \frac{(a; q)_n (q/b; q)_m}{(q; q)_n (q; q)_m}\left(\frac{b}{a}\right)^m z^{n-m}$$

$$= \sum_{n-m\geq 0} \frac{(a; q)_n (q/b; q)_m}{(q; q)_n (q; q)_m}\left(\frac{b}{a}\right)^m z^{n-m} + \sum_{m-n>0} \frac{(a; q)_n (q/b; q)_m}{(q; q)_n (q; q)_m}\left(\frac{b}{a}\right)^m z^{n-m}.$$

The first sum equals

$$\sum_{t\geq 0} z^t \sum_{m\geq 0} \frac{(a; q)_{m+t}(q/b; q)_m}{(q; q)_{m+t}(q; q)_m}\left(\frac{b}{a}\right)^m$$

$$= \sum_{t\geq 0} \frac{(a, q)_t}{(q; q)_t} z^t \sum_{m\geq 0} \frac{(aq^t; q)_m(q/b; q)_m}{(q^{t+1}; q)_m(q; q)_m}\left(\frac{b}{a}\right)^m = \sum_{t\geq 0} \frac{(a, q)_t}{(q; q)_t} z^t \frac{(q/a, q^t b; q)_\infty}{(q^{t+1}, b/a; q)_\infty}$$

$$= \frac{(q/a, b; q)_\infty}{(q, b/a; q)_\infty} \sum_{t\geq 0} \frac{(a; q)_t}{(b; q)_t} z^t,$$

where we have used (4.8) with the replacements $a \to aq^t$, $b \to q/b$ and $c \to q^{t+1}$.

By similar reasoning, the second sum may be written as

$$\sum_{t\geq 1}\left(\frac{b}{az}\right)^t \sum_{n\geq 0} \frac{(a; q)_n (q/b; q)_{n+t}}{(q; q)_n (q; q)_{n+t}}\left(\frac{b}{a}\right)^n$$

$$= \sum_{t\geq 1} \frac{(q/b; q)_t}{(q; q)_t}\left(\frac{b}{az}\right)^t \sum_{n\geq 0} \frac{(a; q)_n (q^{t+1}/b; q)_n}{(q; q)_n (q^{t+1}; q)_n}\left(\frac{b}{a}\right)^n$$

$$= \sum_{t\geq 1} \frac{(q/b; q)_t}{(q; q)_t}\left(\frac{b}{az}\right)^t \frac{(q^{t+1}/a, b; q)_\infty}{(q^{t+1}, b/a; q)_\infty} = \frac{(q/a, b; q)_\infty}{(q, b/a; q)_\infty} \sum_{t\geq 1} \frac{(q/b; q)_t}{(q/a; q)_t}\left(\frac{b}{az}\right)^t,$$

where we have again used (4.8), this time with the replacements $a \to a$, $b \to q^{t+1}/b$ and $c \to q^{t+1}$. The result now follows. $\qquad\square$

7.1   Constant term results

In [24], Andrews used a constant term method involving (7.2) to prove eight
identities of Rogers [217, 219], identities which were found independently by
Rodney Baxter in connection with his solution of the Hard Hexagon Model.
The basic idea in this constant term approach is that, assuming convergence,

$$\sum_{n=0}^{\infty} a_n b_n = \text{coefficient of } z^0 \text{ in } \sum_{k=-\infty}^{\infty} a_k z^k \sum_{m=0}^{\infty} \frac{b_m}{z^m}.$$

If the infinite series have representations as infinite products, which in turn
can be manipulated and re-expanded as different infinite series, it may lead
to an identity between infinite series. We illustrate the method with one of
the examples of Andrews ([24, (4.5)]).

Corollary 7.1.

$$\sum_{n=0}^{\infty} \frac{q^{n(3n-1)/2}}{(q;q)_n (q;q^2)_n} = \frac{1}{(q;q^2)_\infty} \sum_{n=0}^{\infty} \frac{q^{2n^2}}{(q^2;q^2)_n}. \tag{7.5}$$

Remark: Andrews took (7.5) one step further, by using one of the
Rogers–Ramanujan identities (10.17) to write the series on the right side
as an infinite product.

Proof.

$$\sum_{n=0}^{\infty} \frac{q^{n(3n-1)/2}}{(q;q)_n (q;q^2)_n} = \text{coeff. of } z^0 \text{ in } \sum_{n=-\infty}^{\infty} \frac{q^{n^2} z^n}{(q;q^2)_n} \sum_{m=0}^{\infty} \frac{q^{m(m-1)/2} z^{-m}}{(q;q)_m}$$

$$= \text{coeff. of } z^0 \text{ in } \frac{(q^2, -qz, -q/z; q^2)_\infty}{(q, -1/z; q^2)_\infty} (-1/z; q)_\infty \quad \text{(by (7.2) and (3.4))}$$

$$= \frac{1}{(q;q^2)_\infty} \text{ coeff. of } z^0 \text{ in } (q^2, -qz, -q/z; q^2)_\infty (-q/z; q^2)_\infty$$

$$= \frac{1}{(q;q^2)_\infty} \text{ coeff. of } z^0 \text{ in } \sum_{n=-\infty}^{\infty} z^n q^{n^2} \sum_{m=0}^{\infty} \frac{q^{m^2} z^{-m}}{(q^2;q^2)_m} \quad \text{(by (6.1) and (3.4))}$$

$$= \frac{1}{(q;q^2)_\infty} \sum_{n=0}^{\infty} \frac{q^{2n^2}}{(q^2;q^2)_n}.$$

$$\square$$

Some other applications of the constant term method to deriving identities are given in the exercises (see problems 7.2 and 7.3).

7.2   Vanishing coefficients

Andrews and Bressoud [44], extending results of Richmond and Szekeres [216], used Ramanujan's $_1\psi_1$ summation formula (7.1) to prove the following result on vanishing coefficients.

Theorem 7.2. (*Andrews and Bressoud*) *If $1 \leq r < k$ are relatively prime integers of opposite parity and*

$$\frac{(q^r, q^{2k-r}; q^{2k})_\infty}{(q^{k-r}, q^{k+r}; q^{2k})_\infty} =: \sum_{n=0}^\infty \phi_n q^n, \tag{7.6}$$

then $\phi_{kn+r(k-r+1)/2}$ is always zero.

Proof. In (7.1), replace q with q^{2k}, set $z = q^{k+r}$, $a = q^{-k}$ and $b = q^k$ to get

$$\sum_{n=-\infty}^\infty \frac{1-q^{-k}}{1-q^{2nk-k}} q^{n(k+r)} = \frac{(q^{2k}, q^{2k}, q^r, q^{2k-r}; q^{2k})_\infty}{(q^k, q^{3k}, q^{k+r}, q^{k-r}; q^k)_\infty}$$

$$\implies \frac{-(q^k, q^k; q^k)_\infty}{q^k(q^{2k}, q^{2k}; q^{2k})_\infty} \sum_{n=-\infty}^\infty \frac{q^{n(k+r)}}{1-q^{2nk-k}} = \frac{(q^r, q^{2k-r}; q^{2k})_\infty}{(q^{k+r}, q^{k-r}; q^k)_\infty}.$$

On the left side, the product preceding the infinite series is a function of q^k, so it is sufficient to show that if the infinite series is expanded as an infinite series in q,

$$\sum_{n=-\infty}^\infty \frac{q^{n(k+r)}}{1-q^{2nk-k}} = \sum_{n=0}^\infty \frac{q^{n(k+r)}}{1-q^{2nk-k}} - \sum_{n=1}^\infty \frac{q^{n(k-r)+k}}{1-q^{2nk+k}} =: \sum_{m=0}^\infty d_m q^m,$$

then $d_{kn+r(k-r+1)/2}$ is identically zero for all n.

Since the two terms $1/(1 - q^{2nk-k})$ and $1/(1 - q^{2nk+k})$ may be expanded as power series in q^k, terms of the form $q^{kn+r(k-r+1)/2}$ can arise if and only if the numerator $q^{n(k+r)}$ or $q^{n(k-r)+k}$ is of this same form. All such terms (and only those) arise in the first sum if n is replaced with $nk + (k-r+1)/2$, and all such terms (and only those) arise in the second sum if n is replaced with $nk - (k-r+1)/2$. Thus to complete the proof, it is sufficient to show that

$$\sum_{n=0}^\infty \frac{q^{(nk+(k-r+1)/2)(k+r)}}{1-q^{2(nk+(k-r+1)/2)k-k}} - \sum_{n=1}^\infty \frac{q^{(nk-(k-r+1)/2)(k-r)+k}}{1-q^{2(nk-(k-r+1)/2)k+k}}$$

$$= \sum_{n=0}^\infty \frac{q^{(k-r+1)(k+r)/2} q^{kn(k+r)}}{1-q^{2k^2n+k(k-r)}} - \sum_{n=1}^\infty \frac{q^{k-(k-r)(k-r+1)/2} q^{kn(k-r)}}{1-q^{2nk^2-k(k-r)}} = 0.$$

However, this follows easily, since

$$\sum_{n=1}^{\infty} \frac{q^{k-(k-r)(k-r+1)/2}q^{kn(k-r)}}{1-q^{2nk^2-k(k-r)}}$$

$$= \sum_{n=1}^{\infty} q^{k-(k-r)(k-r+1)/2}q^{kn(k-r)} \sum_{m=0}^{\infty} q^{m(2nk^2-k(k-r))}$$

$$= \sum_{m=0}^{\infty} q^{k-(k-r)(k-r+1)/2}q^{-mk(k-r)+2mk^2+k(k-r)} \sum_{n=0}^{\infty} q^{n[2mk^2+k(k-r)]}$$

$$= \sum_{m=0}^{\infty} \frac{q^{(k+r)(k-r+1)/2}q^{km(k+r)}}{1-q^{2mk^2+k(k-r)}}.$$

□

Remark: These results were further extended by Alladi and Gordon [7] and Mc Laughlin [195] (see exercise 7.7).

7.3   Dissections of quotients of infinite products

The following dissection of a quotient of infinite products is an easy consequence of the $_1\psi_1$ summation formula (7.1).

Corollary 7.2. *For each positive integer k, if $|q| < |x| < 1$, then*

$$\frac{(q,q,xy,q/xy;q)_\infty}{(y,q/y,x,q/x;q)_\infty} = \sum_{r=0}^{k-1} x^r \frac{(q^k,q^k,x^kyq^r,q^{k-r}/x^ky;q^k)_\infty}{(yq^r,q^{k-r}/y,x^k,q^k/x^k;q^k)_\infty}. \tag{7.7}$$

Proof. Replace a, b and z with y, yq and x, respectively, in (7.1) to get

$$\sum_{n=-\infty}^{\infty} \frac{x^n}{1-yq^n} = \frac{(q,q,xy,q/xy;q)_\infty}{(y,q/y,x,q/x;q)_\infty}. \tag{7.8}$$

Now set

$$\sum_{n=-\infty}^{\infty} \frac{x^n}{1-yq^n} = \sum_{r=0}^{k-1} x^r \sum_{n=-\infty}^{\infty} \frac{(x^k)^n}{1-yq^r(q^k)^n}$$

and apply (7.8) to each of the inner sums, and (7.7) follows.   □

In [95], Chan specialized this even further in order to prove six identities of Hirschhorn and Sellers (private communication from Hirschhorn to the author of [95]), identities which were dissections of quotients of theta functions. Replace q with q^k and set $y = -1$ in (7.7), and rearrange slightly to get

$$\frac{(-x, -q^k/x; q^k)_\infty}{(x, q^k/x; q^k)_\infty}$$

$$= 2\frac{(-q^k, -q^k; q^k)_\infty}{(q^k, q^k; q^k)_\infty} \sum_{r=0}^{k-1} x^r \frac{(q^{k^2}, q^{k^2}, -x^k q^{kr}, -q^{k(k-r)}/x^k; q^{k^2})_\infty}{(-q^{kr}, -q^{k(k-r)}, x^k, q^{k^2}/x^k; q^{k^2})_\infty}. \quad (7.9)$$

We state and prove one the identities in Chan's paper [95], to give an example of an application.

Corollary 7.3. *For $|q| < 1$ there holds*

$$\sum_{n=-\infty}^{\infty} (-1)^n q^{(3n^2+n)/2} \Big/ \sum_{n=-\infty}^{\infty} q^{(3n^2+n)/2}$$

$$= \frac{1}{(q^3, q^3, q^6, q^{12}, q^{15}, q^{15}; q^{18})_\infty} - \frac{2q}{(q^3, q^6, q^9, q^9, q^{12}, q^{15}; q^{18})_\infty}. \quad (7.10)$$

Proof. By the Jacobi triple product identity, the left side of (6.3) equals

$$\frac{(q, q^2, q^3; q^3)_\infty}{(-q, -q^2, q^3; q^3)_\infty} = \frac{(q, q^2; q^3)_\infty}{(-q, -q^2; q^3)_\infty}.$$

Now apply (7.9), with $k = 3$ and $x = -q$, and the result follows after some of the elementary infinite product manipulations of the type described in Chapter 2. ☐

Exercises

7.1 Derive the q-binomial theorem (3.1), the Jacobi triple product identity (6.1) and the identity at (7.2) from Ramanujan's $_1\psi_1$ formula (7.1).

7.2 By using constant term methods or otherwise, prove, for $|q/c| < 1$, that

$$\sum_{n=-\infty}^{\infty} \frac{(1-a^2)(c;q)_{2n}}{(1-a^2q^{2n})_n(d;q)_{2n}} \left(\frac{q}{c}\right)^{2n}$$

$$= \frac{(-q, -q, qa/c, d/a; q)_\infty}{(-qa, -q/a, d, q/c; q)_\infty} \sum_{n=-\infty}^{\infty} \frac{(1-a)(c/a; q)_n}{(1-aq^n)_n(d/a; q)_n} \left(\frac{-q}{c}\right)^n$$

$$= \frac{(-q, -q, -qa/c, -d/a; q)_\infty}{(qa, q/a, d, q/c; q)_\infty} \sum_{n=-\infty}^{\infty} \frac{(1+a)(-c/a; q)_n}{(1+aq^n)_n(-d/a; q)_n} \left(\frac{-q}{c}\right)^n. \quad (7.11)$$

7.3 In a manner similar to that used to prove question 7.2, or otherwise, prove for $|q/a|, |b/a| < 1$, that

$$\sum_{n=-\infty}^{\infty} \frac{(1-c)(a^2; q^2)_n}{(1-cq^{2n})_n(b^2; q^2)_n} \left(\frac{q}{a}\right)^{2n}$$

$$= \frac{(q, -q, qc/a, -b/a, b/c; q)_\infty}{(-b, -q/a, cq, q/c, b/a; q)_\infty} \sum_{n=-\infty}^{\infty} \frac{(a, cq/b; q)_n}{(b, cq/a; q)_n} \left(\frac{-b}{a}\right)^n. \quad (7.12)$$

Furthermore, if a is replaced with $-a$ and/or b is replaced with $-b$, the value of the right side of (7.12) is unchanged.

7.4 (Andrews and Askey [33], Askey [50]) Prove Ramanujan's $_1\psi_1$ sum (7.1) as follows:
(i) Define $f(b)$ by

$$\sum_{n=-\infty}^{\infty} \frac{(a; q)_n}{(b; q)_n} z^n = \frac{(a; q)_\infty}{(b; q)_\infty} \sum_{n=-\infty}^{\infty} \frac{(bq^n; q)_\infty}{(aq^n; q)_\infty} z^n =: \frac{(a; q)_\infty}{(b; q)_\infty} f(b), \quad (7.13)$$

and show

$$f(b) = \frac{1 - b/a}{1 - b/az} f(bq).$$

(ii) Deduce that, for each positive integer n,

$$f(b) = \frac{(b/a; q)_n}{(b/az; q)_n} f(bq^n) = \frac{(b/a; q)_\infty}{(b/az; q)_\infty} f(0). \quad (7.14)$$

(iii) Evaluate $f(q)$ from (7.13) by substituting $b = q$, and then evaluate $f(0)$ from (7.14) by also replacing b with q, and substituting for $f(q)$.
(iv) Once $f(0)$ has been determined, find $f(b)$ and substitute in (7.13) to complete the proof.

7.5 Let k be a positive integer and set $\omega = \exp(2\pi i/k)$. Show that if $|b/a| < |z| < 1$ and $|q| < 1$, then

$$\sum_{n=-\infty}^{\infty} \frac{(a; q)_{kn}}{(b; q)_{kn}} z^{kn} = \frac{1}{k} \sum_{i=1}^{k} \frac{(q, b/a, a\omega^i z, q/a\omega^i z; q)_\infty}{(b, q/a, \omega^i z, b/a\omega^i z; q)_\infty}. \quad (7.15)$$

7.6 (Ismail [148]) Prove Ramanujan's $_1\psi_1$ sum (7.1) as follows:

(i) Conclude from (7.3) that the series in (7.1) is an analytic function of b provided $|z|, |b/az| < 1$.

(ii) Show that when $b = q^m$, m any positive integer, both sides of (7.1) are equal to

$$\frac{(q, q/za; q)_{m-1}(az; q)_\infty}{(q/a; q)_{m-1}(z; q)_\infty},$$

and thus conclude by the identity theorem that (7.1) holds generally.

7.7 (Mc Laughlin [195]) Suppose $k > 1$, $m > 1$ are positive integers, and let $r = sm + t$, for some integers s and t, where $0 \le s < k$, $1 \le t < m$ and r and k are relatively prime. Let the sequence $\{c_n\}$ be defined by

$$\frac{(q^{r-tk}, q^{mk-(r-tk)}; q^{mk})_\infty}{(q^r, q^{mk-r}; q^{mk})_\infty} =: \sum_{n=0}^\infty c_n q^n. \tag{7.16}$$

Show that $c_{kn-rs} = 0$ for all n.

7.8 Prove that if $|q| < 1$, then

$$\sum_{n=-\infty}^\infty q^{(3n^2+n)/2} \Big/ \sum_{n=-\infty}^\infty (-1)^n q^{(3n^2+n)/2}$$

$$= \frac{1}{(q^3, q^{15}; q^{18})_\infty^6 (q^6, q^{12}; q^{18})_\infty^3} + \frac{2q}{(q^3, q^{15}; q^{18})_\infty^5 (q^6, q^{12}; q^{18})_\infty^3 (q^9; q^{18})_\infty^2}$$

$$+ \frac{4q^2}{(q^3, q^{15}; q^{18})_\infty^4 (q^6, q^{12}; q^{18})_\infty^3 (q^9; q^{18})_\infty^4}. \tag{7.17}$$

7.9 Without appealing to (7.8), prove that if $|q| < |x|, |y| < 1$, then

$$\sum_{n=-\infty}^\infty \frac{x^n}{1 - yq^n} = \sum_{n=-\infty}^\infty \frac{y^n}{1 - xq^n}.$$

7.10 Prove that if $|\alpha|, |\beta| < 1$, then

(i) $$1 + 2\sum_{n=1}^\infty \frac{\alpha^n + \beta^n}{1 + (\alpha\beta)^n} = 1 - 2\sum_{n=1}^\infty (-1)^n \left(\frac{\alpha^{n-1}\beta^n}{1 - \alpha^{n-1}\beta^n} + \frac{\alpha^n \beta^{n-1}}{1 - \alpha^n \beta^{n-1}} \right)$$

$$= \frac{(-\alpha, -\beta, \alpha\beta, \alpha\beta; \alpha\beta)_\infty}{(\alpha, \beta, -\alpha\beta, -\alpha\beta; \alpha\beta)_\infty}, \tag{7.18}$$

(ii) $$\sum_{n=1}^\infty \frac{\alpha^{2n-1} - \beta^{2n-1}}{1 - (\alpha\beta)^{2n-1}} = \alpha \frac{(\beta/\alpha, \alpha^3\beta, \alpha^2\beta^2, \alpha^2\beta^2; \alpha^2\beta^2)_\infty}{(\alpha^2, \beta^2, \alpha\beta, \alpha\beta; \alpha^2\beta^2)_\infty}. \tag{7.19}$$

7.11 (Bailey [62]) By using constant term methods or otherwise, prove that if $|bd/ac| < |z| < 1$, then

(i) $\displaystyle\sum_{n=-\infty}^{\infty} \frac{(a, c; q)_n}{(b, d; q)_n} z^n = \frac{(az, cz, qb/acz, qd/acz; q)_\infty}{(b, d, q/a, q/c; q)_\infty}$

$$\times \sum_{n=-\infty}^{\infty} \frac{(acz/b, acz/d; q)_n}{(az, cz; q)_n} \left(\frac{bd}{acz}\right)^n.$$

$$(7.20)$$

If, in addition $|b/a|, |d/c| < 1$, then prove that

(ii) $\displaystyle\sum_{n=-\infty}^{\infty} \frac{(a, c; q)_n}{(b, d; q)_n} z^n$

$$= \frac{(b/a, d/c, az, qb/acz; q)_\infty}{(b, q/c, z, bd/caz; q)_\infty} \sum_{n=-\infty}^{\infty} \frac{(a, acz/b; q)_n}{(az, d; q)_n} \left(\frac{b}{a}\right)^n. \quad (7.21)$$

7.12 Prove that if $|b/a|, |q| < 1$ and n is any integer, then

$$\sum_{k=-\infty}^{\infty} \frac{(q^{1-n}/b, a; q)_k}{(q^{1-n}/a, b; q)_k} \left(\frac{-b}{a}\right)^k = \begin{cases} \frac{(q, b/a, -b, -q/a; q)_\infty}{(-q, -b/a, b, q/a; q)_\infty} \frac{(a^2, b, bq; q^2)_{n/2}}{(b^2, a, aq; q^2)_{n/2}}, & n \text{ even,} \\ 0, & n \text{ odd.} \end{cases}$$

$$(7.22)$$

Chapter 8

Bailey's $6\psi_6$ Summation

Bailey's [56] $6\psi_6$ identity at (8.1) is arguably the most important summation formula for bilateral basic hypergeometric series. For example, as will be seen in a later chapter, it was used by Slater [236,237] to derive Bailey pairs and, from these, her list of 130 identities of the Rogers–Ramanujan type.

Subsequent proofs were given by Slater and Lakin [238] (two proofs), Andrews [17], Askey and Ismail [52], Askey [51], Schlosser [225], Jouhet and Schlosser [164], and Chu [104]. The simplest proof is possibly that of Askey and Ismail [52], which we reproduce below. We will return to some of the other proofs in later chapters.

Theorem 8.1. *If* $|qa^2/(bcde)| < 1$ *and none of the denominators in* (8.1) *vanish, then*

$$
\sum_{n=-\infty}^{\infty} \frac{(q\sqrt{a}, -q\sqrt{a}, b, c, d, e; q)_n}{(\sqrt{a}, -\sqrt{a}, aq/b, aq/c, aq/d, aq/e; q)_n} \left(\frac{qa^2}{bcde} \right)^n
$$
$$
= \frac{(aq, aq/bc, aq/bd, aq/be, aq/cd, aq/ce, aq/de, q, q/a; q)_\infty}{(aq/b, aq/c, aq/d, aq/e, q/b, q/c, q/d, q/e, qa^2/bcde; q)_\infty}. \quad (8.1)
$$

Proof. Set $e = a/x$ so that it is necessary to prove

$$
\sum_{n=-\infty}^{\infty} \frac{(q\sqrt{a}, -q\sqrt{a}, b, c, d, a/x; q)_n}{(\sqrt{a}, -\sqrt{a}, aq/b, aq/c, aq/d, xq; q)_n} \left(\frac{qax}{bcd} \right)^n
$$
$$
= \frac{(aq, aq/bc, aq/bd, xq/b, aq/cd, xq/c, xq/d, q, q/a; q)_\infty}{(aq/b, aq/c, aq/d, xq, q/b, q/c, q/d, xq/a, qax/bcd; q)_\infty} \quad (8.2)
$$

if $|qax/(bcd)| < 1$. By considering the denominators of the infinite product and the convergence of the bilateral series, it can be seen that both sides of

(8.2) are analytic functions of x provide $|x| < \min\{|bcd/aq|, |a/q|, 1/|q|\}$. It will be shown that both sides of (8.2) agree for $x = q^m$, $m = 0, 1, 2, \ldots$, so that (8.2) holds generally by the identity theorem, thus also proving (8.1).

If $x = q^m$, then by (2.5),

$$\frac{1}{(xq; q)_{-n}} = \frac{1}{(q^{m+1}; q)_{-n}} = \frac{(q^{-m}; q)_n (-1)^n}{q^{n(n-1)/2 - mn}} = 0, \text{ if } n > m,$$

so that the series in (8.2) terminates below at $n = -m$, and equals

$$\sum_{n=-m}^{\infty} \frac{(1 - aq^{2n})(b, c, d, aq^{-m}; q)_n}{(1-a)(aq/b, aq/c, aq/d, q^{m+1}; q)_n} \left(\frac{q^{m+1}a}{bcd}\right)^n$$

$$= \sum_{n=0}^{\infty} \frac{(1 - aq^{2n-2m})(b, c, d, aq^{-m}; q)_{n-m}}{(1-a)(aq/b, aq/c, aq/d, q^{m+1}; q)_{n-m}} \left(\frac{q^{m+1}a}{bcd}\right)^{n-m}$$

$$= \frac{(1 - aq^{-2m})(b, c, d, aq^{-m}; q)_{-m}}{(1-a)(aq/b, aq/c, aq/d, q^{m+1}; q)_{-m}} \left(\frac{bcd}{q^{m+1}a}\right)^m$$

$$\times \sum_{n=0}^{\infty} \frac{(1 - aq^{-2m+2n})(bq^{-m}, cq^{-m}, dq^{-m}, aq^{-2m}; q)_n}{(1 - aq^{-2m})(aq^{1-m}/b, aq^{1-m}/c, aq^{1-m}/d, q; q)_n} \left(\frac{q^{m+1}a}{bcd}\right)^n.$$

This last series may be summed using the $_6\phi_5$ summation formula at (5.6), so that the left side of (8.2) equals

$$\frac{(1 - aq^{-2m})(b, c, d, aq^{-m}; q)_{-m}}{(1-a)(aq/b, aq/c, aq/d, q^{m+1}; q)_{-m}} \left(\frac{bcd}{q^{m+1}a}\right)^m$$

$$\times \frac{(aq^{1-2m}, aq/bc, aq/bd, aq/cd; q)_\infty}{(aq^{1-m}/b, aq^{1-m}/c, aq^{1-m}/d, aq^{1+m}/bcd; q)_\infty}$$

$$= \frac{(aq, aq/bc, aq/bd, aq/cd; q)_\infty}{(aq/b, aq/c, aq/d, aq^{1+m}/bcd; q)_\infty} \qquad \text{(by (2.11))}$$

$$\times \frac{(aq^{-2m}; q)_{2m}(b, c, d, aq^{-m}; q)_{-m}}{(q^{m+1}; q)_{-m}} \left(\frac{bcd}{q^{m+1}a}\right)^m$$

$$= \frac{(aq, aq/bc, aq/bd, aq/cd; q)_\infty}{(aq/b, aq/c, aq/d, aq^{1+m}/bcd; q)_\infty} \frac{(q, q/a; q)_m}{(q/b, q/c, q/d; q)_m},$$

where the last equality follows after applying (2.5) and (2.24). This last expression is precisely what is obtained upon substituting $x = q^m$ into the infinite product on the right side of (8.2), thus completing the proof, by the remarks above. $\qquad \square$

Remark: The bilateral series in the theorem may be written, using (2.5), as

$$\sum_{n=0}^{\infty} \frac{(1 - aq^{2n})(b, c, d, e; q)_n}{(1 - a)(aq/b, aq/c, aq/d, aq/e; q)_n} \left(\frac{qa^2}{bcde} \right)^n$$

$$+ \sum_{n=1}^{\infty} \frac{(1 - q^{2n}/a)(b/a, c/a, d/a, e/a; q)_n}{(1 - 1/a)(q/b, q/c, q/d, q/e; q)_n} \left(\frac{qa^2}{bcde} \right)^n. \quad (8.3)$$

This also shows that in certain circumstances two unilateral series may be combined into a bilateral series (see Q8.9).

Corollary 8.1. *(i) If $|qa^{3/2}/(bcd)| < 1$, then*

$$\sum_{n=-\infty}^{\infty} \frac{(-q\sqrt{a}, b, c, d; q)_n}{(-\sqrt{a}, aq/b, aq/c, aq/d; q)_n} \left(\frac{qa^{3/2}}{bcd} \right)^n$$

$$= \frac{(aq, aq/bc, aq/bd, \sqrt{a}q/b, aq/cd, \sqrt{a}q/c, \sqrt{a}q/d, q, q/a; q)_\infty}{(aq/b, aq/c, aq/d, \sqrt{a}q, q/b, q/c, q/d, q/\sqrt{a}, qa^{3/2}/bcd; q)_\infty}. \quad (8.4)$$

(ii) If $|qa/(bc)| < 1$, then

$$\sum_{n=-\infty}^{\infty} \frac{(b, c; q)_n}{(aq/b, aq/c; q)_n} \left(\frac{-qa}{bc} \right)^n$$

$$= \frac{(aq/bc; q)_\infty (aq^2/b^2, aq^2/c^2, q^2, aq, q/a; q^2)_\infty}{(aq/b, aq/c, q/b, q/c, -qa/bc; q)_\infty}. \quad (8.5)$$

Proof. For (8.4), set $e = \sqrt{a}$ in (8.1), and then set $d = -\sqrt{a}$ and simplify to get (8.5). □

Remark: The two identities above are bilateral extensions of unilateral identities, as setting $d = a$ in (8.4) recovers the q-Dixon sum (5.7), and setting $c = a$ in (8.5) recovers the Bailey–Daum sum (4.10).

Exercises

8.1 By starting with (5.12), find an alternative expression for the first series at (8.3). By employing a change of parameters, find an alternative expression for the second series at (8.3), and hence show that

$$\left(\frac{a}{e},\frac{q}{e},\frac{a^2q}{bcd},\frac{aq^2}{bcd};q\right)_\infty \sum_{k=1}^\infty \frac{\left(1-\dfrac{a^2q^{2k+1}}{bcd}\right)\left(e,\dfrac{aq}{bc},\dfrac{aq}{bd},\dfrac{aq}{cd};q\right)_k}{\left(1-\dfrac{a^2q}{bcd}\right)\left(\dfrac{aq}{b},\dfrac{aq}{c},\dfrac{aq}{d},\dfrac{a^2q^2}{bcde};q\right)_k}\left(\frac{a}{e}\right)^k$$

$$+\left(\frac{1}{e},\frac{aq}{e},\frac{aq}{bcd},\frac{a^2q^2}{bcd};q\right)_\infty \sum_{k=1}^\infty \frac{\left(1-\dfrac{aq^{2k+1}}{bcd}\right)\left(\dfrac{e}{a},\dfrac{aq}{bc},\dfrac{aq}{bd},\dfrac{aq}{cd};q\right)_k}{\left(1-\dfrac{aq}{bcd}\right)\left(\dfrac{q}{b},\dfrac{q}{c},\dfrac{q}{d},\dfrac{a^2q^2}{bcde};q\right)_k}\left(\frac{1}{e}\right)^k$$

$$=\frac{\left(\dfrac{aq}{bc},\dfrac{aq}{bd},\dfrac{aq}{be},\dfrac{aq}{cd},\dfrac{aq}{ce},\dfrac{aq}{de},q,a,\dfrac{q}{a},\dfrac{a^2q^2}{bcd},\dfrac{aq^2}{bcd};q\right)_\infty}{\left(\dfrac{aq}{b},\dfrac{aq}{c},\dfrac{aq}{d},\dfrac{q}{b},\dfrac{q}{c},\dfrac{q}{d},\dfrac{a^2q^2}{bcde};q\right)_\infty}.$$

8.2 (Andrews [17]) Show that The Jacobi triple product identity (6.1)

$$\sum_{n=-\infty}^{\infty}(-z)^n q^{n^2}=(zq,q/z,q^2;q^2)_\infty$$

follows from (8.1).

8.3 Prove that (5.6) follows from (8.1) and hence show (Andrews [17]) that

$$1+4\sum_{n=1}^{\infty}\frac{(-1)^n q^{n(n+1)/2}}{1+q^n}=\left(\frac{(q;q)_\infty}{(-q;q)_\infty}\right)^2.$$

8.4 (Andrews [17]) By similarly employing (5.6), show that

$$1+8\sum_{n=1}^{\infty}\frac{(-q)^n}{(1+q^n)^2}=\left(\frac{(q;q)_\infty}{(-q;q)_\infty}\right)^4. \tag{8.6}$$

8.5 (Andrews [17]) (i) Prove that for $N\geq 1$,

$$\lim_{z\to 1}\frac{(-q,-1;q)_n^N - z^L(-1/z,-zq;q)_n^N}{1-z}$$

$$=(-1,-q;q)_n^N\left(L-\frac{N}{2}+\frac{Nq^n}{1+q^n}\right).$$

(ii) By using part (i) and (8.1), or otherwise, show that

$$1+16\sum_{n=1}^{\infty}\frac{-q^n+4q^{2n}-q^{3n}}{(1+q^n)^4}=\left(\frac{(q;q)_\infty}{(-q;q)_\infty}\right)^8. \tag{8.7}$$

Remark: Andrews [17] used the identities in questions 8.3–8.5 to prove certain arithmetic identities involving the representation of positive integers as sums of squares.

8.6 (Andrews [17]) Prove that

$$\sum_{n=0}^{\infty}\left\{\frac{q^{5n+1}}{(1-q^{5n+1})^2}-\frac{q^{5n+2}}{(1-q^{5n+2})^2}-\frac{q^{5n+3}}{(1-q^{5n+3})^2}+\frac{q^{5n+4}}{(1-q^{5n+4})^2}\right\}$$

$$=\sum_{n=-\infty}^{\infty}\frac{(1-q^2)(1+q^{5n+2})(1-q^{5n+2})q^{5n+1}}{(1-q^{5n+1})^2(1-q^{5n+3})^2}$$

$$=\frac{q(1-q^2)(1-q^4)}{(1-q)^2(1-q^3)^2}\sum_{n=-\infty}^{\infty}\frac{(q^7,-q^7,q,q,q^3,q^3;q^5)_n\,q^{5n}}{(q^2,-q^2,q^8,q^8,q^6,q^6;q^5)_n}$$

$$=q\frac{(q^5;q^5)_{\infty}^5}{(q;q)_{\infty}}.$$

Remarks: Andrews [17] used the result in Question 8.6 to prove the Ramanujan congruence $p(5n+4)\equiv 0(\mathrm{mod}\ 5)$. The exponent "2" on the $(1-q^{5n+1})$ factor in the denominator of the second expression above is missing in [17, (3.46)].

8.7 (Andrews [17]) Derive the quintuple product identity (6.19),

$$\sum_{n=-\infty}^{\infty}(-1)^n q^{n(3n-1)/2}z^{3n}(1+zq^n)=(-z,-q/z,q;q)_{\infty}(qz^2,q/z^2;q^2)_{\infty},$$

from (8.1).

8.8 (i) Prove the following special case of Watson's [257] identity (5.10):

$$\sum_{k=0}^{\infty}\frac{1-aq^{2k}}{1-a}\frac{(a,c,d,e;q)_k}{(aq/c,aq/d,aq/e,q;q)_k}\left(\frac{a^2q}{cde}\right)^k q^{k^2}$$

$$=\frac{(aq,aq/de;q)_{\infty}}{(aq/d,aq/e;q)_{\infty}}\sum_{j=0}^{\infty}\frac{(d,e;q)_j}{(aq/c,q;q)_j}\left(\frac{aq}{de}\right)^j.\quad (8.8)$$

(ii) Use the identity in (i) and the method used in the proof of Theorem 8.1 to prove the following result of Bailey [62]:

$$\sum_{n=-\infty}^{\infty}\frac{(e,f;q)_n}{(aq/c,aq/d;q)_n}\left(\frac{qa}{ef}\right)^n=\frac{(q/c,q/d,aq/e,aq/f;q)_{\infty}}{(aq,q/a,aq/cd,aq/ef;q)_{\infty}}$$

$$\times \sum_{n=-\infty}^{\infty} \frac{(1 - aq^{2n})(c, d, e, f; q)_n}{(1 - a)(aq/c, aq/d, aq/e, aq/f; q)_n} \left(\frac{qa^3}{cdef}\right)^n q^{n^2}. \quad (8.9)$$

Specifically, show that the result in part(i) implies that, if the replacement $c = aq/\gamma$ is made in (8.9), then (8.9) holds for $\gamma = q^{m+1}$, $m \geq 0$.
(This proof is also sketched in [39, pp. 55–56].)

8.9 (Bailey [63]) By specializing the parameter a in (5.6) and combining two unilateral series into a bilateral series, show

(i) $\quad \displaystyle\sum_{n=-\infty}^{\infty} \frac{(b, c, d; q)_n}{(q/b, q/c, q/d; q)_n} \left(\frac{q}{bcd}\right)^n = \frac{(q, q/bc, q/bd, q/cd; q)_\infty}{(q/b, q/c, q/d, q/bcd; q)_\infty}, \quad (8.10)$

(ii) $\quad \displaystyle\sum_{n=-\infty}^{\infty} \frac{(b, c, d; q)_n}{(q^2/b, q^2/c, q^2/d; q)_n} \left(\frac{q^2}{bcd}\right)^n = \frac{(q, q^2/bc, q^2/bd, q^2/cd; q)_\infty}{(q^2/b, q^2/c, q^2/d, q^2/bcd; q)_\infty}.$

$$(8.11)$$

8.10 (Bailey [63]) By setting $b = c = d = q^{-N}$ in the identities in Q8.9, prove that for each non-negative integer N,

(i) $\quad \displaystyle\sum_{n=-N}^{N} (-1)^n \begin{bmatrix} 2N \\ N+n \end{bmatrix}_q^3 q^{n(3n+1)/2} = \frac{(q; q)_{3N}}{(q; q)_N^3},$

(ii) $\quad \displaystyle\sum_{n=-N-1}^{N} (-1)^n \begin{bmatrix} 2N+1 \\ N+n+1 \end{bmatrix}_q^3 q^{n(3n+1)/2} = \frac{(q; q)_{3N+1}}{(q; q)_N^3}.$

8.11 Prove that

$$\frac{(q/e, q/f; q)_\infty}{(aq/e, aq/f; q)_\infty} \sum_{n=-\infty}^{\infty} \frac{(e, f; q)_n}{(aq/c, aq/d; q)_n} \left(\frac{qa}{ef}\right)^n$$

$$= \frac{(q/c, q/d; q)_\infty}{(aq/c, aq/d; q)_\infty} \sum_{n=-\infty}^{\infty} \frac{(c, d; q)_n}{(aq/e, aq/f; q)_n} \left(\frac{qa}{cd}\right)^n. \quad (8.12)$$

Chapter 9

The Rogers–Fine Identity

The Rogers–Fine identity (9.1) was first prove by Rogers [219], and rediscovered by Fine [115], who investigated several implications of the identity. Most of (9.1) was proved also by Starcher [246]. Ramanujan stated an extension of (9.1) — see [69, pp. 16–17], where it is clear from the proof there that the Rogers–Fine identity follows from Watson's [257] transformation (5.10). The identity was proved combinatorially by Andrews [15], and many special cases and implications were proved combinatorially by Berndt and Yee [75].

As remarked, the identity is really a special case of (5.10), but the identity is important in its own right for its many applications to q-series and the theory of partitions. This derivation of (9.1) from (5.10) is left as an exercise (see Q9.1), as it is a more interesting path to follow the proof of Fine [115].

Theorem 9.1. *If $|\tau|, |q| < 1$ and none of the denominators in (9.1) vanish, then*

$$\sum_{n=0}^{\infty} \frac{(\alpha; q)_n}{(\beta; q)_n} \tau^n = \sum_{n=0}^{\infty} \frac{(1 - \alpha\tau q^{2n})(\alpha, \alpha\tau q/\beta; q)_n \beta^n \tau^n q^{n^2-n}}{(\beta; q)_n (\tau; q)_{n+1}}. \tag{9.1}$$

Proof. Define

$$F(a, b; t) := \sum_{n=0}^{\infty} \frac{(aq; q)_n}{(bq; q)_n} t^n. \tag{9.2}$$

Then

$$F(a, b; t) = 1 + \frac{(1 - aq)}{(1 - bq)} t \sum_{n=1}^{\infty} \frac{(aq^2; q)_{n-1}}{(bq^2; q)_{n-1}} t^{n-1} \tag{9.3}$$

71

$$= 1 + \frac{(1-aq)}{(1-bq)} t \sum_{n=0}^{\infty} \frac{(aq^2;q)_n}{(bq^2;q)_n} t^n = 1 + \frac{(1-aq)}{(1-bq)} t F(aq, bq; t).$$

Next, for ease of notation define

$$A_n := \frac{(aq;q)_n}{(bq;q)_n} \implies (1 - bq^{n+1})A_{n+1} = (1 - aq^{n+1})A_n$$

$$\implies \sum_{n=0}^{\infty} (1 - bq^{n+1})A_{n+1}t^{n+1} = \sum_{n=0}^{\infty} (1 - aq^{n+1})A_n t^{n+1}$$

$$\implies F(a,b;t;q) - 1 - bF(a,b;tq;q) + b = tF(a,b;t;q) - aqtF(a,b;tq;q)$$

$$\implies F(a,b;t) = \frac{1-b}{1-t} + \frac{b - aqt}{1-t} F(a,b;tq). \tag{9.4}$$

By composing (9.3) and (9.4) we get

$$F(a,b;t) = \frac{1 - atq}{1-t} + \frac{(1-aq)(1-atq/b)btq}{(1-bq)(1-t)} F(aq, bq; tq). \tag{9.5}$$

The iteration of this last transformation (noting that $F(aq^N, bq^N; tq^N) \to 1$ as $N \to \infty$) gives

$$F(a,b;t) = \sum_{n=0}^{\infty} \frac{(1 - atq^{2n+1})(aq, atq/b;q)_n b^n t^n q^{n^2}}{(bq;q)_n(t;q)_{n+1}}. \tag{9.6}$$

Upon making the replacements $a \to \alpha/q$, $b \to \beta/q$ and $t \to \tau$, (9.1) follows.
□

We next consider some applications.

9.1 False theta series identities

From (6.9) and (6.12), the theta series $\psi(q) = \sum_{n=0}^{\infty} q^{n(n+1)/2}$ has a representation has an infinite product, by the Jacobi triple product identity. The *false theta series* $\sum_{n=0}^{\infty} (-1)^n q^{n(n+1)/2}$ does not have such a representation, but false theta series may be found in many identities stated by Ramanujan in the Lost Notebook [213], some of which were proved in [21] (see also [38, Chapter 9]). We recall some examples.

Corollary 9.1. *If $|q| < 1$ and $a \neq -q^{-2k}$ for k a positive integer, then*

$$\sum_{n=0}^{\infty} \frac{(-aq;q^2)_n}{(-aq^2;q^2)_n} (-aq)^n = \sum_{n=0}^{\infty} (-a)^n q^{n(n+1)/2}. \tag{9.7}$$

Proof. In (9.1), replace q with q^2, set $\alpha = \tau = -aq$ and $\beta = -aq^2$, so that the left side of (9.1) becomes the left side of (9.7). After simplification, the right side of (9.1) equals

$$\sum_{n=0}^{\infty}(1 - aq^{2n+1})a^{2n}q^{2n^2-n} = \sum_{n=0}^{\infty}(-a)^{2n}q^{2n^2+n} + \sum_{n=0}^{\infty}(-a)^{2n+1}q^{2n^2+3n+1},$$

which is equal to the right side of (9.7), separated into odd-indexed and even-indexed terms. □

Corollary 9.2. *If $|q| < 1$, then*

$$\sum_{n=0}^{\infty} \frac{(q; q^2)_n}{(-q; q^2)_{n+1}} q^n = \sum_{n=0}^{\infty}(-1)^n q^{2n^2+2n}. \tag{9.8}$$

Proof. In (9.1), replace q with q^2, set $\alpha = \tau = q$ and $\beta = -q^3$, and (9.8) after multiplying both sides by $1/(1 + q)$ and simplifying. □

9.2 Transformation of series

As Fine points out [115, section 26], the right side of (9.1) can be specialized to give a generalized mock theta function, which may be further specialized to give many of the third order mock theta functions. The left side of (9.1) then gives a transformation of this generalized mock theta function.

Corollary 9.3. *Let $|q|, |zq| < 1$, and suppose s, z are such that none of the denominators in (9.9) vanish. Then*

$$1 + \sum_{n=1}^{\infty} \frac{sz^n q^n}{(sq; q)_n} = \sum_{n=0}^{\infty} \frac{s^n z^n q^{n^2}}{(qz, sq; q)_n}. \tag{9.9}$$

Proof. In (9.1), set $\alpha = 0$, replace τ with zq and β with sq^2 and multiply both sides by $1/(1 - sq)$. This gives

$$\sum_{n=0}^{\infty} \frac{z^n q^n}{(sq; q)_{n+1}} = \sum_{n=0}^{\infty} \frac{s^n z^n q^{n^2+2n}}{(sq, zq; q)_{n+1}}. \tag{9.10}$$

Multiply both sides by szq, re-index on each side so that each sum starts at $n = 1$, add 1 to both sides and (9.9) follows. □

As an example, let $s = z = -1$ to obtain the following well-known identity for Ramanujan's third order mock theta function $f(q)$:

$$f(q) := \sum_{n=0}^{\infty} \frac{q^{n^2}}{(-q, -q; q)_n} = 1 + \sum_{n=1}^{\infty} \frac{(-1)^{n-1}q^n}{(-q; q)_n}.$$

As is also well known, setting $s = 1/z$ in (9.9) recovers the two expressions for the rank generating function for the unrestricted partitions of a positive integer (see Chapter 15).

Remark: It is clear from the proof that

$$\sum_{n=0}^{\infty} \frac{s^n z^n q^{n^2}}{(qz, sq; q)_n} = 1 + \frac{szq}{1 - sq} F(0, sq; zq),$$

so that all of the transformations for $F(a, b; t)$ found by Fine [115] also give rise to transformations for the series on the left above, and hence for all of the mock theta functions that derive from this series by specializing s and z.

Several transformations stated by Ramanujan also follow from (9.1).

Corollary 9.4. *(Ramanujan, [213, p. 32])*

$$\sum_{n=0}^{\infty} \frac{(-aq)^n}{(aq; q^2)_{n+1}} = \sum_{n=0}^{\infty} \frac{(-1)^n a^{2n} q^{2n(n+1)}}{(a^2 q^2; q^4)_{n+1}}. \tag{9.11}$$

Proof. In (9.1), set $\alpha = 0$ and replace q with q^2. Then set $\tau = -aq$ and $\beta = aq^3$. Finally, multiply both sides by $1/(1 - aq)$ and combine the q-products in the denominator on the right side.   □

Remark: The proof above of the transformation (9.11) is that given in [38, p. 226].

9.3   Other Fine transformations

Fine [115] stated and proved a number of other transformations for $F(a, b; t)$.

Corollary 9.5. *(Fine [115, p. 13, Eq. (12.2)] If $|t|, |q| < 1$ and none of the denominators in (9.12) vanish, then*

$$\sum_{n=0}^{\infty} \frac{(aq; q)_n}{(bq; q)_n} t^n = \sum_{n=0}^{\infty} \frac{(b/a; q)_n (-at)^n q^{n(n+1)/2}}{(bq; q)_n (t; q)_{n+1}}. \tag{9.12}$$

Proof. One can check that the transformation (Fine [115, p. 3, Eq. (4.3)])

$$F(a, b; t) = \frac{1}{1-t} + \frac{(1 - b/a)(-atq)}{(1 - bq)(1 - t)} F(a, bq; tq) \qquad (9.13)$$

holds, by showing that the coefficient of t^k on each side of the equivalent transformation

$$(1 - t)F(a, b; t) = 1 + \frac{(1 - b/a)(-atq)}{(1 - bq)} F(a, bq; tq)$$

is

$$\frac{(aq; q)_k}{(bq; q)_k} - \frac{(aq; q)_{k-1}}{(bq; q)_{k-1}} = \frac{(b - a)q^k(aq; q)_{k-1}}{(bq; q)_k}, \quad k \geq 1,$$

and equals 1 for $k = 0$.

The result follows upon iteration of (9.13), noting that $F(a, bq^N; tq^N) \to 1$ as $N \to \infty$. $\qquad\square$

Corollary 9.6. *(Fine [115, p. 12, Eq. (11.1–11.2)] If $|t|, |q| < 1$ and none of the denominators in (9.12) vanish, then*

$$\sum_{n=0}^{\infty} \frac{(aq; q)_n}{(bq; q)_n} t^n = (1 - 1/b)aq \sum_{n=0}^{\infty} \frac{(aq, atq/b; q)_n q^n}{(aq/b; q)_{n+1}}$$

$$+ \frac{(aq, atq/b; q)_\infty}{(aq/b; q)_\infty} \sum_{n=0}^{\infty} \frac{t^n}{(bq; q)_n}. \qquad (9.14)$$

Proof. The proof is similar to that of (9.12). That the transformation (Fine [115, p. 3, Eq. (4.5)])

$$F(a, b; t) = \frac{(1 - 1/b)aq}{1 - aq/b} + \frac{(1 - aq)(1 - atq/b)}{(1 - aq/b)} F(aq, b; t) \qquad (9.15)$$

holds follows upon multiplying both sides by $b - aq$ and then showing that the coefficient of t^k on each side of the resulting expression is

$$\frac{(b - aq)(aq; q)_k}{(bq; q)_k}, \quad k \geq 0.$$

If (9.15) is iterated N times, the result is

$$\sum_{n=0}^{\infty} \frac{(aq; q)_n}{(bq; q)_n} t^n = (1 - 1/b)aq \sum_{n=0}^{N-1} \frac{(aq, atq/b; q)_n q^n}{(aq/b; q)_{n+1}}$$

$$+ \frac{(aq, atq/b; q)_N}{(aq/b; q)_N} \sum_{n=0}^{\infty} \frac{(aq^N; q)_n t^n}{(bq; q)_n},$$

and (9.14) follows upon letting $N \to \infty$, noting that $F(aq^N, b; t) \to F(0, b; t)$ as $N \to \infty$. □

Exercises

9.1 Show that the Rogers–Fine identity (9.1) follows from Watson's transformation (5.10).

9.2 (i) By using the Rogers–Fine identity (9.1) or otherwise, derive the identity (for $|q|, |t| < 1$)

$$\sum_{n=0}^{\infty} \frac{t^n}{1 - xq^n} = \sum_{n=0}^{\infty} \frac{1 - xtq^{2n}}{(1 - xq^n)(1 - tq^n)} x^n t^n q^{n^2}, \tag{9.16}$$

and hence prove Clausen's [106] identity

$$\sum_{n=1}^{\infty} \frac{q^n}{1 - q^n} = \sum_{n=1}^{\infty} \frac{1 + q^n}{1 - q^n} q^{n^2}.$$

(ii) Show that if $|q| < |t| < 1$, then (9.16) may be extended to a bilateral identity (Agarwal [3, Eq. (6.16)], with $\mu = 1$)

$$\sum_{n=-\infty}^{\infty} \frac{t^n}{1 - xq^n} = \sum_{n=-\infty}^{\infty} \frac{1 - xtq^{2n}}{(1 - xq^n)(1 - tq^n)} x^n t^n q^{n^2}. \tag{9.17}$$

(iii) Prove that if $|q|, |t|, |x| < 1$, then

$$\sum_{n=0}^{\infty} \frac{xt^n q^n}{(1 - xq^n)^2} = \sum_{n=0}^{\infty} \frac{nx^n}{1 - tq^n}, \tag{9.18}$$

and if $|q| < |t|, |x| < 1$, then

$$\sum_{n=-\infty}^{\infty} \frac{xt^n q^n}{(1 - xq^n)^2} = \sum_{n=-\infty}^{\infty} \frac{nx^n}{1 - tq^n}. \tag{9.19}$$

Remark: It is not difficult to show (by expanding each denominator as a power series) that

$$\sum_{n=1}^{\infty} \frac{q^n}{1 - q^n} = \sum_{n=1}^{\infty} d(n)q^n,$$

where $d(n)$ denotes the number of positive integer divisors of n.

9.3 By employing (9.1) or otherwise, prove the identity of Ramanujan [213] (see also [38, **Entry 9.5.1**])

$$\sum_{n=0}^{\infty} \frac{(-1)^n a^{2n} q^{n^2+n}}{(-aq;q^2)_{n+1}} = \sum_{n=0}^{\infty} a^{3n} q^{3n^2+2n}(1 - aq^{2n+1}).$$

Remark: An identity equivalent to this was proved by Warnaar [255, p. 388].

9.4 By employing (9.1) or otherwise, prove that if $|b|, |t| < 1$, then

$$(1 - t)\sum_{n=0}^{\infty} \frac{(ab;q)_n}{(bq;q)_n} t^n = (1 - b)\sum_{n=0}^{\infty} \frac{(at;q)_n}{(tq;q)_n} b^n. \tag{9.20}$$

9.5 (i) Use Abel's Theorem to conclude from Q9.4 that

$$\lim_{t \to 1-} (1 - t)\sum_{n=0}^{\infty} \frac{(ab;q)_n}{(bq;q)_n} t^n = \frac{(ab;q)_\infty}{(bq;q)_\infty} = (1 - b)\sum_{n=0}^{\infty} \frac{(a;q)_n}{(q;q)_n} b^n.$$

(Note that the second equality is equivalent to the q-binomial theorem.
(ii) (Fine [115, p. 15, Eq. (14.2)]) Show from part (i) that

$$\sum_{n=0}^{\infty} \frac{(1 - abq^{2n})(ab, a; q)_n b^n q^{n^2}}{(bq, q; q)_n} = \frac{(ab;q)_\infty}{(bq;q)_\infty}.$$

9.6 Iterate the transformation (9.4) to prove that

$$\sum_{n=0}^{\infty} \frac{(aq;q)_n}{(bq;q)_n} t^n = (1 - b)\sum_{n=0}^{\infty} \frac{(atq/b;q)_n}{(t;q)_{n+1}} b^n.$$

Note that this transformation is equivalent to that in Q9.4.

9.7 (Ramanujan, [213, p. 36]) Prove that

$$\sum_{n=0}^{\infty} \frac{(-1)^n b^{2n} q^{n^2}}{(b^2 q^2; q^2)_n} = 1 + b\sum_{n=1}^{\infty} \frac{(-b)^n q^n}{(bq;q)_n}.$$

9.8 Prove that

$$\sum_{n=0}^{\infty} \frac{(\alpha;q)_n}{(\tau;q)_{n+1}} \tau^n = \sum_{n=0}^{\infty} \frac{(\alpha;q)_n^2(1 - \alpha\tau q^{2n})\tau^{2n} q^{n^2}}{(\tau;q)_{n+1}^2}.$$

9.9 (i) (Fine [115, p. 3, Eq. (4.4)]) Prove that

$$F(a,b;t) = \frac{b}{b-at} + \frac{(b-a)t}{(1-bq)(b-at)}F(a,bq;t).$$

(ii) (Fine [115, p. 5, Eqs. (7.1)–(7.2)]) Hence show that

$$\sum_{n=0}^{\infty} \frac{(aq;q)_n}{(bq;q)_n}t^n + \frac{b/at}{1-b/at}\sum_{n=0}^{\infty} \frac{(b/a;q)_n q^n}{(bq,bq/at;q)_n}$$

$$= \frac{(b/a;q)_\infty}{(bq,b/at;q)_\infty}\sum_{n=0}^{\infty}(aq;q)_n t^n.$$

(iii) (Fine [115, p. 5, Eq. (7.3)]) Hence show that

$$\sum_{n=0}^{\infty} \frac{(t;q)_n q^n}{(bq,q;q)_n} = \frac{(t;q)_\infty}{(bq,q;q)_\infty}\sum_{n=0}^{\infty}(bq/t;q)_n t^n.$$

9.10 (i) (Fine [115, p. 3, Eq. (4.6)]) By composing the last equality at (9.4) and (9.15), show that

$$\frac{1-t}{1-b}F(a,b;t) = 1 - \frac{b-atq}{b-aq}aq$$

$$+ \frac{(1-aq)(b-atq)(b-atq^2)}{(b-aq)(1-tq)}\left\{\frac{1-tq}{1-b}F(aq,b;tq)\right\}.$$

(ii) (Fine [115, p. 14, Eq. (13.1)]) By iterating the transformation in part (i), show that

$$\frac{1-t}{1-b}F(a,b;t) = \sum_{n=0}^{\infty} \frac{(aq;q)_n(atq/b;q)_{2n}}{(tq,aq/b;q)_n}b^n$$

$$- aq\sum_{n=0}^{\infty} \frac{(aq;q)_n(atq/b;q)_{2n+1}}{(tq;q)_n(aq/b;q)_{n+1}}(bq)^n.$$

(iii) (Fine [115, p. 14, Eq. (13.2)]) By using the result in Q9.4, and making a suitable change of variables, prove

$$F(a,b;t) = \sum_{n=0}^{\infty} \frac{(atq/b;q)_n(aq;q)_{2n}}{(bq,aq/b;q)_n}t^n$$

$$- \frac{atq}{b}\sum_{n=0}^{\infty} \frac{(atq/b;q)_n(aq;q)_{2n+1}}{(bq;q)_n(aq/b;q)_{n+1}}(tq)^n.$$

(iv) (Fine [115, p. 15, Eq. (13.5)]) By taking a suitable limit in part (i), show that

$$\frac{(aq;q)_\infty}{(b;q)_\infty} = \sum_{n=0}^{\infty} \frac{(aq;q)_n(aq/b;q)_{2n}}{(q,aq/b;q)_n} b^n - aq \sum_{n=0}^{\infty} \frac{(aq;q)_n(aq/b;q)_{2n+1}}{(q;q)_n(aq/b;q)_{n+1}} (bq)^n.$$

Chapter 10

Bailey Pairs

One of the most effective methods for developing basic hypergeometric identities involves the use of *Bailey Pairs*. Many series-product identities that derive from Bailey pairs are *not* special cases of more general series-product identities such has Bailey's $_6\psi_6$ identity. Instead, Bailey pairs initially give rise to series-series identities, after which various parameters need to be specialized so that the Jacobi triple identity (6.1), quintuple product identity (6.19), Bailey's $_6\psi_6$ identity (8.1), or other identity, may be used to convert one of the infinite series to an infinite product.

Since the topic is of such importance to the subject of basic hypergeometric series, some time will be spent developing the material. Most of the Bailey pairs (see below for an explanation of the term) that presently exist were found by Slater [236,237], who was primarily interested in finding new identities of Rogers–Ramanujan type, and employed Bailey's $_6\psi_6$ identity (8.1) to develop these pairs. When the authors in [197] retraced Slater's steps to see if the same methods would uncover any new and interesting identities when applied to Chu's [103] $_{10}\psi_{10}$ extension of Bailey's $_6\psi_6$ summation formula, it was discovered that many of Slater's Bailey pairs were actually special cases of more general Bailey pairs containing free parameters. Slater could certainly have discovered these more general pairs, had not her primary interest, as mentioned above, been the discovery of new infinite series — product identities.

The concept of what is now termed a Bailey pair was initiated by Bailey in two papers [58,61], and while he stated a terminating version of (10.6) below, like Slater in her two papers [236,237], Bailey was primarily interested in *infinite* series — product identities, and did not investigate the possibilities of this terminating transformation. Consequently he missed possibly its most important property, namely, that it could be used to derive a new Bailey pair from an existing Bailey pair, and that this process could be iterated to produce a *Bailey chain*, an infinite sequence of Bailey pairs. This property was discovered by Andrews [22], and we examine it in a subsequent chapter.

In this chapter we will outline how Slater's methods allow these more general Bailey pairs to be derived, and will then specialize the parameters to show how they lead to various interesting identities, including identities of Rogers–Ramanujan type. We do not consider all the Bailey pairs found by Slater [236,237], and it is recommended that the reader who is interested in acquiring more details read the papers of Slater and also the paper [197].

10.1 The Bailey transform and Bailey pairs

To describe the method of Bailey pairs, we first describe the *Bailey transform* (see [61]).

Lemma 10.1. *Subject to suitable convergence conditions, if*

$$\beta_n = \sum_{r=0}^{n} \alpha_r U_{n-r} V_{n+r} \tag{10.1}$$

and

$$\gamma_n = \sum_{r=n}^{\infty} \delta_r U_{r-n} V_{r+n}, \tag{10.2}$$

then

$$\sum_{n=0}^{\infty} \alpha_n \gamma_n = \sum_{n=0}^{\infty} \beta_n \delta_n. \tag{10.3}$$

Proof. We assume that the sequences are such that all the infinite series converge, and that the change in the order of summation below is justified.

$$\sum_{n=0}^{\infty} \alpha_n \gamma_n = \sum_{n=0}^{\infty} \alpha_n \sum_{r=n}^{\infty} \delta_r U_{r-n} V_{r+n}$$

$$= \sum_{r=0}^{\infty} \delta_r \sum_{n=0}^{r} U_{r-n} V_{r+n} \alpha_n$$

$$= \sum_{r=0}^{\infty} \delta_r \beta_r.$$

$\square$

The interest lies in finding choices for the sequences $\{U_n\}$, $\{V_n\}$, $\{\alpha_n\}$ and $\{\delta_n\}$ such that the expressions for β_n and γ_n have closed form, and thus that (10.3) is an equation relating single series, rather than double- or iterated series.

Bailey [61, pp. 2–3] takes

$$U_n = \frac{1}{(q;q)_n}, V_n = \frac{1}{(aq;q)_n} \text{ and } \delta_n = (\rho_1, \rho_2; q)_n \left(\frac{aq}{\rho_1 \rho_2}\right)^n \tag{10.4}$$

to obtain the following general transformation.

Theorem 10.1. *Subject to suitable convergence conditions, if*

$$\beta_n(a, q) = \sum_{r=0}^{n} \frac{\alpha_r(a, q)}{(q;q)_{n-r}(aq;q)_{n+r}}, \tag{10.5}$$

then

$$\sum_{n=0}^{\infty} (\rho_1, \rho_2; q)_n \left(\frac{aq}{\rho_1 \rho_2}\right)^n \beta_n(a, q)$$

$$= \frac{(aq/\rho_1, aq/\rho_2; q)_\infty}{(aq, aq/\rho_1\rho_2; q)_\infty} \sum_{n=0}^{\infty} \frac{(\rho_1, \rho_2; q)_n}{(aq/\rho_1, aq/\rho_2; q)_n} \left(\frac{aq}{\rho_1 \rho_2}\right)^n \alpha_n(a, q). \tag{10.6}$$

Proof. With the choices for U_n, V_n and δ_n above,

$$\gamma_n = \sum_{r=n}^{\infty} \delta_r U_{r-n} V_{r+n} = \sum_{r=0}^{\infty} \delta_{r+n} U_r V_{r+2n}$$

$$= \sum_{r=0}^{\infty} \frac{(\rho_1, \rho_2; q)_{r+n}}{(q;q)_r (aq;q)_{r+2n}} \left(\frac{aq}{\rho_1 \rho_2}\right)^{r+n}$$

$$= \frac{(\rho_1, \rho_2; q)_n}{(aq;q)_{2n}} \left(\frac{aq}{\rho_1 \rho_2}\right)^n \sum_{r=0}^{\infty} \frac{(\rho_1 q^n, \rho_2 q^n; q)_r}{(q, aq^{2n+1}; q)_r} \left(\frac{aq}{\rho_1 \rho_2}\right)^r$$

$$= \frac{(\rho_1, \rho_2; q)_n}{(aq;q)_{2n}} \left(\frac{aq}{\rho_1 \rho_2}\right)^n \frac{(aq^{n+1}/\rho_1, aq^{n+1}/\rho_2; q)_\infty}{(aq^{2n+1}, aq/\rho_1\rho_2; q)_\infty}$$

$$= \frac{(aq/\rho_1, aq/\rho_2; q)_\infty}{(aq, aq/\rho_1\rho_2; q)_\infty} \frac{(\rho_1, \rho_2; q)_n}{(aq/\rho_1, aq/\rho_2; q)_n} \left(\frac{aq}{\rho_1 \rho_2}\right)^n, \tag{10.7}$$

where the next to last sum is by Heine's q-Gauss summation formula (4.8). The result is now a consequence of Lemma 10.1. $\qquad\square$

Remark: We call a pair of sequences $(\alpha_n, \beta_n) = (\alpha_n(a, q), \beta_n(a, q))$ that satisfy (10.5) a *Bailey pair with respect to a*, and the pair $(\delta_n, \gamma_n) = (\delta_n(a, q), \gamma_n(a, q))$ that satisfies (10.2) (with $U_n = 1/(q; q)_n$, $V_n = 1/(aq; q)_n$) a *conjugate Bailey pair with respect to a*. Note that unless otherwise specified, $\alpha_0 = \beta_0 = 1$, and in particular that this convention takes precedence over any formula $\alpha_n = f(n)$ for which $f(0) \neq 1$.

The following particular cases of Theorem 10.1 are useful for deriving various types of identity, including many identities of Rogers–Ramanujan type (Slater [237] made frequent use of some of these special cases).

Corollary 10.1. *If (α_n, β_n) is a Bailey pair with respect to a, then*

$$\sum_{n=0}^{\infty} a^n q^{n^2} \beta_n(a, q) = \frac{1}{(aq)_\infty} \sum_{r=0}^{\infty} a^r q^{r^2} \alpha_r(a, q), \tag{10.8}$$

$$\sum_{n=0}^{\infty} a^n q^{n^2} (-q; q^2)_n \beta_n(a, q^2) = \frac{(-aq; q^2)_\infty}{(aq^2; q^2)_\infty} \sum_{r=0}^{\infty} \frac{a^r q^{r^2} (-q; q^2)_r}{(-aq; q^2)_r} \alpha_r(a, q^2), \tag{10.9}$$

$$\sum_{n=0}^{\infty} (-a)^n q^{n^2} (q; q^2)_n \beta_n(a, q^2) = \frac{(aq; q^2)_\infty}{(aq^2; q^2)_\infty} \sum_{r=0}^{\infty} \frac{(-a)^r q^{r^2} (q; q^2)_r}{(aq; q^2)_r} \alpha_r(a, q^2), \tag{10.10}$$

$$\sum_{n=0}^{\infty} a^n q^{n(n+1)/2} (-1)_n \beta_n(a, q) = \frac{(-aq)_\infty}{(aq)_\infty} \sum_{r=0}^{\infty} \frac{a^r q^{r(r+1)/2} (-1)_r}{(-aq)_r} \alpha_r(a, q), \tag{10.11}$$

$$\sum_{n=0}^{\infty} a^n q^{n(n-1)/2} (-q)_n \beta_n(a, q) = \frac{(-a)_\infty}{(aq)_\infty} \sum_{r=0}^{\infty} \frac{a^r q^{r(r-1)/2} (-q)_r}{(-a)_r} \alpha_r(a, q), \tag{10.12}$$

$$\sum_{n=0}^{\infty} (-a)^n q^{n^2} (aq; q^2)_n \beta_n(a^2, q^2)$$

$$= \frac{(aq; q^2)_\infty}{(a^2 q^2; q^2)_\infty} \sum_{r=0}^{\infty} (-a)^r q^{r^2} \alpha_r(a^2, q^2), \tag{10.13}$$

$$\sum_{n=0}^{\infty} a^n q^{n^2} (-aq; q^2)_n \beta_n(a^2, q^2)$$

$$= \frac{(-aq; q^2)_\infty}{(a^2 q^2; q^2)_\infty} \sum_{r=0}^{\infty} a^r q^{r^2} \alpha_r(a^2, q^2), \tag{10.14}$$

$$\sum_{n=0}^{\infty} (aq; q^2)_n \, (-1)^n \, \beta_n(a, q) = \frac{1}{(aq^2; q^2)_\infty (-1; q)_\infty} \sum_{n=0}^{\infty} (-1)^n \, \alpha_n(a, q).$$

(10.15)

Proof. For (10.8), let $\rho_1, \rho_2 \to \infty$ in (10.6). The case (10.9) follows after replacing q with q^2, letting $\rho_1 \to \infty$ and setting $\rho_2 = -q$, and (10.10) is similar, except for setting $\rho_2 = q$. Similarly, (10.11) is a consequence of letting $\rho_1 \to \infty$ and setting $\rho_2 = -1$, and the proof of (10.12) is identical to that of (10.11), apart from setting $\rho_2 = -q$. For (10.13), let $\rho_1 \to \infty$ and set $\rho_2 = \sqrt{aq}$, and then replace a with a^2 and q with q^2, and (10.14) is similar, apart from setting $\rho_2 = -\sqrt{aq}$. Finally, (10.15) follows after setting $\rho_1 = \sqrt{aq}$ and $\rho_2 = -\sqrt{aq}$. $\qquad\square$

Remarks: The transformation at (10.15) has not been investigated much in the literature, possibly for a number of reasons. Firstly the series involving β_n will not converge, if $\beta_n \not\to 0$ as $n \to \infty$. Secondly, even if this series does converge and the series involving the α_n does converge to an infinite product, the power of q on the series side may not be quadratic in the exponent, and such identities may not be regarded as being of Rogers–Ramanujan–Slater type. However, this transformation was examined by the authors in [197], where some interesting applications were given. The transformation at (10.8) is known as the "weak Bailey lemma". Also, it may seem that the transformation at (10.14) may be derived trivially from that at (10.13), by replacing q with $-q$, but this is true only if a is independent of q. As may be seen from Q10.13, they can produce quite different identities when $a = q$.

The initial approach to finding Bailey pairs in many cases is to transform existing summation formulae so that they have the form at (10.5), and the pair (α_n, β_n) is then extracted. This was the method use by Slater [236,237], working with Bailey's ${}_6\psi_6$ summation formula (8.1) and the ${}_6\phi_5$ summation formula (5.5). As an example, we follow Slater [236] and derive a quite general Bailey pair from the latter formula.

Lemma 10.2. *The pair of sequences* (α_n, β_n) *defined by*

$$\alpha_n = \frac{(1 - aq^{2n})(a, b, c; q)_n}{(1 - a)(aq/b, aq/c, q; q)_n} \left(\frac{-aq}{bc} \right)^n q^{(n^2 - n)/2},$$

(10.16)

$$\beta_n = \frac{(aq/bc; q)_n}{(aq/b, aq/c, q; q)_n},$$

constitute a Bailey pair with respect to a.

Proof. In (5.5), use (2.14) to replace the $(q^{-n}; q)_k$ factor, and then divide both sides by $(aq, q; q)_n$. This gives

$$\sum_{k=0}^{n} \frac{1 - aq^{2k}}{1 - a} \frac{(a, b, c; q)_k}{(aq/b, aq/c, q; q)_k (q; q)_{n-k} (aq; q)_{n+k}} \left(\frac{-aq}{bc} \right)^k q^{k(k-1)/2}$$
$$= \frac{(aq/bc; q)_n}{(aq/b, aq/c, q; q)_n},$$

and (10.16) is now seen to hold by inspection. $\square$

Remark: The identity obtained by inserting the pair at (10.16) into (10.6), namely,

$$\sum_{n=0}^{\infty} \frac{(\rho_1, \rho_2, aq/bc; q)_n}{(aq/b, aq/c, q; q)_n} \left(\frac{aq}{\rho_1 \rho_2} \right)^n = \frac{(aq/\rho_1, aq/\rho_2; q)_\infty}{(aq, aq/\rho_1 \rho_2; q)_\infty}$$
$$\times \sum_{n=0}^{\infty} \frac{(1 - aq^{2n})(a, b, c, \rho_1, \rho_2; q)_n}{(1 - a)(aq/b, aq/c, aq/\rho_1, aq/\rho_2, q; q)} \left(\frac{-a^2 q^2}{bc\rho_1 \rho_2} \right)^n q^{(n^2-n)/2},$$

is equivalent to the special case $n \to \infty$ of Watson's [257] transformation (5.10) for a terminating $_8\phi_7$ series. Thirty of Slater's [236] Bailey pairs $(B(1) - B(4), F(1) - F(4), E(3), E(6), E(7), H(1) - H(19))$ are special cases of the pair at (10.16), and twenty three of the identities in the list of one hundred and thirty identities in her 1952 paper [237] derive from some of these pairs, so that these twenty three identities (those numbered **S.1, S.2, S.6 - S.14, S.18, S.23, S.24, S.27 - S.30, S.34** and **S.36 - S.39**, where we here and subsequently denote the identity labelled "(n)" in Slater's paper [237] as **S.n**) including, as will be seen below, the Rogers–Ramanujan identities, are ultimately special cases of Watson's [257] transformation at (5.10).

10.2 A proof of the Rogers–Ramanujan identities

We are now in a position to give the proof of the famous Rogers–Ramanujan identities outlined by Slater [237].

Theorem 10.2. *If* $|q| < 1$, *then*

$$\sum_{n=0}^{\infty} \frac{q^{n^2}}{(q; q)_n} = \frac{1}{(q, q^4; q^5)_\infty}; \tag{10.17}$$

$$\sum_{n=0}^{\infty} \frac{q^{n(n+1)}}{(q;q)_n} = \frac{1}{(q^2,q^3;q^5)_\infty}. \tag{10.18}$$

Proof. Let $b, c \to \infty$ in (10.16), and substitute the resulting Bailey pair in (10.8) to get

$$\sum_{n=0}^{\infty} \frac{a^n q^{n^2}}{(q;q)_n} = \frac{1}{(aq;q)_\infty} \sum_{n=0}^{\infty} a^n q^{n^2} \frac{(1-aq^{2n})(a;q)_n}{(1-a)(q;q)} (-aq)^n q^{(3n^2-3n)/2}.$$

For (10.17), let $a \to 1$ and simplify to get

$$\sum_{n=0}^{\infty} \frac{q^{n^2}}{(q;q)_n} = \frac{1}{(q;q)_\infty} \left(1 + \sum_{n=1}^{\infty}(1+q^n)(-1)^n q^{(5n^2-n)/2} \right)$$

$$= \frac{1}{(q;q)_\infty} \sum_{n=-\infty}^{\infty} (-1)^n q^{(5n^2-n)/2} = \frac{(q^2,q^3,q^5;q^5)_\infty}{(q;q)_\infty} = \frac{1}{(q,q^4;q^5)_\infty},$$

where the next-to-last equality use the Jacobi triple product identity (6.1). The proof of (10.18) follows similarly, upon setting $a = q$. □

10.3   Example: mod 3 Bailey pairs

Before giving further examples of identities derived through inserting Bailey pairs in one of the transformations in Corollary 10.1, we first give an illustration of how Slater [236] used Bailey's $_6\psi_6$ summation formula (8.1) to derive Bailey pairs. The Bailey pairs in the next corollary, and in questions 10.1–10.3 were given in [197], and some of the **A** and **J** pairs of Slater [236, 237] are special cases. However, the more general pairs could easily have been found by Slater, by the same methods she used to find the more specialized pairs (her primary purpose was to find new identities of Rogers–Ramanujan type, as stated earlier).

Corollary 10.2. *The sequences* (α_n, β_n) *below are Bailey pairs with respect to a, where*

$$\alpha_{3r} = \frac{1-aq^{6r}}{1-a} \frac{(a;q^3)_r}{(q^3;q^3)_r} q^{\frac{9r^2-3r}{2}} (-a)^r, \qquad \alpha_{3r\pm 1} = 0, \tag{10.19}$$

$$\beta_n = \frac{(a;q^3)_n}{(a;q)_{2n}(q;q)_n};$$

$$\alpha_{3r} = \frac{\left(aq;q^3\right)_r}{\left(q^3;q^3\right)_r} q^{\frac{9r^2-r}{2}} (-a)^r, \qquad\qquad \alpha_{3r-1} = 0, \qquad\qquad (10.20)$$

$$\alpha_{3r+1} = \frac{\left(aq;q^3\right)_r}{\left(q^3;q^3\right)_r} q^{\frac{9r^2+11r}{2}+1} (-a)^{r+1}, \qquad \beta_n = \frac{(aq;q^3)_n}{(aq;q)_{2n}(q;q)_n};$$

$$\alpha_{3r} = \frac{\left(aq;q^3\right)_r}{\left(q^3;q^3\right)_r} q^{\frac{9r^2+5r}{2}} (-a)^r, \qquad\qquad \alpha_{3r-1} = 0, \qquad\qquad (10.21)$$

$$\alpha_{3r+1} = -\frac{\left(aq;q^3\right)_r}{\left(q^3;q^3\right)_r} q^{\frac{9r^2+5r}{2}} (-a)^r, \qquad \beta_n = \frac{(aq;q^3)_n q^n}{(aq;q)_{2n}(q;q)_n}.$$

Proof. Replace q with q^3 in (8.1) and then set $b = q^{-n}$, $c = q^{1-n}$ and $d = q^{2-n}$ so that the series terminates above. This gives, after simplifying,

$$\sum_{k=-\infty}^{n/3} \frac{(1-aq^{6k})(q^{-n};q)_{3k}(e;q^3)_k}{(1-a)(aq^{n+1};q)_{3k}(aq^3/e;q^3)_k} \left(\frac{q^{3n}a^2}{e}\right)^k$$

$$= \frac{(a,q^3/a,aq/e,aq^2/e;q^3)_\infty}{(q,q^2,q^3/e,a^2/e;q^3)_\infty} \frac{(aq,q;q)_n(a^2/e;q^3)_n}{(a;q)_{2n}(aq/e;q)_n}. \qquad (10.22)$$

Set $e = a$ so that all the terms with $k < 0$ vanish, apply (2.14) to the term $(q^{-n};q)_{3k}$, and divide both sides by $(aq,q;q)_n$ to get

$$\sum_{k=0}^{n/3} \frac{(1-aq^{6k})(a;q^3)_k(-a)^k q^{(9k^2-3k)/2}}{(1-a)(q^3;q^3)_k(aq;q)_{n+3k}(q;q)_{n-3k}} = \frac{(a;q^3)_n}{(a;q)_{2n}(q;q)_n}, \qquad (10.23)$$

and (10.19) follows from the definition of a Bailey pair. (Note that Slater's pair $J(1)$ from [237] follows upon setting $a = 1$.)

Replace a with aq in (10.23), and (10.20) follows from the identity

$$\frac{(1-aq^{6r+1})q^{(9r^2-r)/2}}{(aq;q)_{n+3r+1}(q;q)_{n-3r}} = \frac{q^{(9r^2-r)/2}}{(aq;q)_{n+3r}(q;q)_{n-3r}} - \frac{a\,q^{(9r^2+11r)/2+1}}{(aq;q)_{n+3r+1}(q;q)_{n-3r-1}},$$

and (10.21) follows from the identity

$$\frac{(1-aq^{6r+1})q^{(9r^2-r)/2}}{(aq;q)_{n+3r+1}(q;q)_{n-3r}} = \frac{q^{-n}q^{(9r^2+5r)/2}}{(aq;q)_{n+3r}(q;q)_{n-3r}} - \frac{q^{-n}q^{(9r^2+5r)/2}}{(aq;q)_{n+3r+1}(q;q)_{n-3r-1}}.$$

$\square$

Each of these Bailey pairs gives rises to a transformation between basic hypergeometric series upon substitution into (10.6). For example, the pair at (10.19) leads to the following identity.

Corollary 10.3. *If* $|q|, |aq/\rho_1\rho_2| < 1$*, then*

$$\sum_{n=0}^{\infty} \frac{(\rho_1, \rho_2; q)_n (a; q^3)_n}{(a; q)_{2n}(q; q)_n} \left(\frac{aq}{\rho_1\rho_2}\right)^n = \frac{(aq/\rho_1, aq/\rho_2; q)_\infty}{(aq, aq/\rho_1\rho_2; q)_\infty}$$

$$\times \sum_{n=0}^{\infty} \frac{(1 - aq^{6n})(\rho_1, \rho_2; q)_{3n}(a; q^3)_n a^{4n} q^{9n^2/2 + 3n/2}}{(1 - a)(aq/\rho_1, aq/\rho_2; q)_{3n}(q^3; q^3)_n} \left(\frac{-1}{\rho_1\rho_2}\right)^{3n}. \quad (10.24)$$

The transformations in Corollary 10.1 may also be applied to the Bailey pairs in Corollary 10.2.

Corollary 10.4. *The following identities hold.*

$$\sum_{n=0}^{\infty} \frac{(q^3; q^3)_n q^{n^2+3n}}{(q; q)_{2n+2}(q; q)_n} = \frac{(q^3, q^{24}, q^{27}; q^{27})_\infty}{(q; q)_\infty}. \quad (10.25)$$

$$1 + \sum_{n=1}^{\infty} \frac{(-q; q^2)_n (q^6; q^6)_{n-1} q^{n^2}}{(q^2; q^2)_{2n-1}(q^2; q^2)_n} = \frac{(-q; q^2)_\infty (q^{15}, q^{21}, q^{36}; q^{36})_\infty}{(q^2; q^2)_\infty}. \quad (10.26)$$

$$1 + \sum_{n=1}^{\infty} \frac{(-1; q)_n (q^3; q^3)_{n-1} q^{n(n+1)/2}}{(q; q)_{2n-1}(q; q)_n} = \frac{(-q; q)_\infty (q^9, q^9, q^{18}; q^{18})_\infty}{(q; q)_\infty}. \quad (10.27)$$

$$\sum_{n=0}^{\infty} \frac{(q^3; q^3)_n (-q)^n}{(q^2; q^2)_{n+1}(q; q)_n} = \frac{(q; q^2)_\infty}{(q^2; q^2)_\infty} \frac{(q^{18}; q^{18})_\infty}{(q^9; q^{18})_\infty}. \quad (10.28)$$

Proof. To get (10.25), let $a = q^2$ in (10.21), substitute the resulting pair in (10.8), combine the two series into a single bi-lateral series (shifting the summation index by 1 in one series), and use the Jacobi triple product identity (6.1) to sum the resulting series on the right.

The identity at (10.26) follows similarly, by letting $a \to 1$ in (10.19), substituting the resulting pair into (10.9), and again using the Jacobi triple product identity to sum the resulting series on the right. The identity at (10.27) follows upon substituting the same pair into (10.11) and dealing similarly with the resulting series on the right.

Finally, to get (10.28), let $a = q^2$ in (10.21), substitute the resulting pair in (10.15), use (6.12) to sum the resulting series, and simplify. □

Remark: The identities at (10.26) and (10.27) are originally due to Slater [237, **S.114, S.78**], (10.25) appears in Bailey's paper [59, p. 434, (B1)] (where Bailey attributes it to Dyson), and (10.28) first appeared in [197, Eq. (3.13)].

The derivation of some of the other Bailey pairs employed by Slater [236, 237] are given as exercises below. All of the mod 3, mod 2 and mod 4 pairs listed in Appendix V may be derived using methods similar to those employed above and in the exercises following.

Exercises

10.1 (i) By starting with (10.22) and setting $a = q$ (instead of $e = a$), prove that

$$\sum_{k=-(n+1)/3}^{n/3} \frac{(1 - q^{6k+1})(e; q^3)_k (-1)^k q^{(9k^2+k)/2}}{(q^4/e; q^3)_k (q; q)_{n+3k+1} (q; q)_{n-3k} \, e^k} = \frac{(q^2/e; q^3)_n}{(q; q)_{2n} (q^2/e; q)_n}. \quad (10.29)$$

(ii) By separating the above sum into two sums, those with $k \geq 0$ and $k < 0$, and for $k < 0$ using (2.5), show that

$$\sum_{k=0}^{n/3} \frac{(1 - q^{6k+1})(e; q^3)_k (-1)^k q^{(9k^2+k)/2}}{(q^4/e; q^3)_k (q; q)_{n+3k+1} (q; q)_{n-3k} \, e^k}$$

$$+ \sum_{k=1}^{(n+1)/3} \frac{(1 - q^{-6k+1})(e/q; q^3)_k (-1)^k q^{(9k^2+7k)/2}}{(q^3/e; q^3)_k (q; q)_{n-3k+1} (q; q)_{n+3k} \, e^k} = \frac{(q^2/e; q^3)_n}{(q; q)_{2n} (q^2/e; q)_n}. \quad (10.30)$$

(iii) Finally, by applying the identities

$$(1 - q^{6k+1}) q^{(9k^2+k)/2} = q^{(9k^2+k)/2} (1 - q^{n+3k+1})$$
$$- q^{(9k^2+13k)/2+1} (1 - q^{n-3k}),$$

$$(1 - q^{-6k+1}) q^{(9k^2+7k)/2} = q^{(9k^2+7k)/2} (1 - q^{n-3k+1})$$
$$- q^{(9k^2-5k)/2+1} (1 - q^{n+3k}),$$

to the series in (10.30), derive the Bailey pair (with respect to $a = 1$)

$$\alpha_{3k} = (-1)^k \left(\frac{q^{(9k^2+k)/2}(e; q^3)_k}{(q^4/e; q^3)_k e^k} + \frac{q^{(9k^2+7k)/2}(e/q; q^3)_k}{(q^3/e; q^3)_k e^k} \right), \quad (10.31)$$

$$\alpha_{3k+1} = \frac{(-1)^{k+1} q^{(9k^2+13k)/2+1}(e;q^3)_k}{(q^4/e;q^3)_k e^k},$$

$$\alpha_{3k-1} = \frac{(-1)^{k+1} q^{(9k^2-5k)/2+1}(e/q;q^3)_k}{(q^3/e;q^3)_k e^k}, \qquad \beta_n = \frac{(q^2/e;q^3)_n}{(q;q)_{2n}(q^2/e;q)_n}.$$

10.2 Instead of the identities in part (iii) of Q10.1, apply the identities

$$(1 - q^{6k+1})q^{(9k^2+k)/2} = q^{(9k^2+7k)/2-n}((1 - q^{n+3k+1}) - (1 - q^{n-3k})),$$
$$(1 - q^{-6k+1})q^{(9k^2+7k)/2} = q^{(9k^2+k)/2-n}((1 - q^{n-3k+1}) - (1 - q^{n+3k})),$$

to the series in (10.30) to derive the Bailey pair (with respect to $a = 1$)

$$\alpha_{3k} = (-1)^k \left(\frac{q^{9k^2/2+7k/2}(e;q^3)_k}{(q^4/e;q^3)_k e^k} + \frac{q^{9k^2/2+k/2}(e/q;q^3)_k}{(q^3/e;q^3)_k e^k} \right), \qquad (10.32)$$

$$\alpha_{3k+1} = \frac{(-1)^{k+1} q^{9k^2/2+7k/2}(e;q^3)_k}{(q^4/e;q^3)_k e^k},$$

$$\alpha_{3k-1} = \frac{(-1)^{k+1} q^{9k^2/2+k/2}(e/q;q^3)_k}{(q^3/e;q^3)_k e^k}, \qquad \beta_n = \frac{q^n(q^2/e;q^3)_n}{(q;q)_{2n}(q^2/e;q)_n}.$$

10.3 By slightly modifying (10.30), prove that (α_n, β_n) is a Bailey pair (with respect to $a = q$), where

$$\alpha_{3k} = (-1)^k \frac{1 - q^{6k+1}}{1 - q} q^{\frac{9k^2+k}{2}} \frac{(e;q^3)_k}{(q^4/e;q^3)_k} e^k, \qquad \alpha_{3k+1} = 0, \qquad (10.33)$$

$$\alpha_{3k-1} = (-1)^{k+1} \frac{1 - q^{6k-1}}{1 - q} q^{\frac{9k^2-5k}{2}+1} \frac{(e/q;q^3)_k}{(q^3/e;q^3)_k} e^k, \qquad \beta_n = \frac{(q^2/e;q^3)_n}{(q;q)_{2n}(q^2/e;q)_n}.$$

10.4 (Slater [237, **S.95**]) Use the $e = 0$ case of the Bailey pair at (10.31) in conjunction with the transformation at (10.9) and the quintuple product identity (6.21) to prove the identity

$$\sum_{n=0}^{\infty} \frac{(-q;q^2)_n q^{3n^2-2n}}{(q^2;q^2)_{2n}} = (q^3, q^7, q^{10}; q^{10})_\infty (q^4, q^{16}; q^{20})_\infty \frac{(-q;q^2)_\infty}{(q^2;q^2)_\infty}.$$

10.5 Specialize the pair at (10.32) and insert in (10.15) to prove the identity

$$1 + \sum_{n=1}^{\infty} \frac{(q;q^2)_n (q^3;q^3)_{n-1}(-q)^n}{(q;q)_{2n}(q;q)_{n-1}} = \frac{(q;q^2)_\infty}{(q^2;q^2)_\infty}(-q^3, -q^6, q^9; q^9)_\infty.$$

10.6 Prove Ramanujan's identity [213, p. 33]

$$\sum_{n=0}^{\infty} \frac{(q/z;q)_n(z;q)_n q^{n^2}}{(q;q)_{2n}} = \frac{(qz,q^2/z,q^3;q^3)_\infty}{(q;q)_\infty}, \qquad (10.34)$$

by letting $e \to 0$ in the Bailey pair at (10.33), inserting the resulting pair into (10.5)(after replacing ρ_1 with z and ρ_2 with q/z), using (2.8) to simplify the ratio of q-products in the resulting series, using (2.5) to combine the two series into a single bi-lateral series, and finally using a limiting case of Bailey's $_6\psi_6$ summation formula to convert the bi-lateral series to an infinite product.

10.7 (i) Mirror the proof in Corollary 10.2 by replacing q with q^2 in (8.1) and then setting $b = q^{-n}$, $c = q^{1-n}$ and $e = a$ to derive the identity

$$\sum_{k=0}^{n/2} \frac{(1 - aq^{4k})(a,d;q^2)_k\, a^k q^{2k^2}}{(1-a)(aq^2/d,q^2;q^2)_k(aq;q)_{n+2k}(q;q)_{n-2k}\, d^k} = \frac{(aq/d;q^2)_n}{(aq;q^2)_n(aq/d,q;q)_n}. \qquad (10.35)$$

(ii) ([197, Eq. (2.47)]) Hence derive the Bailey pair, with respect to a, with $\alpha_{2k-1} = 0$,

$$\alpha_{2k} = \frac{1 - aq^{4k}}{1-a} \frac{(a,d;q^2)_k\, q^{2k^2} a^k}{\left(aq^2/d,q^2;q^2\right)_k d^k}, \qquad \beta_n = \frac{(aq/d;q^2)_n}{(aq;q^2)_n(aq/d,q;q)_n}. \qquad (10.36)$$

10.8 Replace a with aq in (10.35) and use the identities

$$(1 - aq^{4r+1})q^{2r^2+r} = q^{2r^2+r}(1 - aq^{n+2r+1}) - aq^{2r^2+5r+1}(1 - q^{n-2r}),$$

$$(1 - aq^{4r+1})q^{2r^2+r} = q^{2r^2+3r-n}((1 - aq^{n+2r+1}) - (1 - q^{n-2r})),$$

to derive the Bailey pairs, with respect to a ([197, Eqs. (2.48), (2.49)]),

$$\alpha_{2r} = \frac{(aq,d;q^2)_r\, q^{2r^2+r} a^r}{(aq^3/d,q^2;q^2)_r\, d^r}, \qquad \alpha_{2r+1} = -\frac{(aq,d;q^2)_r\, q^{2r^2+5r+1} a^{r+1}}{(aq^3/d,q^2;q^2)_r\, d^r}, \qquad (10.37)$$

$$\beta_n = \frac{(aq^2/d;q^2)_n}{(aq^2;q^2)_n(aq^2/d,q;q)_n}.$$

$$\alpha_{2r} = \frac{(aq,d;q^2)_r\, q^{2r^2+3r} a^r}{(aq^3/d,q^2;q^2)_r\, d^r}, \qquad \alpha_{2r+1} = -\frac{(aq,d;q^2)_r\, q^{2r^2+3r} a^r}{(aq^3/d,q^2;q^2)_r\, d^r}, \qquad (10.38)$$

$$\beta_n = \frac{(aq^2/d;q^2)_n q^n}{(aq^2;q^2)_n(aq^2/d,q;q)_n}.$$

10.9 (i) Use the Bailey pair corresponding to the special case $a = q^2$ and $d = q$ of the general Bailey pair (10.36) in conjunction with one of the transformations in Corollary 10.1 derive the following identity of Slater [237, **S.56**]:

$$\sum_{n=0}^{\infty} \frac{(-q;q)_n q^{n(n+2)}}{(q;q^2)_{n+1}(q;q)_{n+1}} = \frac{(-q,-q^{11},q^{12};q^{12})_\infty}{(q;q)_\infty}.$$

(ii) Similarly use the special case $a = q^2$ and $d = -q^2$ of the general Bailey pair (10.36) to prove the identity [237, **S.35**]

$$\sum_{n=0}^{\infty} \frac{(-q;q^2)_n(-q;q)_n q^{n(n+3)/2}}{(q;q)_{2n+1}} = \frac{(q,q^7,q^8;q^8)_\infty(-q;q)_\infty}{(q;q)_\infty}.$$

(iii) Use the special case $a \to 1$ and $d \to \infty$ of the same general Bailey pair to prove the identity [237, **S.61**]

$$\sum_{n=0}^{\infty} \frac{q^{n^2}}{(q;q^2)_n(q;q)_n} = \frac{(q^6,q^8,q^{14};q^{14})_\infty}{(q;q)_\infty}.$$

10.10 (i) Replace q with q^4 and then set $b = q^{-n}$, $c = q^{1-n}$, $d = q^{2-n}$ and $e = q^{3-n}$ in Bailey's ${}_6\psi_6$ sum, and then rearrange the resulting identity to get

$$\sum_{k=-\infty}^{n/4} \frac{1 - aq^{8k}}{1 - a} \frac{q^{8r^2-4r}a^{2r}}{(aq;q)_{n+4k}(q;q)_{n-4k}} = \frac{(q^4/a,a/q,a,aq;q^4)_\infty}{(q,q^2,q^3,a^2/q^2;q^4)_\infty} \frac{(-a/q;q^2)_n}{(a;q)_{2n}}.$$

$$(10.39)$$

(ii) Let $a \to q$ in (10.39), separate the resulting series on the left into terms with $k \geq 0$ and $k < 0$ (and replacing k with $-k$ in the latter), and use the identities

$$(1 - q^{8r+1})q^{8r^2-2r} = q^{8r^2-2r}(1 - q^{n+4r+1}) - q^{8r^2+6r+1}(1 - q^{n-4r}),$$
$$(1 - q^{-8r+1})q^{8r^2+2r} = q^{8r^2+2r}(1 - q^{n-4r+1}) - q^{8r^2-6r+1}(1 - q^{n+4r}),$$

to derive the "mod 4" Bailey pair, with respect to $a = 1$, from Slater's **K** table ([236, p. 471]),

$$\alpha_{4r} = q^{8r^2-2r} + q^{8r^2+2r}, \qquad \alpha_{4r+1} = -q^{8r^2+6r+1}, \qquad (10.40)$$

$$\alpha_{4r-2} = 0, \qquad \alpha_{4r-1} = -q^{8r^2-6r+1}, \qquad \beta_n = \frac{(-q^2;q^2)_{n-1}}{(q;q)_{2n}}.$$

(iii) Use the identities

$$(1 - q^{8r+1})q^{8r^2-2r} = q^{-n}q^{8r^2+2r}((1 - q^{n+4r+1}) - (1 - q^{n-4r})),$$
$$(1 - q^{-8r+1})q^{8r^2+2r} = q^{-n}q^{8r^2-2r}((1 - q^{n-4r+1}) - (1 - q^{n+4r})),$$

instead to prove another of Slater's mod 4 Bailey pairs, also with respect to $a = 1$ ([236, p. 471]),

$$\alpha_{4r} = q^{8r^2+2r} + q^{8r^2-2r}, \qquad \alpha_{4r+1} = -q^{8r^2+2r}, \tag{10.41}$$

$$\alpha_{4r-1} = -q^{8r^2-2r}, \qquad \alpha_{4r-2} = 0, \qquad \beta_n = \frac{q^n\,(-q^2; q^2)_{n-1}}{(q; q)_{2n}}.$$

10.11 Use the Bailey pair at (10.41) and various parts of Corollary 10.1 to prove the following identities (the first three are due to Slater, [237, **S.104, S.120, S.127**]):

$$1 + \sum_{n=1}^{\infty} \frac{(-q; q)_n(-q^2; q^2)_{n-1}q^{n(n+1)/2}}{(q; q)_{2n}}$$
$$= \frac{(-q^{16}, -q^{16}, q^{32}; q^{32}) - q(-q^8, -q^{24}, q^{32}; q^{32})_\infty}{(q; q)_\infty(q; q^2)_\infty}; \tag{10.42}$$

$$1 + \sum_{n=1}^{\infty} \frac{(-q^2; q^2)_{n-1}q^{n(n+1)}}{(q; q)_{2n}}$$
$$= \frac{(-q^{22}, -q^{26}, q^{48}; q^{48})_\infty - q(-q^{14}, -q^{34}, q^{48}; q^{48})_\infty}{(q; q)_\infty}; \tag{10.43}$$

$$1 + \sum_{n=1}^{\infty} \frac{(-q; q^2)_n(-q^4; q^4)_{n-1}q^{n(n+2)}}{(q^2; q^2)_{2n}}$$
$$= \frac{(-q^{28}, -q^{36}, q^{64}; q^{64})_\infty - q(-q^{20}, -q^{44}, q^{64}; q^{64})_\infty}{(q; q^2)_\infty(q^4; q^4)_\infty}; \tag{10.44}$$

$$1 + \sum_{n=1}^{\infty} \frac{(q; q^2)_n(-q^2; q^2)_{n-1}(-q)^n}{(q; q)_{2n}} = \frac{(-q^6, -q^{10}, q^{16}; q^{16})_\infty}{(-q; q)_\infty(q^2; q^2)_\infty}. \tag{10.45}$$

10.12 By considering yet another special case of (10.6), show that if $\sum_{n=0}^{\infty}(\sqrt{aq}; q)_n^2\beta_n(a, q)$ converges, then

$$\sum_{n=0}^{\infty} \alpha_n(a, q) = 0.$$

Remark: It may be possible to give a more general result than that in Q10.12 by other methods, but we do not pursue that here.

10.13 By considering the Bailey pair derived from that at (10.16) by setting $a = q$, $b = -q$ and $c \to \infty$ in conjunction with the transformations at (10.13) and (10.14), derive the identities

$$\sum_{n=0}^{\infty} \frac{(-1)^n q^{n^2+n}}{(-q^2; q^2)_n} = \sum_{n=0}^{\infty} q^{3n^2+n} - \sum_{n=1}^{\infty} q^{3n^2-n},$$

$$\sum_{n=0}^{\infty} \frac{q^{n^2+n}}{(q^2; q^2)_n} = (-q^2; q^2)_\infty.$$

Remark: Observe that the left side of the first identity is a false theta series, in that if the "−" sign between the two series were replaced by "+", the two series could be combine into an infinite product via the Jacobi triple product identity (6.1). The second identity is of course a special case of the q-binomial theorem (3.1).

10.14 (i) By specializing the parameters in (10.16), derive the Bailey pairs ([236, page 468, **F(1)**, **F(2)**]):

$$\alpha_n = q^{n^2}(q^{n/2} + q^{-n/2}), \qquad \beta_n = \frac{1}{(q^{1/2}, q; q)_n} \qquad (a = 1), \qquad (10.46)$$

$$\alpha_n = q^{n^2+n/2}\frac{1 + q^{n+1/2}}{1 + q^{1/2}}, \qquad \beta_n = \frac{1}{(q^{3/2}, q; q)_n} \qquad (a = q). \qquad (10.47)$$

ii) By employing these Bailey pairs in various parts of Corollary 10.1, derive the following identities:

$$\sum_{n=0}^{\infty} \frac{q^{2n^2}}{(q; q)_{2n}} = \frac{(-q^3, -q^5, q^8; q^8)_\infty}{(q^2; q^2)_\infty}; \qquad \text{(S.39)} \quad (10.48)$$

$$\sum_{n=0}^{\infty} \frac{q^{2n^2+2n}}{(q; q)_{2n+1}} = \frac{(-q, -q^7, q^8; q^8)_\infty}{(q^2; q^2)_\infty}; \qquad \text{(S.38)} \quad (10.49)$$

$$\sum_{n=0}^{\infty} \frac{q^{n^2}(-q; q^2)_n}{(q; q)_{2n}} = \frac{(-q^2, -q^4, q^6; q^6)_\infty(-q; q^2)_\infty}{(q^2; q^2)_\infty}; \qquad \text{(S.29)} \quad (10.50)$$

$$\sum_{n=0}^{\infty} \frac{q^{n^2+n}(-q^2; q^2)_n}{(q; q)_{2n+1}} = \frac{(-q, -q^5, q^6; q^6)_\infty(-q^2; q^2)_\infty}{(q^2; q^2)_\infty}; \qquad \text{(S.28)} \quad (10.51)$$

$$\sum_{n=0}^{\infty} \frac{q^{n^2+n}(-1; q^2)_n}{(q; q)_{2n}} = \frac{(-q^3, -q^3, q^6; q^6)_\infty(-q^2; q^2)_\infty}{(q^2; q^2)_\infty}. \qquad \text{(S.48)} \quad (10.52)$$

Chapter 11

Bailey Chains

As mentioned in the previous chapter, Bailey [61] stated a finite version of (10.6), but was not interested in its implications, and thus he missed that this finite transformation could be used to produce new Bailey pairs from existing Bailey pairs, and that this process could be iterated to produce an infinite sequence of Bailey pairs, or a *Bailey chain*. This property was discovered by Andrews [22], and we now describe Andrews' discovery. Recall that a Bailey pair with respect to a is defined at (10.5).

Lemma 11.1. *If* (α_n, β_n) *is a Bailey pair with respect to* a, *and* N *is a non-negative integer, then*

$$\sum_{n=0}^{N} \frac{(\rho_1, \rho_2, q^{-N}; q)_n \, q^n}{(\rho_1 \rho_2 q^{-N}/a; q)_n} \beta_n$$

$$= \frac{(aq/\rho_1, aq/\rho_2; q)_N}{(aq, aq/\rho_1\rho_2; q)_N} \sum_{n=0}^{N} \frac{(\rho_1, \rho_2, q^{-N}; q)_n \, q^{-n(n-3)/2}}{(aq/\rho_1, aq/\rho_2, aq^{1+N}; q)_n} \left(\frac{-aq^N}{\rho_1 \rho_2} \right)^n \alpha_n. \quad (11.1)$$

Proof. The result will follow by (10.3), if it can be shown that (δ_n, γ_n) is a conjugate Bailey pair with respect to a, where

$$\delta_n = \frac{(\rho_1, \rho_2, q^{-N}; q)_n \, q^n}{(\rho_1 \rho_2 q^{-N}/a; q)_n},$$

$$\gamma_n = \frac{(aq/\rho_1, aq/\rho_2; q)_N}{(aq, aq/\rho_1\rho_2; q)_N} \frac{(\rho_1, \rho_2, q^{-N}; q)_n \, q^{-n(n-3)/2}}{(aq/\rho_1, aq/\rho_2, aq^{1+N}; q)_n} \left(\frac{-aq^N}{\rho_1 \rho_2} \right)^n.$$

However, this is straightforward. With U_n and V_n as at (10.4),

$$\sum_{r=n}^{\infty} \delta_r U_{r-n} V_{r+n} = \sum_{r=0}^{\infty} \delta_{r+n} U_r V_{r+2n}.$$

95

$$= \sum_{r=0}^{\infty} \frac{(\rho_1, \rho_2, q^{-N}; q)_{r+n} q^{r+n}}{(\rho_1 \rho_2 q^{-N}/a; q)_{r+n}(q; q)_r (aq; q)_{r+2n}}$$

$$= \frac{(\rho_1, \rho_2, q^{-N}; q)_n q^n}{(\rho_1 \rho_2 q^{-N}/a; q)_n (aq; q)_{2n}} \sum_{r=0}^{N-n} \frac{(\rho_1 q^n, \rho_2 q^n, q^{-(N-n)}; q)_r q^r}{(\rho_1 \rho_2 q^{n-N}/a, aq^{2n+1}, q; q)_r}$$

$$= \frac{(\rho_1, \rho_2, q^{-N}; q)_n q^n}{(\rho_1 \rho_2 q^{-N}/a; q)_n (aq; q)_{2n}} \frac{(aq^{n+1}/\rho_1, aq^{n+1}/\rho_2; q)_{N-n}}{(aq^{2n+1}, aq/\rho_1 \rho_2; q)_{N-n}}$$

$$= \frac{(aq/\rho_1, aq/\rho_2; q)_N}{(aq, aq/\rho_1 \rho_2; q)_N} \frac{(\rho_1, \rho_2, q^{-N}; q)_n q^{nN-n(n-1)/2}}{(aq/\rho_1, aq/\rho_2, aq^{N+1}; q)_n} \left(\frac{-aq}{\rho_1 \rho_2} \right)^n = \gamma_n,$$

where the second last equality is by the q-Pfaff–Saalschütz sum (4.5), and the next-to-last equality is by elementary q-product manipulations of the type at (2.11) and (2.12).   □

Note that letting $N \to \infty$ recovers (10.6), so that any identity deriving from the latter transformation is now seen to have a finite analogue. As an example of this transformation formula (11.1), we see that inserting the pair at (10.16) and replacing ρ_1 with d and ρ_2 with e immediately gives Watson's [257] transformation (5.10) for a terminating $_8\phi_7$ series.

As a second example, the Bailey pair at (10.19) implies that

$$\sum_{n=0}^{N} \frac{(\rho_1, \rho_2, q^{-N}; q)_n (a; q^3)_n \, q^n}{(\rho_1 \rho_2 q^{-N}/a, q; q)_n (a; q)_{2n}} = \frac{(aq/\rho_1, aq/\rho_2; q)_N}{(aq, aq/\rho_1 \rho_2; q)_N}$$

$$\times \sum_{n=0}^{N/3} \frac{1 - aq^{6n}}{1 - a} \frac{(a; q^3)_n}{(q^3; q^3)_n} \frac{(\rho_1, \rho_2, q^{-N}; q)_{3n} a^{4n}}{(aq/\rho_1, aq/\rho_2, aq^{1+N}; q)_{3n}} \left(\frac{q^{N+1}}{\rho_1 \rho_2} \right)^{3n} \quad (11.2)$$

holds for all non-negative integers N.

We next describe how Andrews modified this transformation to provide a mechanism to derive new Bailey pairs from existing pairs. Andrews mentioned this process in [51], and gave details of the proof in [24].

Theorem 11.1. *(Bailey's Lemma) If (α_n, β_n) is a Bailey pair with respect to a, then so is (α_n', β_n'), where*

$$\alpha_n' = \frac{(\rho_1, \rho_2; q)_n}{(aq/\rho_1, aq/\rho_2; q)_n} \left(\frac{aq}{\rho_1 \rho_2} \right)^n \alpha_n, \quad (11.3)$$

$$\beta_n' = \frac{1}{(aq/\rho_1, aq/\rho_2; q)_n} \sum_{j=0}^{n} \frac{(\rho_1, \rho_2; q)_j (aq/\rho_1 \rho_2; q)_{n-j}}{(q; q)_{n-j}} \left(\frac{aq}{\rho_1 \rho_2} \right)^j \beta_j.$$

Proof. By the definition of β'_n and (2.14),

$$
\begin{aligned}
\beta'_n &= \frac{(aq/\rho_1\rho_2; q)_n}{(aq/\rho_1, aq/\rho_2, q; q)_n} \sum_{j=0}^{n} \frac{(\rho_1, \rho_2, q^{-n}; q)_j}{(\rho_1\rho_2 q^{1-n}/a; q)_j} q^j \beta_j \\
&= \frac{1}{(aq, q; q)_n} \sum_{j=0}^{n} \frac{(\rho_1, \rho_2, q^{-n}; q)_j \, q^{-j(j-3)/2}}{(aq/\rho_1, aq/\rho_2, aq^{1+n}; q)_j} \left(\frac{-aq^n}{\rho_1\rho_2}\right)^j \alpha_j \text{ (by (11.1))} \\
&= \sum_{j=0}^{n} \frac{(\rho_1, \rho_2; q)_j}{(aq/\rho_1, aq/\rho_2; q)_j (aq; q)_{n+j} (q; q)_{n-j}} \left(\frac{aq}{\rho_1\rho_2}\right)^j \alpha_j \text{ (by (2.13))} \\
&= \sum_{j=0}^{n} \frac{\alpha'_j(a, q)}{(q; q)_{n-j} (aq; q)_{n+j}}.
\end{aligned}
$$

$\square$

Clearly this process may iterated an arbitrary number times, to produce a *Bailey chain*:

$$
(\alpha_n, \beta_n) \to (\alpha'_n, \beta'_n) \to (\alpha_n", \beta_n") \to \cdots \to (\alpha_n^{(k)}, \beta_n^{(k)}) \to \cdots
$$

with the pair of parameters (ρ_1, ρ_2) being replaced with a different pair of parameters $(\rho_1^{(i)}, \rho_2^{(i)})$ at the i-th step, and if the iteration is stopped after k steps, the final Bailey pair may then be substituted into (11.1) with one final pair of new parameters, $(\rho_1^{(k+1)}, \rho_2^{(k+1)})$, being used there. We will not state this multi-sum transformation in its full generality here, the reader may easily derive it (or view it in [22, Theorem 1]), but instead will concentrate on various special cases.

For later use, we record some special cases of (11.3) here.

Corollary 11.1. *If (α_n, β_n) is a Bailey pair with respect to a, then so are*

$$
\alpha_n^{(1)} = a^n q^{n^2} \alpha_n, \tag{11.4}
$$

$$
\beta_n^{(1)} = \sum_{j=0}^{n} \frac{a^j q^{j^2}}{(q; q)_{n-j}} \beta_j;
$$

$$
\alpha_n^{(2)} = \frac{(-\sqrt{q}; q)_n (a\sqrt{q})^n q^{n(n-1)/2}}{(-a\sqrt{q}; q)_n} \alpha_n, \tag{11.5}
$$

$$
\beta_n^{(2)} = \frac{1}{(-a\sqrt{q}; q)_n} \sum_{j=0}^{n} \frac{(-\sqrt{q}; q)_j (a\sqrt{q})^j q^{j(j-1)/2}}{(q; q)_{n-j}} \beta_j;
$$

$$\alpha_n^{(3)} = \frac{(\sqrt{q};q)_n(-a\sqrt{q})^n q^{n(n-1)/2}}{(a\sqrt{q};q)_n}\alpha_n, \tag{11.6}$$

$$\beta_n^{(3)} = \frac{1}{(a\sqrt{q};q)_n}\sum_{j=0}^{n}\frac{(\sqrt{q};q)_j(-a\sqrt{q})^j q^{j(j-1)/2}}{(q;q)_{n-j}}\beta_j;$$

$$\alpha_n^{(4)} = \frac{(-1;q)_n a^n q^{n(n+1)/2}}{(-aq;q)_n}\alpha_n, \tag{11.7}$$

$$\beta_n^{(4)} = \frac{1}{(-aq;q)_n}\sum_{j=0}^{n}\frac{(-1;q)_j a^j q^{j(j+1)/2}}{(q;q)_{n-j}}\beta_j;$$

$$\alpha_n^{(5)} = \frac{(-q;q)_n a^n q^{n(n-1)/2}}{(-a;q)_n}\alpha_n, \tag{11.8}$$

$$\beta_n^{(5)} = \frac{1}{(-a;q)_n}\sum_{j=0}^{n}\frac{(-q;q)_j a^j q^{j(j-1)/2}}{(q;q)_{n-j}}\beta_j;$$

$$\alpha_n^{(6)} = (-\sqrt{aq})^n q^{n(n-1)/2}\alpha_n, \tag{11.9}$$

$$\beta_n^{(6)} = \frac{1}{(\sqrt{aq};q)_n}\sum_{j=0}^{n}\frac{(\sqrt{aq};q)_j(-\sqrt{aq})^j q^{j(j-1)/2}}{(q;q)_{n-j}}\beta_j;$$

$$\alpha_n^{(7)} = (\sqrt{aq})^n q^{n(n-1)/2}\alpha_n, \tag{11.10}$$

$$\beta_n^{(7)} = \frac{1}{(-\sqrt{aq};q)_n}\sum_{j=0}^{n}\frac{(-\sqrt{aq};q)_j(\sqrt{aq})^j q^{j(j-1)/2}}{(q;q)_{n-j}}\beta_j.$$

Proof. That the stated pairs are indeed Bailey pairs are a consequence, respectively, of the following specializations in (11.3)

(11.4)	$\rho_1, \rho_2 \to \infty,$
(11.5)	$\rho_1 \to \infty, \rho_2 = -\sqrt{q},$
(11.6)	$\rho_1 \to \infty, \rho_2 = \sqrt{q},$
(11.7)	$\rho_1 \to \infty, \rho_2 = -1,$
(11.8)	$\rho_1 \to \infty, \rho_2 = -q,$
(11.9)	$\rho_1 \to \infty, \rho_2 = \sqrt{aq},$
(11.10)	$\rho_1 \to \infty, \rho_2 = -\sqrt{aq}.$

$\square$

11.1 The Bailey lattice

As remarked after the proof of Theorem 11.1, the process may be iterated, with a different pair of parameters $(\rho_1^{(i)}, \rho_2^{(i)})$ being used at the i-th step, and since the parameters $(\rho_1^{(i)}, \rho_2^{(i)})$ may vary continuously, we may think of a single Bailey pair giving rise to a "continuum" of Bailey pairs through the repeated application of Theorem 11.1. If instead, however, we restricted to the various special cases of Corollary 11.1, then at each step there is a *finite* number of directions to go to arrive at a new Bailey pair, so that a single Bailey pair may generate an infinite *lattice* of Bailey pairs, the Bailey lattice. Moreover, there is a certain amount of structured relationship between the lattice "steps" — it is easy to see, for example, that two steps along the path determined by (11.10) is equivalent to one step along the path determined by (11.4). Finally, we should also make clear that the lattice steps indicated by Corollary 11.1 are not the only ones possible — see, for example, Lovejoy [185].

A natural question of course is why lose the fully generality of the "Bailey continuum" for the special cases of the Bailey lattice. One reason is that the special cases let us generate multi-sum infinite product identities. As will be seen below when considering multi-sum identities of Rogers–Ramanujan type, to make the "α" side summable it will be convenient to specialize a (or to choose initial Bailey pairs relative to a particular value of a) so that the q-products in the expressions for the $\alpha_n^{(i)}$ vanish or cancel, leaving only pure powers of q and a, so that the Jacobi triple product identity (6.1) may be used. For example, setting $a = 1$ in (11.5) gives $\alpha_n^{(2)} = (a\sqrt{q})^n q^{n(n-1)/2} \alpha_n$.

Thus, for example, by starting with a Bailey pair relative to $a = 1$ with the aim of generating a multi-sum identity of extended Rogers–Ramanujan type (multi-sum equals infinite product), at each stage any one of the transformations (11.4), (11.5) or (11.9) may be used to walk along the lattice (note that (11.5) and (11.10) coincide for $a = 1$) and then, after an arbitrary finite number of steps, this final Bailey pair may be substituted into any one of (10.8), (10.9), (10.13) and possibly (if the β side converges) (10.15). In all cases, the single sum on the α side will have as its n-th term a power of q (in which the exponent will be quadratic in n) times the original α_n, and may thus be summable via the Jacobi triple product identity (6.1).

Note that at each stage the new $\alpha_n^{(i)}$ is formed by multiplying the current $\alpha_n^{(i-1)}$ by some factor, so that the final $\alpha_n^{(i)}$ depends only on the steps along the lattice, not on the order in which the steps are taken. This implies the final $\beta_n^{(i)}$ is also independent of the order in which the summations are performed, which is not obvious from looking at the sums. In fact, certain transformation are implied by the fact that the order of summations may be switched. One may check, for example, with two iterations of (11.3), with the parameters (ρ_1, ρ_2) replaced with another pair of parameters at the second step, that the implied transformation is the $_4\phi_3$ transformation (4.7) of Sears.

To give an example of the process of iteration in Theorem 11.1, we start with the "unit" Bailey pair

$$\alpha_n = \frac{1 - aq^{2n}}{1 - a} \frac{(a; q)_n}{(q; q)_n} (-1)^n q^{n(n-1)/2}, \tag{11.11}$$

$$\beta_n = \delta_{n,0} = \begin{cases} 1 & n = 0 \\ 0 & n > 0. \end{cases}$$

That this is a Bailey pair may be easily seen to follow from (5.3), after employing (2.14). The insertion of this pair in (11.1) leads to an identity equivalent to the $_6\phi_5$ summation formula at (5.5), while inserting the same pair in (11.3) produces the pair (10.16), and, as noted above, inserting this pair in (11.1) gives Watson's [257] transformation (5.10) for a terminating $_8\phi_7$ series. Further iteration will lead to transformations with a single series on one side and nested multiple series on the other.

11.2 Polynomial versions of identities of Rogers–Ramanujan–Slater type

As noted above, letting $N \to \infty$ in (11.1) recovers (10.6), so any infinite identity deriving from (10.6) or one of the special cases in Corollary 10.1 has a finite analogue containing the integer parameter N, and the infinite identity is recovered upon letting $N \to \infty$. For example, the pair used to derive the Rogers–Ramanujan identity (10.17) in Theorem 10.2 was the pair (relative to $a = 1$)

$$\alpha_n = (-1)^n q^{(3n^2 - n)/2} (1 + q^n), \tag{11.12}$$

$$\beta_n = \frac{1}{(q; q)_n},$$

and this pair was substituted into (10.8). If the same pair is substituted into the corresponding special case of (11.1) (which results from letting $\rho_1, \rho_2 \to \infty$, using (2.14), and finally cancelling a factor of $(q; q)_N$)

$$\sum_{k=0}^{N} \frac{a^k q^{k^2}}{(q; q)_{N-k}} \beta_k = \sum_{k=0}^{N} \frac{a^k q^{k^2}}{(aq; q)_{N+k} (q; q)_{N-k}} \alpha_k \tag{11.13}$$

and a is set equal to 1, then we get that

$$\sum_{k=0}^{N} \frac{q^{k^2}}{(q;q)_{N-k}(q;q)_k} = \frac{1}{(q;q)_N^2} + \sum_{k=1}^{N} \frac{q^{(5k^2-k)/2}(1+q^k)(-1)^k}{(q;q)_{N+k}(q;q)_{N-k}}$$

$$= \sum_{k=-N}^{N} \frac{q^{(5k^2-k)/2}(-1)^k}{(q;q)_{N+k}(q;q)_{N-k}} \tag{11.14}$$

holds for all non-negative integers N, and (10.17) is recovered upon letting $N \to \infty$.

Remark: Both sides of (11.14) can be converted to polynomials in q by multiplying across by $(q;q)_{2N}$ to get

$$\frac{(q;q)_{2N}}{(q;q)_N} \sum_{k=0}^{N} \begin{bmatrix} N \\ k \end{bmatrix}_q q^{k^2} = \sum_{k=-N}^{N} \begin{bmatrix} 2N \\ N-k \end{bmatrix}_q q^{(5k^2-k)/2}(-1)^k. \tag{11.15}$$

A polynomial version of (10.18) may also be derived, and indeed polynomial versions of all 130 identities in Slater's paper [237] may be similarly produced, as has been done by Werley in [261]. Polynomial analogues of Slater identities are not unique, and a different set of polynomials, produced by a different method, was given by Sills [233], and sporadic examples have been given elsewhere, for example by Bressoud [84] (see Q12.4).

11.3 Multiple series Rogers–Ramanujan type identities

When Andrews developed the concept of a Bailey chain in [51], the principal purpose in the paper was to show that all 130 of the Slater's identities in [237] could be embedded in infinite families of multi-sum identities. Some examples of multi-sum identities had appeared elsewhere, for example, Andrews [18], Bressoud [83], Jain and Verma [161], Milne [204] and Paule [209]. To illustrate the method, we consider the Bailey chain (11.4). Upon applying the transformations in this chain $k - 1$ times and then substituting in (10.8), we get the following theorem, due to Andrews [51, p. 273, Theorem 2]).

Theorem 11.2. *If (α_n, β_n) is a Bailey pair with respect to a and k is a positive integer, then*

$$\frac{1}{(aq;q)_\infty} \sum_{n=0}^{\infty} a^{kn} q^{kn^2} \alpha_n$$

$$= \sum_{n_k \geq n_{k-1} \geq \cdots \geq n_1 \geq 0} \frac{a^{n_1 + \cdots + n_k} q^{n_1^2 + \cdots + n_k^2} \beta_{n_1}}{(q;q)_{n_k-n_{k-1}} (q;q)_{n_{k-1}-n_{k-2}} \cdots (q;q)_{n_2-n_1}}. \quad (11.16)$$

Remark: The use of a different part of Corollary 10.1 instead of (10.8) will produce another multi-sum transformation similar to that at (11.16). Inserting particular Bailey pairs may result in the left side of (11.16) being summable by the Jacobi triple product identity (6.1) or the quintuple product identity (6.21).

Corollary 11.2.

$$\sum_{n_{k-1} \geq n_{k-2} \geq \cdots \geq n_1 \geq 0} \frac{q^{n_1^2 + \cdots + n_{k-1}^2}}{(q;q)_{n_{k-1}-n_{k-2}} (q;q)_{n_{k-2}-n_{k-3}} \cdots (q;q)_{n_2-n_1} (q;q)_{n_1}}$$

$$= \frac{(q^k, q^{k+1}, q^{2k+1}; q^{2k+1})_\infty}{(q;q)_\infty}. \quad (11.17)$$

$$\sum_{n_{k-1} \geq n_{k-2} \geq \cdots \geq n_1 \geq 0} \frac{q^{n_1^2 + \cdots + n_{k-1}^2 + n_1 + \cdots + n_{k-1}}}{(q;q)_{n_{k-1}-n_{k-2}} (q;q)_{n_{k-2}-n_{k-3}} \cdots (q;q)_{n_2-n_1} (q;q)_{n_1}}$$

$$= \frac{(q, q^{2k}, q^{2k+1}; q^{2k+1})_\infty}{(q;q)_\infty}. \quad (11.18)$$

Proof. See Q11.5.   □

Notice that setting $k = 2$ in the identities above results in the Rogers–Ramanujan identities (Theorem 10.2).

11.4   The Andrews–Gordon identities

The identities in Corollary 11.2 are special cases of a more general identity, first proved combinatorially by Gordon [128], and subsequently proved analytically by Andrews [18]. To derive these using the Bailey lattice requires a trick, namely to change a Bailey pair with respect to $a = q$ into a Bailey pair with respect to $a = 1$ at a certain point on the path along the lattice. We will use the idea given in Warnaar's proof [254]. As part of his proof, Warnaar states and proves two quite general theorems ([254, Theorems 3.1,

3.2]), each of which enables a Bailey pair relative to a to be derived from a Bailey pair relative to $b = aq^N$. Instead we will use a simpler result (Lemma 13.1, Eq. (13.3)), which enables a Bailey pair relative to a to be derived from a Bailey pair relative to aq. To avoid confusion, we recall before starting the proof the convention that $\alpha_0 = \beta_0 = 1$, and that any formula that might appear to indicate otherwise is thus valid only for $n > 0$.

Theorem 11.3. *Let $k \geq 2$ and $1 \leq i \leq k$ be integers, and let $|q| < 1$. Then*

$$
\sum_{n_{k-1} \geq n_{k-2} \geq \cdots \geq n_1 \geq 0} \frac{q^{n_1^2 + \cdots + n_{k-1}^2 + n_1 + \cdots + n_{k-i}}}{(q;q)_{n_{k-1} - n_{k-2}} (q;q)_{n_{k-2} - n_{k-3}} \cdots (q;q)_{n_2 - n_1} (q;q)_{n_1}}
$$
$$
= \frac{(q^i, q^{2k+1-i}, q^{2k+1}; q^{2k+1})_\infty}{(q;q)_\infty}. \quad (11.19)
$$

Remark: In the case $i = k$, we follow the usual convention and regard the empty sum $n_1 + \cdots + n_{k-i} = n_1 + \cdots + n_0$ as having the value 0.

Proof. Start with the Bailey pair, with respect to $a = q$ ($b, c \to \infty$, $a = q$ in (10.16))

$$
\alpha_n^{(0)} = \frac{1 - q^{2n+1}}{1 - q} (-1)^n q^{n(3n+1)/2}, \qquad \beta_n^{(0)} = \frac{1}{(q;q)_n}, \quad (11.20)
$$

and apply the transformation (11.4) $k - i$ times to get the Bailey pair, again with respect to $a = q$,

$$
\alpha_n^{(k-i)} = q^{(k-i)(n^2+n)} \frac{1 - q^{2n+1}}{1 - q} (-1)^n q^{n(3n+1)/2},
$$
$$
\beta_n^{(k-i)} =
$$
$$
\sum_{n \geq n_{k-i} \geq n_{k-i-1} \geq \cdots \geq n_1 \geq 0} \frac{q^{n_1^2 + \cdots + n_{k-i}^2 + n_1 + \cdots + n_{k-i}}}{(q;q)_{n - n_{k-i}} (q;q)_{n_{k-i} - n_{k-i-1}} \cdots (q;q)_{n_2 - n_1} (q;q)_{n_1}}.
$$

Next use (13.3) to derive, after a little algebra, the Bailey pair, with respect to $a = 1$,

$$
\alpha_n^{(k-i)*} = (-1)^n q^{(3/2-i+k)n^2} \left(q^{(1/2-i+k)n} + q^{(-1/2+i-k)n} \right),
$$
$$
\beta_n^{(k-i)*} =
$$
$$
\sum_{n \geq n_{k-i} \geq n_{k-i-1} \geq \cdots \geq n_1 \geq 0} \frac{q^{n_1^2 + \cdots + n_{k-i}^2 + n_1 + \cdots + n_{k-i}}}{(q;q)_{n - n_{k-i}} (q;q)_{n_{k-i} - n_{k-i-1}} \cdots (q;q)_{n_2 - n_1} (q;q)_{n_1}}.
$$

Apply (11.4) a further $i - 2$ times (now with $a = 1$) to get the Bailey pair with respect to $a = 1$

$$\alpha_n^{(k-2)} = (-1)^n q^{(-1/2+k)n^2} \left(q^{(1/2-i+k)n} + q^{(-1/2+i-k)n} \right),$$

$$\beta_n^{(k-2)} =$$

$$\sum_{n \geq n_{k-2} \geq n_{k-3} \geq \cdots \geq n_1 \geq 0} \frac{q^{n_1^2 + \cdots + n_{k-2}^2 + n_1 + \cdots + n_{k-i}}}{(q;q)_{n - n_{k-2}}(q;q)_{n_{k-2} - n_{k-3}} \cdots (q;q)_{n_2 - n_1}(q;q)_{n_1}}.$$

This final pair is inserted in (10.8) (with $a = 1$), and (11.19) follows upon using the Jacobi triple product identity (6.1) on the series containing the α_n.   □

Remark: The reader is asked to fill in some of the algebraic details of the proof in Q11.6.

By following the same steps as in Theorem 11.3, but with an arbitrary Bailey pair, one can derive the following result (stated and proved in [4, Corollary 4.2], with a replaced with a/q).

Theorem 11.4. *Let $k \geq 2$ and $1 \leq i \leq k$ be integers, and let $|q| < 1$. Let $(\alpha_n(a,q), \beta_n(a,q))$ be a Bailey pair. Then, assuming convergence,*

$$\sum_{n_{k-1} \geq n_{k-2} \geq \cdots \geq n_1 \geq 0} \frac{q^{n_1^2 + \cdots + n_{k-1}^2 + n_1 + \cdots + n_{k-i}} a^{n_1 + n_2 + \cdots + n_{k-1}} \beta_{n_1}(aq;q)}{(q;q)_{n_{k-1} - n_{k-2}}(q;q)_{n_{k-2} - n_{k-3}} \cdots (q;q)_{n_2 - n_1}}$$

$$= \frac{1}{(aq;q)_\infty} \left[1 + \sum_{r=0}^{\infty} q^{(k-1)r^2}(1 - aq) \right.$$

$$\times \left. \left(\frac{a^{(k-1)r} q^{r(k-i)} \alpha_r(aq,q)}{1 - aq^{2r+1}} - \frac{a^{i+(k-1)(r-1)} q^{r(i-k+2)-1} \alpha_{r-1}(aq,q)}{1 - aq^{2r-1}} \right) \right].$$

$$\tag{11.21}$$

Proof. See Q11.7.   □

Remark: In order to use (11.21) to derive identities of the Andrews–Gordon type via the Jacobi triple product identity (6.1), $\alpha_r(aq, q)$ has to have a factor $1 - aq^{2r+1}$ to cancel the factor of this form in the denominator above, so that this theorem may not lead to Andrews–Gordon type extensions of all the identities on the Slater list.

The following Andrews–Gordon type identities follow as special cases.

Corollary 11.3. *Let $k \geq 2$ and $1 \leq i \leq k$ be integers, and let $|q| < 1$. Then*

$$\sum_{n_{k-1} \geq n_{k-2} \geq \cdots \geq n_1 \geq 0} \frac{q^{n_1^2 + \cdots + n_{k-1}^2 + 3n_1 + \cdots + 3n_{k-i} + 2n_{k-i+1} + \cdots + 2n_{k-1}}}{(q;q)_{n_{k-1} - n_{k-2}}(q;q)_{n_{k-2} - n_{k-3}} \cdots (q;q)_{n_2 - n_1}}$$
$$\times \frac{(q^3; q^3)_{n_1}}{(q;q)_{2n_1 + 2}(q;q)_{n_1}} = \frac{(q^{3i}, q^{18k-9-3i}, q^{18k-9}; q^{18k-9})_\infty}{(q;q)_\infty}. \quad (11.22)$$

$$\sum_{n_{k-1} \geq n_{k-2} \geq \cdots \geq n_1 \geq 0} \frac{q^{n_1^2 + \cdots + n_{k-1}^2 + 2n_1 + \cdots + 2n_{k-i} + n_{k-i+1} + \cdots + n_{k-1}}}{(q;q)_{n_{k-1} - n_{k-2}}(q;q)_{n_{k-2} - n_{k-3}} \cdots (q;q)_{n_2 - n_1}}$$
$$\times \frac{1}{(q;q^2)_{n_1 + 1}(q;q)_{n_1}} = \frac{(q^{2i}, q^{8k-2-2i}, q^{8k-2}; q^{8k-2})_\infty}{(q;q)_\infty}. \quad (11.23)$$

$$\sum_{n_{k-1} \geq n_{k-2} \geq \cdots \geq n_1 \geq 0} \frac{q^{n_1^2 + \cdots + n_{k-1}^2 + n_1 + \cdots + n_{k-i}}}{(q;q)_{n_{k-1} - n_{k-2}}(q;q)_{n_{k-2} - n_{k-3}} \cdots (q;q)_{n_2 - n_1}(q^2; q^2)_{n_1}}$$
$$= \frac{(q^i, q^{2k-i}, q^{2k}; q^{2k})_\infty}{(q;q)_\infty}. \quad (11.24)$$

Proof. See Q11.8. □

Remark: Note that the cases $k = 2$ and $i = 1, 2$ of (11.22) result in two of the Bailey–Dyson mod 27 identities ([58, p. 434], see also [237, (90), (91)]), and the cases $k = 2$ and $i = 1, 2$ of (11.23) give two of Slater's mod 14 identities ([237, (59), (60)]).

Remark: Many other multi-sum identities are stated and proved by the authors in [85].

Exercises

11.1 (i) (An Alternative proof of (10.34) — Andrews [28, p. 9, Eq. (4.6)]) Modify the statement at (6.2) to show that (α_n, β_n) is a Bailey pair with respect to $a = 1$, where

$$\alpha_n = (-1)^n (z^n q^{n(n-1)/2} + z^{-n} q^{n(n+1)/2}), \quad (11.25)$$
$$\beta_n = \frac{(z, q/z; q)_n}{(q;q)_{2n}}.$$

(ii) Hence prove Ramanujan's identity (10.34) (see [213, p. 33]).

(iii) Extend this identity by showing, for each positive integer k, that

$$\sum_{n_k \geq n_{k-1} \geq \cdots \geq n_1 \geq 0} \frac{q^{n_1^2 + \cdots + n_k^2}(z, q/z; q)_{n_1}}{(q;q)_{n_k - n_{k-1}}(q;q)_{n_{k-1} - n_{k-2}} \cdots (q;q)_{n_2 - n_1}(q;q)_{2n_1}}$$
$$= \frac{(q^k z, q^{k+1}/z, q^{2k+1}; q^{2k+1})_\infty}{(q;q)_\infty}. \quad (11.26)$$

11.2 (i) By using the same Bailey pair as in Q11.1, but a different part of Corollary 10.1, prove another identity of Ramanujan ([213, p. 33])

$$\sum_{n=0}^\infty \frac{(-q, z, q^2/z; q^2)_n q^{n^2}}{(q^2; q^2)_{2n}} = \frac{(qz, q^3/z, q^4; q^4)_\infty (-q; q^2)_\infty}{(q^2; q^2)_\infty}. \quad (11.27)$$

(ii) By considering the comment following Theorem 11.2 together with the transformation from Corollary 10.1 used in part (i), extend the identity in part (i) by showing that for each positive integer $k \geq 1$, that

$$\sum_{n_k \geq n_{k-1} \geq \cdots \geq n_1 \geq 0} \frac{q^{n_k^2 + 2(n_{k-1}^2 + \cdots + n_1^2)}(-q; q^2)_{n_k}(z, q^2/z; q^2)_{n_1}}{(q^2; q^2)_{n_k - n_{k-1}}(q^2; q^2)_{n_{k-1} - n_{k-2}} \cdots (q^2; q^2)_{n_2 - n_1}(q^2; q^2)_{2n_1}}$$
$$= \frac{(q^{2k-1}z, q^{2k+1}/z, q^{4k}; q^{4k})_\infty (-q; q^2)_\infty}{(q^2; q^2)_\infty}. \quad (11.28)$$

11.3 (i) By replacing N with $2n + 1$ in (3.5), and following similar steps to those indicated in the derivation of (6.2) and in question 11.1, derive the Bailey pair (α_n, β_n), with respect to $a = q$, where

$$\alpha_n = \frac{q^{n(n+1)/2}\left((-z)^{-n} + (-z)^{n+1}\right)}{1 - z}, \quad \beta_n = \frac{(qz, q/z; q)_n}{(q^2; q)_{2n}}. \quad (11.29)$$

(ii) Hence prove the companion identity to Ramanujan's identity in Q11.1 (see [197, Eq. (1.5)], with z replaced with q/z)

$$\sum_{n=0}^\infty \frac{(q/z; q)_n (z; q)_{n+1} q^{n^2+n}}{(q;q)_{2n+1}} = \frac{(z, q^3/z, q^3; q^3)_\infty}{(q;q)_\infty}. \quad (11.30)$$

(iii) Extend this identity by showing, for each positive integer k, that

$$\sum_{n_k \geq n_{k-1} \geq \cdots \geq n_1 \geq 0} \frac{q^{n_1^2 + \cdots + n_k^2 + n_1 + \cdots + n_k}(z; q)_{n_1+1}(q/z; q)_{n_1}}{(q;q)_{n_k - n_{k-1}}(q;q)_{n_{k-1} - n_{k-2}} \cdots (q;q)_{n_2 - n_1}(q;q)_{2n_1+1}}$$
$$= \frac{(z, q^{2k+1}/z, q^{2k+1}; q^{2k+1})_\infty}{(q;q)_\infty}. \quad (11.31)$$

11.4 (i) By using the same Bailey pair as in Q11.3 and a different part of Corollary 10.1, prove the identity

$$\sum_{n=0}^{\infty} \frac{(-q, q/z; q)_n (z; q)_{n+1} q^{(n^2+n)/2}}{(q; q)_{2n+1}} = \frac{(z, q^2/z, q^2; q^2)_\infty (-q; q)_\infty}{(q; q)_\infty}. \quad (11.32)$$

(ii) Extend this identity by showing, for each positive integer k, that

$$\sum_{n_k \geq n_{k-1} \geq \cdots \geq n_1 \geq 0} \frac{q^{(n_1^2 + \cdots + n_k^2 + n_1 + \cdots + n_k)/2} (z; q)_{n_1+1} (-q, q/z; q)_{n_1}}{(q; q)_{n_k - n_{k-1}} (q; q)_{n_{k-1} - n_{k-2}} \cdots (q; q)_{n_2 - n_1} (q; q)_{2n_1+1}}$$

$$= \frac{(z, q^{k+1}/z, q^{k+1}; q^{k+1})_\infty (-q; q)_\infty}{(q; q)_\infty}. \quad (11.33)$$

11.5 By letting $b, c \to \infty$ in (10.16) and then substituting the Bailey pairs deriving from the cases $a = 1$ and $a = q$, respectively, into the $k - 1$ case of (11.16), prove the identities in Corollary 11.2.

11.6 Complete the proof of Theorem 11.3 by filling the algebraic details that were just indicated in the proof.

11.7 By following the steps in the proof of Theorem 11.3, but with the Bailey pair at (11.20) (with respect to $a = q$) replaced with an arbitrary Bailey pair $(\alpha_n(aq, q), \beta_n(aq, q))$ (with respect to aq), prove Theorem 11.4.

11.8 (i) By inserting the Bailey pair (see [197, Eq. (2.26)]) with respect to a defined by

$$\alpha_{3r} = \frac{1 - aq^{6r}}{1 - a} \frac{(a; q^3)_r}{(q^3; q^3)_r} q^{\frac{9r^2 - 3r}{2}} (-a)^r, \qquad \alpha_{3r \pm 1} = 0, \quad (11.34)$$

$$\beta_n = \frac{(a; q^3)_n}{(a; q)_{2n} (q; q)_n},$$

into (11.21), prove the summation formula (11.22).

(ii) Use the Bailey pair with respect to a defined by

$$\alpha_{2r} = \frac{1 - aq^{4r}}{1 - a} \frac{(a; q^2)_r}{(q^2; q^2)_r} q^{3r^2 - r} (-a)^r, \qquad \alpha_{2r+1} = 0, \quad (11.35)$$

$$\beta_n = \frac{1}{(aq; q^2)_n (q; q)_n},$$

to prove (11.23).

(iii) Likewise use the Bailey pair, with respect to $a = q$

$$\alpha_n = \frac{1 - q^{2n+1}}{1 - q}(-1)^n q^{n^2}, \qquad \beta_n = \frac{1}{(q^2; q^2)_n}, \qquad (11.36)$$

to prove (11.24).

11.9 Use the Bailey pair at (11.20) and the method of section 11.2 to prove, for each positive integer N, that

$$\frac{(q; q)_{2N+1}}{(q; q)_N} \sum_{k=0}^{N} \begin{bmatrix} N \\ k \end{bmatrix}_q q^{k^2+k} = \sum_{k=-N-1}^{N} \begin{bmatrix} 2N + 1 \\ N - k \end{bmatrix}_q q^{(5k^2+3k)/2}(-1)^k. \quad (11.37)$$

11.10 By mirroring the proof of Theorem 11.4 with different parts of Corollaries 10.1 and 11.1, prove that if $k \geq 2$ and $1 \leq i \leq k$ are integers, $|q| < 1$, and $(\alpha_n(a, q), \beta_n(a, q))$ is a Bailey pair, then, assuming convergence (and that $n_k = 0$ if it appears (when $i = 1$)),

$$\sum_{n_{k-1} \geq n_{k-2} \geq \cdots \geq n_1 \geq 0} \frac{q^{n_1^2 + \cdots + n_{k-1}^2 + n_1 + \cdots + n_{k-i}} a^{n_1 + n_2 + \cdots + n_{k-1}}}{(q^2; q^2)_{n_{k-1}-n_{k-2}}(q^2; q^2)_{n_{k-2}-n_{k-3}} \cdots (q^2; q^2)_{n_2-n_1}}$$

$$\times \frac{(-aq; q^2)_{n_{k-i+1}}(-aq^2; q^2)_{n_1}\beta_{n_1}(a^2q^2, q^2)}{(-aq^2; q^2)_{n_{k-i+1}}}$$

$$= \frac{(-aq^{1+\delta_{i,1}}; q^2)_\infty}{(a^2q^2; q^2)_\infty}\left[1 + \sum_{r=1}^{\infty} q^{(k-1)r^2}(1 - a^2q^2)\times\right.$$

$$\left.\left(\frac{a^{(k-1)r}q^{r(k-i)}\alpha_r(a^2q^2, q^2)}{1 - a^2q^{4r+2}} - \frac{a^{i+(k-1)r-k+2}q^{r(i-k+4)-2}\alpha_{r-1}(a^2q^2, q^2)}{1 - a^2q^{4r-2}}\right)\right].$$

$$(11.38)$$

11.11 By inserting the Bailey pairs from Q11.8 into (11.38), prove that each of the following identities holds for each pair of integers (i, k), with $k \geq 2$ and $1 \leq i \leq k$ (as before, $n_k = 0$ if it appears (when $i = 1$)):

$$\sum_{n_{k-1} \geq n_{k-2} \geq \cdots \geq n_1 \geq 0} \frac{q^{n_1^2 + \cdots + n_{k-1}^2 + 3n_1 + \cdots + 3n_{k-i} + 2n_{k-i+1} + \cdots + 2n_{k-1}}}{(q^2; q^2)_{n_{k-1}-n_{k-2}}(q^2; q^2)_{n_{k-2}-n_{k-3}} \cdots (q^2; q^2)_{n_2-n_1}}$$

$$\times \frac{(-q; q^2)_{n_{k-i+1}+1}(q^6; q^6)_{n_1}(-q^2; q^2)_{n_1+1}}{(-q^2; q^2)_{n_{k-i+1}+1}(q^2; q^2)_{2n_1+2}(q^2; q^2)_{n_1}}$$

$$= \frac{(1 + \delta_{i,1}q)(-q^{1+\delta_{i,1}}; q^2)_\infty (q^{3i+3}, q^{18k-3-3i}, q^{18k}; q^{18k})_\infty}{(1 + \delta_{i,1}q^2)(q^2; q^2)_\infty}; \quad (11.39)$$

$$\sum_{n_{k-1}\geq n_{k-2}\geq\cdots\geq n_1\geq 0} \frac{q^{n_1^2+\cdots+n_{k-1}^2+2n_1+\cdots+2n_{k-i}+n_{k-i+1}+\cdots+n_{k-1}}}{(q^2;q^2)_{n_{k-1}-n_{k-2}}(q^2;q^2)_{n_{k-2}-n_{k-3}}\cdots(q^2;q^2)_{n_2-n_1}}$$

$$\times\frac{(-q^2;q^2)_{n_{k-i+1}}(-q;q^2)_{n_1+1}}{(-q;q^2)_{n_{k-i+1}+1}(q^2;q^4)_{n_1+1}(q^2;q^2)_{n_1}}$$

$$=\frac{(-q^{2+\delta_{i,1}};q^2)_\infty(q^{2i+2},q^{8k+2-2i},q^{8k+4};q^{8k+4})_\infty}{(q^2;q^2)_\infty};\quad(11.40)$$

$$\sum_{n_{k-1}\geq n_{k-2}\geq\cdots\geq n_1\geq 0} \frac{q^{n_1^2+\cdots+n_{k-1}^2+n_1+\cdots+n_{k-i}}}{(q^2;q^2)_{n_{k-1}-n_{k-2}}(q^2;q^2)_{n_{k-2}-n_{k-3}}\cdots(q^2;q^2)_{n_2-n_1}}$$

$$\times\frac{(-q;q^2)_{n_{k-i+1}}(-q^2;q^2)_{n_1}}{(-q^2;q^2)_{n_{k-i+1}}(q^4;q^4)_{n_1}}$$

$$=\frac{(-q^{1+\delta_{i,1}};q^2)_\infty(q^{i+1},q^{2k+1-i},q^{2k+2};q^{2k+2})_\infty}{(q^2;q^2)_\infty}.\quad(11.41)$$

Remark: The cases $k=2$ and $i=1,2$ of (11.39) correspond, respectively to Slater's identities (76) (after replacing q^2 with q) and (115) in [237], and the cases $k=2$ and $i=1,2$ of (11.40) correspond, respectively to identities of Rogers [219, p. 330 (3), 2nd Eq.] and Slater [237, (45)] (after replacing q^2 with q).

11.12 (Andrews [22, Theorem 5] Let $e\to\infty$ in the Bailey pair at (10.32), and insert the resulting pair into (11.16) to show (after using the quintuple product identity (6.21)), for all integers $k\geq 1$, that

$$\sum_{n_k\geq n_{k-1}\geq\cdots\geq n_1\geq 0} \frac{q^{n_1^2+\cdots+n_k^2+n_1}}{(q;q)_{n_k-n_{k-1}}(q;q)_{n_{k-1}-n_{k-2}}\cdots(q;q)_{n_2-n_1}(q;q)_{2n_1}}$$

$$=\frac{(q^k,q^{5k+4},q^{6k+4};q^{6k+4})_\infty(q^{4k+4},q^{8k+4},q^{12k+8})_\infty}{(q;q)_\infty}.\quad(11.42)$$

Note that the case $k=1$ gives an identity of Rogers ([217, p. 332, Eq. (13)], see also [237, Eq. (99)]).

Chapter 12

WP-Bailey Pairs and Chains

The values for U_n, V_n and δ_n given at (10.4) are not the only possible choices for insertion in Lemma 10.1 that lead to interesting consequences. In particular, in this chapter we investigate WP-Bailey pairs and chains, as defined by Andrews [31, Section 6], and the totality of these chains, the WP-Bailey *tree*.

Before considering the WP-Bailey chains found by Andrews and others, we briefly consider the implications of the Bailey tree for single-sum basic hypergeometric identities. Presently, there are at least ten WP-Bailey chains (Warnaar states at the end of [256] that he had found many more WP-Bailey chains, but stated only two). When the unit WP-Bailey pair (12.15) is substituted into each of these ten chains, a new WP-Bailey pair results. A result of Warnaar [256, Lemma 2.1] (see Lemma 13.2 and Corollary 13.1) results in seven new WP-Bailey pairs deriving from these ten (some pairs that arise by this process are duplicates of existing pairs), for a total of seventeen WP-Bailey pairs. When each of these seventeen WP-Bailey pairs is substituted into each of the ten WP-Bailey chains, a basic hypergeometric identity results, thus giving a total of 170 single-sum basic hypergeometric identities — see Sparks [240] for a list of these 170 identities. In fact, there may be more than 170, since it possible to walk along some chains (in particular the first chain of Andrews, by specializing some of the free parameters) for more than one step, in such a way that the new $\beta_n(a,k)$ have closed form. In addition, Andrews and Berkovich [37, Eqs. (3.3)–(3.6)] also gave two pairs which did not arise directly by travelling along a WP-Bailey chain.

We explore the two chains of Andrews and Berkovich [37] in most detail, as these, taken together, are the most productive in producing both new-and known transformation- and summation formulae. The method of proof for the first chain of Andrews is described in detail, while the proofs of eight of the other nine are just sketched, since the method is essentially the same. The method of proof for the chain at (12.35) is different, and is also spelled out in detail.

Before coming to the various WP-Bailey chains, we briefly examine some other choices for U_n, V_n and δ_n that have been considered.

12.1 A transformation of Sears

In [228], Sears proved the transformation stated in the next theorem. We state it using the notation of Andrews and Bowman [43].

Theorem 12.1. *Let $\{\theta_r\}_{r\geq 0}$ denote a sequence of complex numbers. Then, subject to absolute convergence conditions,*

$$\sum_{n=0}^{\infty} \frac{(a,b;q)_n \theta_n}{(e,q;q)_n} = \sum_{s=0}^{\infty} \frac{(e/a,e/b;q)_s}{(e,q;q)_s} \left(\frac{ab}{e}\right)^s \sum_{t=0}^{\infty} \frac{(ab/e;q)_t \theta_{s+t}}{(q;q)_t}. \tag{12.1}$$

Proof. In Lemma 10.1, let

$$U_n = \frac{(ab/e;q)_n}{(q;q)_n}, \quad V_n = 1, \quad \alpha_n = \frac{(e/a,e/b;q)_n}{(e,q;q)_n}\left(\frac{ab}{e}\right)^n, \quad \delta_n = \theta_n.$$

Then, by (10.1),

$$\begin{aligned}
\beta_n &= \sum_{r=0}^{n} \alpha_r U_{n-r} V_{n+r} \\
&= \sum_{r=0}^{n} \frac{(e/a,e/b;q)_r}{(e,q;q)_r} \frac{(ab/e;q)_{n-r}}{(q;q)_{n-r}} \left(\frac{ab}{e}\right)^r \\
&= \frac{(ab/e;q)_n}{(q;q)_n} \sum_{r=0}^{n} \frac{(e/a,e/b;q)_r}{(e,q;q)_r} \frac{(q^{-n};q)_r}{(q^{1-n}e/ab;q)_r} q^r \\
&= \frac{(a,b;q)_n}{(e,q;q)_n},
\end{aligned}$$

where the second equality follows from (2.13), and the final equality from the q-Pfaff–Saalschütz sum (4.5). Similarly, by (10.2),

$$\gamma_n = \sum_{r=n}^{\infty} \delta_r U_{r-n} V_{r+n} = \sum_{r=n}^{\infty} \frac{(ab/e;q)_{r-n}}{(q;q)_{r-n}} \theta_r = \sum_{r=0}^{\infty} \frac{(ab/e;q)_r}{(q;q)_r} \theta_{r+n}.$$

The transformation at (12.1) is now just the statement at (10.3) for these particular choices of the parameters. □

What is desirable is to choose the sequence $\{\theta_r\}_{r \geq 0}$ so that the inner series on the right side of (12.1) becomes summable, reducing the right side to a single series. As an example of a simple application of (12.1), one can easily check that making the choice $\theta_r = z^r$, leads easily to Heine's transformation (4.4), after using the q-binomial theorem (3.1) on the inner sum on the right side. Other examples of applications are given in the exercises.

12.2   WP-Bailey pairs and chains

In [31] Andrews formally extended the definition of a Bailey pair by introducing another parameter k (setting $k = 0$ in the definition at (12.3) below recovers the definition of a regular Bailey pair) and making the choices

$$U_n = \frac{(k/a; q)_n}{(q; q)_n}, \qquad V_n = \frac{(k; q)_n}{(aq; q)_n}, \qquad (12.2)$$

in the Bailey transform (Lemma 10.1) and hence formulated the definition of a WP-Bailey pair, with respect to the parameters a and k, as a pair of sequences $(\alpha_n(a, k), \beta_n(a, k))$ satisfying

$$\beta_n(a, k) = \sum_{j=0}^{n} \frac{(k/a; q)_{n-j}(k; q)_{n+j}}{(q; q)_{n-j}(aq; q)_{n+j}} \alpha_j(a, k). \qquad (12.3)$$

We sometimes write $(\alpha_n(a, k, q), \beta_n(a, k, q))$ to indicate the dependence on q.

In [31], Andrews also described two WP-Bailey chains, thus allowing an infinite tree of WP-Bailey pairs to be generated from a single such pair.

Prior to the paper [31] of Andrews, some particular instances of WP-Bailey pairs (without the WP-Bailey chain mechanism for generating new pairs from existing pairs) had appeared earlier in the literature, although possibly in disguised form.

12.2.1   Bailey's WP-Bailey pairs

In fact, Bailey [59, Eq. (9.3)], briefly considers a pair of sequences (α_n, β_n) related by

$$\beta_n = \sum_{r=0}^{n} \frac{(q^{-n}, kq^n; q)_r}{(aq^{1-n}/k, aq^{n+1}; q)_r} \alpha_r \qquad (12.4)$$

$$= \frac{(aq, q; q)_n}{(k, k/a; q)_n} \sum_{r=0}^{n} \frac{(k/a; q)_{n-r}(k; q)_{n+r}}{(q; q)_{n-r}(aq; q)_{n+r}} \left(\frac{k}{aq} \right)^r \alpha_r.$$

(We have replaced Bailey's x by the more standard q.) Thus, if we take Bailey's pair (α_n, β_n) from above, and define

$$\alpha'_r = \left(\frac{k}{aq} \right)^r \alpha_r, \qquad \beta'_n = \frac{(k, k/a; q)_n}{(aq, q; q)_n} \beta_n,$$

then (α'_r, β'_r) is a WP-Bailey pair. Further, the particular pair Bailey gives following his definition at (12.4) is a disguised form of the WP-Bailey pair at (12.9), first explicitly stated by Singh [234, pp. 112–113] and later by Andrews [31] (and written down explicitly in [37, Eqs. (2.3)–(2.4)]). Although the details in [58] are scant, Bailey then indicates how continuing this line of argument allows him to derive the transformation (5.14) relating two $_{10}\phi_9$ derived previously by him [54] by another method. He also states how a transformation relating a $_{12}\phi_{11}$ series with a $_5\phi_4$ series may be similarly derived.

12.2.2 Bressoud's WP-Bailey pairs

Bressoud [84, p. 215] also considered a particular case of the Bailey transform which also may be interpreted in terms of WP-Bailey pairs, namely

$$U_n = \frac{(\beta; q)_n}{(q; q)_n}, \qquad V_n = \frac{(a\beta; q)_n}{(aq; q)_n}, \qquad \delta_n = r^n. \qquad (12.5)$$

Clearly the substitution $\beta = k/a$ shows that the pairs (α_n, β_n) generated by inserting Bressoud's choices in (10.1) are WP-Bailey pairs. Bressoud started with

$$\alpha_n = \frac{1 - aq^{2n}}{1 - a} \frac{(a, g^{-1}; q)_n}{(agq, q; q)_n} g^n,$$

(we have replaced Bressoud's parameter γ with g, to avoid confusion with the notation in Lemma 10.1) and then showed that

$$\beta_n = \sum_{r=0}^{n} U_{n-r} V_{n+r} \alpha_r = \sum_{r=0}^{n} \frac{(\beta; q)_{n-r}}{(q; q)_{n-r}} \frac{(a\beta; q)_{n+r}}{(aq; q)_{n+r}} \frac{1 - aq^{2r}}{1 - a} \frac{(a, g^{-1}; q)_r}{(agq, q; q)_r} g^r$$

$$= \frac{(a\beta, \beta/g; q)_n}{(agq, q; q)_n} g^n, \qquad (12.6)$$

and

$$
\gamma_n = \sum_{j=n}^{\infty} U_{j-n} V_{n+j} \delta_j = \sum_{j=n}^{\infty} \frac{(\beta;q)_{j-n}}{(q;q)_{j-n}} \frac{(a\beta;q)_{n+j}}{(aq;q)_{n+j}} r^j
$$

$$
= r^n \frac{(a\beta, \beta r; q)_\infty}{(aq, r; q)_\infty} \sum_{j=0}^{\infty} \frac{(q/\beta, r; q)_j}{(\beta r, q; q)_j} \left(a\beta q^{2n}\right)^j . \tag{12.7}
$$

(The proofs are left as exercises — see Q12.3.) By then employing (10.3), Bressoud arrives at the transformation

$$
\sum_{n=0}^{\infty} \frac{(a\beta, \beta/g; q)_n}{(agq, q; q)_n} g^n r^n
$$

$$
= \frac{(a\beta, \beta r; q)_\infty}{(aq, r; q)_\infty} \sum_{n=0}^{\infty} \frac{1 - aq^{2n}}{1 - a} \frac{(a, g^{-1}; q)_n}{(agq, q; q)_n} g^n r^n \sum_{j=0}^{\infty} \frac{(q/\beta, r; q)_j}{(\beta r, q; q)_j} \left(a\beta q^{2n}\right)^j .
$$

$$
\tag{12.8}
$$

As with other similar transformations, the chief interest lies in finding values for the parameters that make the inner series on the right summable, and one particular case is investigated in the exercises. Bressoud also considered another set of values for U_n, V_n and δ_n — see [84] for a description of this case of the Bailey transform and its implications, which included polynomial versions of the Rogers–Selberg mod 7 identities.

12.2.3   Singh's WP-Bailey pair

Singh considered a case of the Bailey transform that also may be represented in terms of WP-Bailey pairs. Since this pair will also make an appearance in our consideration of the work of Andrews [31] and Andrews and Berkovich [37], we will describe Singh's discoveries using the more standard a and k parameters of WP-Bailey pairs.

Lemma 12.1. *(Singh [234, pp. 112–113] The pair of sequences $(\alpha_n(a,k),$ $\beta_n(a,k))$ is a WP-Bailey pair, where*

$$
\alpha_n(a,k) = \frac{(q\sqrt{a}, -q\sqrt{a}, a, \rho_1, \rho_2, a^2 q/k\rho_1\rho_2; q)_n}{(\sqrt{a}, -\sqrt{a}, q, aq/\rho_1, aq/\rho_2, k\rho_1\rho_2/a; q)_n} \left(\frac{k}{a}\right)^n , \tag{12.9}
$$

$$
\beta_n(a,k) = \frac{(k\rho_1/a, k\rho_2/a, k, aq/\rho_1\rho_2; q)_n}{(aq/\rho_1, aq/\rho_2, k\rho_1\rho_2/a, q; q)_n} .
$$

Proof. The proof follows the standard lines — the expression for $\alpha_j(a, k)$ is substituted into (12.3), and some known summation formula (in this case Jackson's $_8\phi_7$ summation formula at (5.11)) is used to sum the right side of (12.3), resulting in the stated expression for $\beta_n(a, k)$. The details otherwise involve just elementary q-product manipulations, so the proof is left as an exercise (see Q12.5). □

Remark: This pair is also produced by substituting the unit WP-Bailey pair in the first WP-Bailey chain of Andrews — see below for details.

Like Bressoud [84], Singh also took $\delta_n = r^n$ in the Bailey transform, to get (this follows simply by replacing β with k/a in (12.7))

$$\gamma_n = \sum_{j=n}^{\infty} U_{j-n} V_{n+j} \delta_j = \sum_{j=n}^{\infty} \frac{(k/a; q)_{j-n}}{(q; q)_{j-n}} \frac{(k; q)_{n+j}}{(aq; q)_{n+j}} r^j$$

$$= r^n \frac{(k, kr/a; q)_\infty}{(aq, r; q)_\infty} \sum_{j=0}^{\infty} \frac{(qa/k, r; q)_j}{(kr/a, q; q)_j} (kq^{2n})^j. \tag{12.10}$$

The Bailey transform (10.3) then gives

$$\sum_{n=0}^{\infty} \frac{(k\rho_1/a, k\rho_2/a, k, aq/\rho_1\rho_2; q)_n}{(aq/\rho_1, aq/\rho_2, k\rho_1\rho_2/a, q; q)_n} r^n = \frac{(k, kr/a; q)_\infty}{(aq, r; q)_\infty} \times$$

$$\sum_{n=0}^{\infty} \frac{(q\sqrt{a}, -q\sqrt{a}, a, \rho_1, \rho_2, a^2q/k\rho_1\rho_2; q)_n k^n r^n}{(\sqrt{a}, -\sqrt{a}, q, aq/\rho_1, aq/\rho_2, k\rho_1\rho_2/a; q)_n a^n} \sum_{j=0}^{\infty} \frac{(qa/k, r; q)_j}{(kr/a, q; q)_j} (kq^{2n})^j. \tag{12.11}$$

The inner series is summable just for $r = 0$ or $r = qa^2/k^2$. The latter value results in the transformation

$$\sum_{n=0}^{\infty} \frac{(k\rho_1/a, k\rho_2/a, k, aq/\rho_1\rho_2; q)_n}{(aq/\rho_1, aq/\rho_2, k\rho_1\rho_2/a, q; q)_n} \left(\frac{qa^2}{k^2}\right)^n = \frac{(qa^2/k, qa/k; q)_\infty}{(aq, qa^2/k^2; q)_\infty} \times$$

$$\sum_{n=0}^{\infty} \frac{(q\sqrt{a}, -q\sqrt{a}, a, \rho_1, \rho_2, a^2q/k\rho_1\rho_2; q)_n (k, q)_{2n}}{(\sqrt{a}, -\sqrt{a}, q, aq/\rho_1, aq/\rho_2, k\rho_1\rho_2/a; q)_n (qa^2/k; q)_{2n}} \left(\frac{qa}{k}\right)^n. \tag{12.12}$$

Singh goes on to consider several special cases of this identity, and derives, amongst other results, a new identity of Rogers–Ramanujan type, and a polynomial analogue of this identity.

12.3   The WP-Bailey chains of Andrews

The significant breakthrough that Andrews made in [31], which distinguished that paper from previous papers on the topic, was that he developed mechanisms from producing new WP-Bailey pairs from existing WP-Bailey pairs (and similarly for regular Bailey pairs), thus creating the first WP-Bailey chains and the first branches in the WP-Bailey lattice (or tree).

In sections 6 and 7 of [31], Andrews stated and proved two WP-Bailey chains. The first of these may be described as follows.

Theorem 12.2. *(Andrews [31]) If $(\alpha_n(a,k), \beta_n(a,k))$ satisfy (12.3), then so does the pair $(\alpha'_n(a,k), \beta'_n(a,k))$, where*

$$\alpha'_n(a,k) = \frac{(\rho_1, \rho_2; q)_n}{(aq/\rho_1, aq/\rho_2; q)_n} \left(\frac{k}{c}\right)^n \alpha_n(a,c), \tag{12.13}$$

$$\beta'_n(a,k) = \frac{(k\rho_1/a, k\rho_2/a; q)_n}{(aq/\rho_1, aq/\rho_2; q)_n}$$

$$\times \sum_{j=0}^{n} \frac{(1 - cq^{2j})(\rho_1, \rho_2; q)_j(k/c; q)_{n-j}(k; q)_{n+j}}{(1 - c)(k\rho_1/a, k\rho_2/a; q)_j(q; q)_{n-j}(qc; q)_{n+j}} \left(\frac{k}{c}\right)^j \beta_j(a,c), \tag{12.14}$$

where $c = k\rho_1\rho_2/aq$.

Proof. Substitute for $\beta_j(a,c)$ in (12.14) using (12.3) (with k replaced with c), change the order of summation and use the fact that for any sequence $\{g_j\}$, $\sum_{j=i}^{n} g_j = \sum_{j=0}^{n-i} g_{i+j}$, to get

$$\beta'_n(a,k) = \frac{(k\rho_1/a, k\rho_2/a; q)_n}{(aq/\rho_1, aq/\rho_2; q)_n} \sum_{j=0}^{n} \frac{(1 - cq^{2j})(\rho_1, \rho_2; q)_j(k/c; q)_{n-j}(k; q)_{n+j}}{(1 - c)(k\rho_1/a, k\rho_2/a; q)_j(q; q)_{n-j}(qc; q)_{n+j}}$$

$$\times \left(\frac{k}{c}\right)^j \sum_{i=0}^{j} \frac{(c/a; q)_{j-i}(c; q)_{j+i}}{(q; q)_{j-i}(aq; q)_{j+i}} \alpha_i(a,c)$$

$$= \frac{(k\rho_1/a, k\rho_2/a; q)_n}{(aq/\rho_1, aq/\rho_2; q)_n} \sum_{i=0}^{n} \sum_{j=0}^{n-i} \frac{(1 - cq^{2j+2i})(\rho_1, \rho_2; q)_{i+j}}{(1 - c)(k\rho_1/a, k\rho_2/a; q)_{i+j}}$$

$$\times \frac{(k/c; q)_{n-i-j}(k; q)_{n+i+j}(c/a; q)_j(c; q)_{j+2i}}{(q; q)_{n-i-j}(qc; q)_{n+i+j}(q; q)_j(aq; q)_{j+2i}} \left(\frac{k}{c}\right)^{j+i} \alpha_i(a,c)$$

$$= \frac{(k\rho_1/a, k\rho_2/a; q)_n}{(aq/\rho_1, aq/\rho_2; q)_n} \sum_{i=0}^{n} \frac{(1 - cq^{2i})(\rho_1, \rho_2; q)_i}{(1 - c)(k\rho_1/a, k\rho_2/a; q)_i}$$

$$\times \frac{(k/c;q)_{n-i}(k;q)_{n+i}(c;q)_{2i}}{(q;q)_{n-i}(qc;q)_{n+i}(aq;q)_{2i}} \left(\frac{k}{c}\right)^i \alpha_i(a,c)$$

$$\times \sum_{j=0}^{n-i} \frac{(1-cq^{2i+2j})(\rho_1 q^i, \rho_2 q^i, kq^{n+i}, c/a, cq^{2i}, q^{-(n-i)};q)_j\, q^j}{(1-cq^{2i})(k\rho_1 q^i/a, k\rho_2 q^i/a, cq^{n+i+1}, aq^{2i+1}, cq^{1-(n-i)}/k, q;q)_j},$$

where (2.13) has been used on the ratio $(k/c;q)_{n-i-j}/(q;q)_{n-i-j}$. Next, the inner series above may be summed using Jackson's $_8\phi_7$ summation formula at (5.11) (with the part of a in the latter sum being played by cq^{2i} in the sum above (also recalling that $c = k\rho_1\rho_2/aq$). Thus,

$$\beta_n'(a,k) = \frac{(k\rho_1/a, k\rho_2/a;q)_n}{(aq/\rho_1, aq/\rho_2;q)_n} \sum_{i=0}^{n} \frac{(1-cq^{2i})(\rho_1,\rho_2;q)_i}{(1-c)(k\rho_1/a, k\rho_2/a;q)_i}$$

$$\times \frac{(k/c;q)_{n-i}(k;q)_{n+i}(c;q)_{2i}}{(q;q)_{n-i}(qc;q)_{n+i}(aq;q)_{2i}} \left(\frac{k}{c}\right)^i \alpha_i(a,c)$$

$$\times \frac{(cq^{2i+1}, cq/\rho_1\rho_2, cq^{1-n}/k\rho_1, cq^{1-n}/k\rho_2;q)_{n-i}}{(k\rho_1 q^i/a, k\rho_2 q^i/a, cq^{1+i-n}/k, cq^{1-n-i}/k\rho_1\rho_2;q)_{n-i}}$$

$$= \sum_{i=0}^{n} \frac{(k/a;q)_{n-i}(k;q)_{n+i}}{(q;q)_{n-i}(aq;q)_{n+i}} \frac{(\rho_1,\rho_2;q)_i}{(aq/\rho_1, aq/\rho_2;q)_i} \left(\frac{k}{c}\right)^i \alpha_i(a,c)$$

$$= \sum_{i=0}^{n} \frac{(k/a;q)_{n-i}(k;q)_{n+i}}{(q;q)_{n-i}(aq;q)_{n+i}} \alpha_i'(a,k),$$

where the next-to-last equality follows from using (2.11) and (2.25) (with n replaced with $n-i$ in the latter). $\qquad\square$

Remark: Notice that letting $k \to 0$ in (12.13) recovers the regular Bailey chain of Andrews at (11.3).

Before considering the second chain of Andrews [31, Section 7], we examine the implications of walking along this chain, starting with the unit WP-Bailey pair at (12.15), which we first need to prove actually is a WP-Bailey pair.

Lemma 12.2. *The pair of sequences* $(\alpha_n(a,k), \beta_n(a,k))$ *form a WP-Bailey pair, where*

$$\alpha_n(a,k) = \frac{(1-aq^{2n})(a,\frac{a}{k};q)_n}{(1-a)(q,kq;q)_n}\left(\frac{k}{a}\right)^n, \qquad (12.15)$$

$$\beta_n(a,k) = \delta_{n,0}.$$

Proof. Upon substituting $\alpha_j(a, k)$ into (12.3)

$$\sum_{j=0}^{n} \frac{(k/a)_{n-j}(k)_{n+j}}{(q)_{n-j}(aq)_{n+j}} \alpha_j(a, k)$$

$$= \frac{(k/a, k; q)_n}{(aq, q; q)_n} \sum_{j=0}^{n} \frac{1 - aq^{2j}}{1-a} \frac{(a, a/k, kq^n, q^{-n}; q)_j \, q^j}{(kq, aq^{n+1}, aq^{1-n}/k, q; q)_j} = \delta_{n,0},$$

where the first equality follows from (2.11) and (2.13), and the second equality by (5.2).   □

Corollary 12.1. *For each non-negative integer* n,

$$\sum_{j=0}^{n} \frac{1 - aq^{2j}}{1-a} \frac{(a, \rho_1, \rho_2, a^2q/k\rho_1\rho_2, kq^n, q^{-n}; q)_j q^j}{(aq/\rho_1, aq/\rho_2, k\rho_1\rho_2/a, aq^{n+1}, aq^{1-n}/k, q; q)_j}$$

$$= \frac{(aq, k\rho_1/a, k\rho_2/a, aq/\rho_1\rho_2; q)_n}{(aq/\rho_1, aq/\rho_2, k\rho_1\rho_2/a, k/a; q)_n}. \quad (12.16)$$

Proof. This follows upon substituting the unit WP-Bailey pair (12.15) into Theorem 12.2 to get the pair at (12.19), setting the two expressions for $\beta'_n(a, k)$ (the one coming from (12.3) and the one coming from (12.14)) equal, and rearranging.   □

Remarks: 1) The identity at (12.16) is equivalent to Jackson's $_8\phi_7$ summation formula at (5.11).

2) We will subsequently derive other basic hypergeometric identities from a new WP-Bailey pair $(\alpha'_n(a, k), \beta'_n(a, k))$ in a similar fashion, namely by equating the expression for $\beta'_n(a, k)$ coming from the WP-Bailey chain with the expression derived by inserting $\alpha'_n(a, k)$ into the defining expression of a WP-Bailey pair (12.3), by simply stating that the pair leads to the identity, without dwelling on the details.

3) At each step along the chain, the parameters ρ_1, ρ_2 may be replaced with a different pair of parameters. As will be seen in the next corollary, the next step along this particular WP-Bailey chain produces Bailey's $_{10}\phi_9$ transformation (5.14) (as demonstrated by Andrews and Berkovich in [37, Eq. (2.9)]).

Corollary 12.2. *For each non-negative integer* n,

$$\sum_{j=0}^{n} \frac{1 - aq^{2j}}{1-a} \frac{(a, \rho_1, \rho_2, \sigma_1, \sigma_2, a^2q/\lambda\rho_1\rho_2, kq^n, q^{-n}; q)_j q^j}{(aq/\rho_1, aq/\rho_2, aq/\sigma_1, aq/\sigma_2, \lambda\rho_1\rho_2/a, aq^{1-n}/k, aq^{n+1}, q; q)_j}$$

$$= \frac{(aq, k\sigma_1/a, k\sigma_2/a, aq/\sigma_1\sigma_2; q)_n}{(aq/\sigma_1, aq/\sigma_2, k\sigma_1\sigma_2/a, k/a; q)_n}$$

$$\sum_{j=0}^{n} \frac{(1 - \lambda q^{2j})(\lambda, \lambda\rho_1/a, \lambda\rho_2/a, \sigma_1, \sigma_2, aq/\rho_1\rho_2, kq^n, q^{-n}; q)_j q^j}{(1 - \lambda)(aq/\rho_1, aq/\rho_2, k\sigma_1/a, k\sigma_2/a, \lambda\rho_1\rho_2/a, \sigma_1\sigma_2 q^{-n}/a, \lambda q^{n+1}, q; q)_j},$$

$$(12.17)$$

where $\lambda = k\sigma_1\sigma_2/aq$.

$$\sum_{j=0}^{n} \frac{\left(q\sqrt{a}, -q\sqrt{a}, iq\sqrt{a}, -iq\sqrt{a}, a, \frac{a^2}{k^2}, kq^n, -kq^n, -q^{-n}, q^{-n}; q\right)_j q^j}{\left(\sqrt{a}, -\sqrt{a}, i\sqrt{a}, -i\sqrt{a}, \frac{k^2 q}{a}, -\frac{aq^{1-n}}{k}, \frac{aq^{1-n}}{k}, -aq^{n+1}, aq^{n+1}, q; q\right)_j}$$

$$= \frac{(a^2 q^2, a^2/k^2; q^2)_n (-k^2 q/a; q)_{2n}}{(k^2/a^2, k^4 q^2/a^2; q^2)_n (-a; q)_{2n}} \left(\frac{k^2}{a^2 q}\right)^n. \quad (12.18)$$

Proof. The WP-Bailey pair that results upon substituting the unit WP-Bailey pair (12.15) into Theorem 12.2 is the pair

$$\alpha_n'(a, k) = \frac{(1 - aq^{2n})(a, \rho_1, \rho_2, a^2 q/k\rho_1\rho_2; q)_n}{(1 - a)(aq/\rho_1, aq/\rho_2, k\rho_1\rho_2/a, q; q)_n} \left(\frac{k}{a}\right)^n, \quad (12.19)$$

$$\beta_n'(a, k) = \frac{(k\rho_1/a, k\rho_2/a, aq/\rho_1\rho_2, k; q)_n}{(aq/\rho_1, aq/\rho_2, k\rho_1\rho_2/a, q; q)_n}.$$

When the above pair is substituted back into Theorem 12.2 (with ρ_1, ρ_2 replaced with σ_1, σ_2), the resulting pair implies (12.17), in the way indicated in Remark 2) above.

For (12.18), set $\rho_1 = iq\sqrt{a}$, $\rho_2 = -iq\sqrt{a}$, $\sigma_1 = -kq^n$ and $\sigma_2 = -q^{-n}$ in (12.17). Then the term $aq/\rho_1\rho_2$ in the series on the right side becomes $1/q$, so that all terms in this series vanish for $j \geq 2$. The result follows after some elementary algebra. $\square$

The second WP-Bailey chain of Andrews may be described as follows.

Theorem 12.3. *(Andrews [31]) If* $(\alpha_n(a, k), \beta_n(a, k))$ *satisfy (12.3), then so does the pair* $(\alpha_n'(a, k), \beta_n'(a, k))$, *where*

$$\alpha_n'(a, k) = \frac{(qa^2/k)_{2n}}{(k)_{2n}} \left(\frac{k^2}{qa^2}\right)^n \alpha_n\left(a, \frac{qa^2}{k}\right), \quad (12.20)$$

$$\beta_n'(a, k) = \sum_{j=0}^{n} \frac{(k^2/qa^2)_{n-j}}{(q)_{n-j}} \left(\frac{k^2}{qa^2}\right)^j \beta_j\left(a, \frac{qa^2}{k}\right). \quad (12.21)$$

Proof. See Q12.7.   □

When the arguments used in the proofs of Corollaries 12.1 and 12.2 are applied to the chain in Theorem 12.3, the following consequences result.

Corollary 12.3. *(i) The pair* $(\alpha'_n(a, k),\ \beta'_n(a, k))$ *is a WP-Bailey pair, where*

$$\alpha'_n(a, k) = \frac{1 - aq^{2n}}{1 - a} \frac{(a, k/aq; q)_n (qa^2/k; q)_{2n}}{(q^2 a^2/k, q; q)_n (k; q)_{2n}} \left(\frac{k}{a}\right)^n, \qquad (12.22)$$

$$\beta'_n(a, k) = \frac{(k^2/qa^2; q)_n}{(q; q)_n}.$$

(ii) For each non-negative integer n,

$$\sum_{j=0}^{n} \frac{(1 - aq^{2j})(a, k/aq, a\sqrt{q/k}, -a\sqrt{q/k}, aq/\sqrt{k}, -aq/\sqrt{k}, kq^n, q^{-n}; q)_j q^j}{(1 - a)(q^2 a^2/k, \sqrt{k}, -\sqrt{k}, \sqrt{kq}, -\sqrt{kq}, aq^{n+1}, aq^{1-n}/k, q; q)_j}$$

$$= \frac{(aq, k^2/a^2 q; q)_n}{(k, k/a; q)_n}. \qquad (12.23)$$

Proof. See Q12.8.   □

Remark: Notice that there is no equivalent of (12.17) for this second chain, since travelling two steps along the chain actually returns the initial WP-Bailey pair. However, as indicated earlier, one can insert a WP-Bailey pair that arises in one chain into another chain, and if the pair at (12.19) is inserted into Theorem 12.3, then the following transformation is a consequence (see remark 2) following Corollary 12.1).

Corollary 12.4. *For each non-negative integer* n, *there holds*

$$\sum_{j=0}^{n} \frac{\left(a, q\sqrt{a}, -q\sqrt{a}, \rho_1, \rho_2, \frac{k}{\rho_1 \rho_2}, a\sqrt{\frac{q}{k}}, -a\sqrt{\frac{q}{k}}, \frac{aq}{\sqrt{k}}, -\frac{aq}{\sqrt{k}}, kq^n, q^{-n}; q\right)_j q^j}{\left(\sqrt{a}, -\sqrt{a}, \frac{aq\rho_1 \rho_2}{k}, \frac{aq}{\rho_1}, \frac{aq}{\rho_2}, \sqrt{k}, -\sqrt{k}, \sqrt{kq}, -\sqrt{kq}, aq^{n+1}, \frac{aq^{1-n}}{k}, q; q\right)_j}$$

$$= \frac{(aq, k^2/a^2 q; q)_n}{(k, k/a; q)_n} \sum_{j=0}^{n} \frac{\left(\frac{a^2 q}{k}, \frac{aq}{\rho_1 \rho_2}, \frac{aq\rho_1}{k}, \frac{aq\rho_2}{k}, q^{-n}; q\right)_j q^j}{\left(\frac{aq\rho_1 \rho_2}{k}, \frac{aq}{\rho_1}, \frac{aq}{\rho_2}, \frac{a^2 q^{2-n}}{k^2}, q; q\right)_j}. \qquad (12.24)$$

Proof. See Q12.9.   □

Remark: This identity is equivalent to a result of Bailey [59, p. 431].

As Andrews and Berkovich noted, it is possible to write both (12.23) and (12.47) as expressions of the form at (12.3), thus deriving two WP-Bailey pairs (actually the first pair derived in this way, the pair at (12.25) below, is just the pair from Corollary 12.3, so that this method just gives a new proof for that pair). These in turn may be substituted back into the chain at (12.13), leading to two basic hypergeometric identities (see [37, Eqs. (3.7) and (3.8)]).

Corollary 12.5. *(Andrews and Berkovich [37]) (i) Each of the pairs* $(\alpha'_n(a,k), \beta'_n(a,k))$ *and* $(\alpha^*_n(a,k),\ \beta^*_n(a,k))$ *are WP-Bailey pairs, where*

$$\alpha'_n(a,k) = \frac{(a, q\sqrt{a}, -q\sqrt{a}, k/aq; q)_n (qa^2/k; q)_{2n}}{(\sqrt{a}, -\sqrt{a}, a^2q^2/k, q; q)_n (k; q)_{2n}} \left(\frac{k}{a}\right)^n, \tag{12.25}$$

$$\beta'_n(a,k) = \frac{(k^2/a^2 q; q)_n}{(q; q)_n};$$

$$\alpha^*_n(a,k) = \frac{(a, q\sqrt{a}, -q\sqrt{a}, a\sqrt{q/k}, -a\sqrt{q/k}, a/\sqrt{k}, -aq/\sqrt{k}, k/a; q)_n}{(\sqrt{a}, -\sqrt{a}, \sqrt{qk}, -\sqrt{qk}, q\sqrt{k}, -\sqrt{k}, a^2q/k, q; q)_n}, \tag{12.26}$$

$$\beta^*_n(a,k) = \frac{(\sqrt{k}, k^2/a^2; q)_n}{(q\sqrt{k}, q; q)_n}.$$

(ii) The following identities hold for each non-negative integer n:

$$\sum_{j=0}^{n} \frac{\left(a, q\sqrt{a}, -q\sqrt{a}, a\sqrt{\frac{q}{c}}, -a\sqrt{\frac{q}{c}}, \frac{aq}{\sqrt{c}}, -\frac{aq}{\sqrt{c}}, \frac{c}{aq}, \rho_1, \rho_2, kq^n, q^{-n}; q\right)_j q^j}{\left(\sqrt{a}, -\sqrt{a}, \sqrt{cq}, -\sqrt{cq}, \sqrt{c}, -\sqrt{c}, \frac{a^2q^2}{c}, \frac{aq}{\rho_1}, \frac{aq}{\rho_2}, \frac{aq^{1-n}}{k}, aq^{n+1}, q; q\right)_j}$$

$$= \frac{\left(\frac{k\rho_1}{a}, \frac{k\rho_2}{a}, \frac{k}{c}, aq; q\right)_n}{\left(\frac{aq}{\rho_1}, \frac{aq}{\rho_2}, \frac{k}{a}, cq; q\right)_n} \sum_{j=0}^{n} \frac{\left(\frac{c^2}{qa^2}, q\sqrt{c}, -q\sqrt{c}, \rho_1, \rho_2, kq^n, q^{-n}; q\right)_j q^j}{\left(\sqrt{c}, -\sqrt{c}, \frac{k\rho_1}{a}, \frac{k\rho_2}{a}, \frac{cq^{1-n}}{k}, cq^{n+1}, q; q\right)_j}, \tag{12.27}$$

$$\sum_{j=0}^{n} \frac{\left(a, q\sqrt{a}, -q\sqrt{a}, a\sqrt{\frac{q}{c}}, -a\sqrt{\frac{q}{c}}, \frac{a}{\sqrt{c}}, -\frac{aq}{\sqrt{c}}, \frac{c}{a}, \rho_1, \rho_2, kq^n, q^{-n}; q\right)_j q^j}{\left(\sqrt{a}, -\sqrt{a}, \sqrt{cq}, -\sqrt{cq}, q\sqrt{c}, -\sqrt{c}, \frac{a^2q}{c}, \frac{aq}{\rho_1}, \frac{aq}{\rho_2}, \frac{aq^{1-n}}{k}, aq^{n+1}, q; q\right)_j}$$

$$= \frac{\left(\frac{k\rho_1}{a}, \frac{k\rho_2}{a}, \frac{k}{c}, aq; q\right)_n}{\left(\frac{aq}{\rho_1}, \frac{aq}{\rho_2}, \frac{k}{a}, cq; q\right)_n} \sum_{j=0}^{n} \frac{\left(\frac{c^2}{a^2}, -q\sqrt{c}, \rho_1, \rho_2, kq^n, q^{-n}; q\right)_j q^j}{\left(-\sqrt{c}, \frac{k\rho_1}{a}, \frac{k\rho_2}{a}, \frac{cq^{1-n}}{k}, cq^{n+1}, q; q\right)_j}, \qquad (12.28)$$

where $c = k\rho_1\rho_2/aq$.

Proof. See Q12.13. □

Remark: The transformation (12.27) may be shown to be equivalent to a formula of Bailey [60], while (12.28) is equivalent to a result of Jain [160].

In general, the expression for the $\beta'_n(a,k)$ at (12.14) cannot be summed at each step, so continuing to move along the chain will produce expressions for the $\beta'_n(a,k)$ that are multi-sums.

12.4 The WP-Bailey chains of Warnaar

In [256], Warnaar stated and proved four WP-Bailey chains. He also proved a number of chains for elliptic hypergeometric series, which we do not consider in this chapter.

Theorem 12.4. *(Warnaar [256]) If $(\alpha_n(a,k), \beta_n(a,k))$ satisfy (12.3), then so do $(\alpha_n^{(i)}(a,k), \beta_n^{(i)}(a,k))$, $i = 1,2,3,4$, where*

$$\alpha_n^{(1)}(a,k) = \frac{1 - \sigma k^{1/2}}{1 - \sigma k^{1/2} q^n} \frac{1 + \sigma a q^n/k^{1/2}}{1 + \sigma a/k^{1/2}} \frac{(a^2/k; q)_{2n}}{(k; q)_{2n}} \left(\frac{k^2}{a^2}\right)^n \alpha_n\left(a, \frac{a^2}{k}\right),$$

$$(12.29)$$

$$\beta_n^{(1)}(a,k) = \frac{1 - \sigma k^{1/2}}{1 - \sigma k^{1/2} q^n} \sum_{j=0}^{n} \frac{1 + \sigma a q^j/k^{1/2}}{1 + \sigma a/k^{1/2}} \frac{(k^2/a^2; q)_{n-j}}{(q; q)_{n-j}} \left(\frac{k^2}{a^2}\right)^j \beta_j\left(a, \frac{a^2}{k}\right),$$

where $\sigma \in \{1, -1\}$ *(as Warnaar [256] indicates, the choices for σ simply reflect that the expression is invariant under simultaneous negation of σ and $k^{1/2}$).*

$$\alpha_n^{(2)}(a,k,q^2) = \alpha_n\left(\sqrt{a}, \frac{k}{q\sqrt{a}}, q\right), \qquad (12.30)$$

$$\beta_n^{(2)}(a,k,q^2) = \frac{\left(\frac{-k}{\sqrt{a}}; q\right)_{2n}}{(-q\sqrt{a}; q)_{2n}} \sum_{j=0}^{n} \frac{\left(1 - \frac{kq^{2j}}{q\sqrt{a}}\right)}{\left(1 - \frac{k}{q\sqrt{a}}\right)} \frac{\left(\frac{aq^2}{k}; q^2\right)_{n-j}}{(q^2; q^2)_{n-j}} \frac{(k; q^2)_{n+j}}{\left(\frac{k^2}{a}; q^2\right)_{n+j}}$$

$$\times \left(\frac{k}{aq}\right)^{n-j} \beta_j\left(\sqrt{a}, \frac{k}{q\sqrt{a}}, q\right).$$

$$\alpha_n^{(3)}(a,k,q^2) = q^{-n}\frac{1+\sqrt{a}q^{2n}}{1+\sqrt{a}}\alpha_n\left(\sqrt{a},\frac{k}{\sqrt{a}},q\right), \tag{12.31}$$

$$\beta_n^{(3)}(a,k,q^2) = q^{-n}\frac{\left(\frac{-kq}{\sqrt{a}};q\right)_{2n}}{(-\sqrt{a};q)_{2n}}\sum_{j=0}^{n}\frac{\left(1-\frac{kq^{2j}}{\sqrt{a}}\right)\left(\frac{a}{k};q^2\right)_{n-j}(k;q^2)_{n+j}}{\left(1-\frac{k}{\sqrt{a}}\right)(q^2;q^2)_{n-j}\left(\frac{k^2q^2}{a};q^2\right)_{n+j}}$$

$$\times\left(\frac{k}{a}\right)^{n-j}\beta_j\left(\sqrt{a},\frac{k}{\sqrt{a}},q\right).$$

$$\alpha_{2n}^{(4)}(a,k,q) = \alpha_n\left(a,\frac{k^2}{a},q^2\right), \qquad \alpha_{2n+1}^{(4)}(a,k,q) = 0 \tag{12.32}$$

$$\beta_n^{(4)}(a,k,q) = \frac{\left(\frac{k^2q}{a};q^2\right)_n}{(aq;q^2)_n}\sum_{j=0}^{\lfloor n/2\rfloor}\frac{\left(1-\frac{k^2q^{4j}}{a}\right)\left(\frac{a}{k};q\right)_{n-2j}(k;q)_{n+2j}}{\left(1-\frac{k^2}{a}\right)(q;q)_{n-2j}\left(\frac{k^2q}{a};q\right)_{n+2j}}$$

$$\times\left(\frac{-k}{a}\right)^{n-2j}\beta_j\left(a,\frac{k^2}{a},q^2\right).$$

Proof. The proof in each case essentially follows the same steps as in the proof of Theorem 12.2, namely, substitute for β_j in the sum for the new $\beta_n^{(i)}(a,k)$ using (12.3), change the order of summation and shift the index on the inner sum so that the summation index starts at 0, use some summation formula on the inner sum, and finally show that expression for $\beta_n^{(i)}(a,k)$ is given by (12.3) (with $\alpha_j(a,k)$ replaced with $\alpha_j^{(i)}(a,k)$). The only real difference in the proof of the various chains involves the summation formula used on the inner sum, so we restrict the proof in each case to indicating which summation formula is used, and leave the remaining details of the proofs as exercises. The summation formulae used are, respectively, the following:

Warnaar's chain	summation formula used in proof
(12.29)	(5.24) (the case $e = d^2q/b$)
(12.30)	(5.11)
(12.31)	(5.25)
(12.32)	(5.11)

□

Just as with the chains of Andrews, one can substitute the unit WP-Bailey pair (12.15) into any of the chains of Warnaar to derive new

WP-Bailey pairs, and hence new basic hypergeometric transformations (see Q12.11), and similarly with any WP-Bailey pair arising from another chain.

12.5   The WP-Bailey chain of Liu and Ma

In [181], Liu and Ma introduce the concept of a general WP-Bailey chain, and showed that the first chain of Andrews and a new WP-Bailey chain due to them follow as special cases. We consider their new chain in this section.

Theorem 12.5. *(Liu and Ma [181]) If $(\alpha_n(a,k), \beta_n(a,k))$ satisfy (12.3), then so do $(\alpha_n^*(a,k), \beta_n^*(a,k))$, where*

$$\alpha_n^*(a,k) = \frac{(a^2 q/k; q^2)_n}{(kq; q^2)_n} \left(\frac{-k}{a}\right)^n \alpha_n\left(a, \frac{a^2}{k}\right), \tag{12.33}$$

$$\beta_n^*(a,k) = \sum_{j=0}^{n/2} \frac{\left(1 - \frac{a^2 q^{2n-4j}}{k}\right)(k; q^2)_{n-j}\left(\frac{k^2}{a^2}; q^2\right)_j}{\left(1 - \frac{a^2}{k}\right)\left(\frac{a^2 q^2}{k}; q^2\right)_{n-j}(q^2; q^2)_j} \left(\frac{-k}{a}\right)^{n-2j} \beta_{n-2j}\left(a, \frac{a^2}{k}\right).$$

Proof. The proof is essentially the same as that for each of the chains listed above: substitute for $\beta_{n-2j}\left(a, a^2/k\right)$ in the expression for $\beta_n^*(a,k)$ above (using (12.3) with k replaced with a^2/k), switch the order of summation (this time it is not necessary to shift the summation index on the inner sum so that it starts at 0, as it already does so), use Jackson's $_8\phi_7$ summation formula (5.11) (with q replaced with q^2), and the result follows after some q-product manipulations.   □

As with other chains, any WP-Bailey pair may be substituted into (12.33) to derive a basic hypergeometric identity (see Q12.12).

12.6   The WP-Bailey chains of Mc Laughlin and Zimmer

Motivated by the general chains of Liu and Ma, Mc Laughlin and Zimmer [201] also investigated general WP-Bailey chains, and discovered three new WP-Bailey chains as particular cases of these general chains. The proofs of two of these proceed as with the proofs of other chains. However, for the third chain (12.35), we have not yet provided a proof of the relevant summation formula, so instead we prove a general WP-Bailey chain, and then show that (12.35) follows as a special case. This will in turn lead to a proof of the afore-mentioned summation formula.

Theorem 12.6. *(Mc Laughlin and Zimmer [201]) If $(\alpha_n(a,k), \beta_n(a,k))$ satisfy (12.3), then so do $(\alpha_n^{(i)}(a,k), \beta_n^{(i)}(a,k))$, $i = 1, 2, 3$, where*

$$\alpha_n^{(1)}(a,k,q) = \frac{\left(\frac{a^2}{k};q\right)_{2n}}{(kq;q)_{2n}} \left(\frac{k^2 q}{a^2}\right)^n \alpha_n\left(a, \frac{a^2}{kq}, q\right), \tag{12.34}$$

$$\beta_n^{(1)}(a,k,q) = \frac{1-k}{1-kq^{2n}} \sum_{j=0}^{n} \frac{\left(1 - \frac{a^2 q^{2j}}{kq}\right)\left(\frac{k^2 q}{a^2};q\right)_{n-j}}{\left(1 - \frac{a^2}{kq}\right)(q;q)_{n-j}} \left(\frac{k^2 q}{a^2}\right)^j \beta_j\left(a, \frac{a^2}{kq}, q\right).$$

$$\alpha_n^{(2)}(a,k,q) = \frac{1+a}{1+aq^{2n}} q^n \alpha_n\left(a^2, \frac{ak}{q}, q^2\right), \tag{12.35}$$

$$\beta_n^{(2)}(a,k,q) = \sum_{j=0}^{n} \frac{\left(1 - \frac{ak}{q}q^{4j}\right)}{\left(1 - \frac{ak}{q}\right)} \frac{(-a;q)_{2j}}{(-k;q)_{2j}} \frac{\left(\frac{qk}{a};q^2\right)_{n-j}}{(q^2;q^2)_{n-j}} \frac{(k^2;q^2)_{n+j}}{(akq;q^2)_{n+j}}$$

$$\times q^j \beta_j\left(a^2, \frac{ak}{q}, q^2\right).$$

$$\alpha_n^{(3)}(a,k,q^2) = \alpha_n\left(\sqrt{a}, \frac{k}{\sqrt{a}}, q\right), \tag{12.36}$$

$$\beta_n^{(3)}(a,k,q^2) = \frac{\left(\frac{-k}{\sqrt{a}};q\right)_{2n}}{(-q\sqrt{a};q)_{2n}} \sum_{j=0}^{n} \frac{\left(1 - \frac{k^2 q^{4j}}{a}\right)}{\left(1 - \frac{k^2}{a}\right)} \frac{\left(\frac{a}{k};q^2\right)_{n-j}}{(q^2;q^2)_{n-j}} \frac{(k;q^2)_{n+j}}{\left(\frac{k^2 q^2}{a};q^2\right)_{n+j}}$$

$$\times \left(\frac{kq}{a}\right)^{n-j} \beta_j\left(\sqrt{a}, \frac{k}{\sqrt{a}}, q\right).$$

Proof of (12.34) *and* (12.36). As with the proofs of other chains in this chapter, the proofs of these chains proceed as follows: substitute for β_j in the sum for the new $\beta_n^{(i)}(a,k)$ using (12.3), change the order of summation and shift the index on the inner sum so that the summation index starts at 0, use some summation formula on the inner sum, and finally show that expression for $\beta_n^{(i)}(a,k)$ is given by (12.3) (with $\alpha_j(a,k)$ replaced with $\alpha_j^{(i)}(a,k)$). This time, the summation formulae used are the following:

chain	summation formula used in proof
(12.34)	(12.46)
(12.35)	(5.25)
(12.36)	(12.18).

$\square$

To prove (12.35), we first prove a general WP-Bailey chain.

Proposition 12.1. *Suppose that $(\alpha_n(a,k,q), \beta_n(a,k,q))$ satisfy (12.3), that g_n is an arbitrary sequence of functions and that e and c are arbitrary constants. Then $(\alpha'_n(a,k,q), \beta'_n(a,k,q))$ also satisfy (12.3), where*

$$\alpha'_n(a,k,q) = g_n \alpha_n(e,c,q^2), \tag{12.37}$$

$$\beta'_n(a,k,q) = \sum_{j=0}^{n} \beta_j(e,c,q^2) \frac{(1-cq^{4j})(k/a;q)_{n-j}(k;q)_{n+j}(eq^2;q^2)_{2j}}{(1-c)(q;q)_{n-j}(aq;q)_{n+j}(cq^2;q^2)_{2j}}$$

$$\times \sum_{r=0}^{n-j} \frac{(1-eq^{4j+4r})(q^{-(n-j)},kq^{n+j};q)_r(e/c,eq^{4j};q^2)_r \, g_{r+j}}{(1-eq^{4j})(aq^{n+j+1},aq^{1-(n-j)}/k;q)_r(cq^{4j+2},q^2;q^2)_r} \left(\frac{qca}{ke}\right)^r.$$

Proof. As previously, if $\alpha'_n(a,k,q)$ is as given, then

$$\beta'_n(a,k,q) = \sum_{r=0}^{n} \frac{(k/a)_{n-r}(k)_{n+r}}{(q)_{n-r}(aq)_{n+r}} g_r \alpha_r(e,c,q^2)$$

$$= \sum_{r=0}^{n} \frac{(k/a)_{n-r}(k)_{n+r}}{(q)_{n-r}(aq)_{n+r}} g_r$$

$$\times \frac{1-eq^{4r}}{1-e} \sum_{j=0}^{r} \frac{1-cq^{4j}}{1-c} \frac{(e/c;q^2)_{r-j}(e;q^2)_{r+j}}{(q^2;q^2)_{r-j}(cq^2;q^2)_{r+j}} \left(\frac{c}{e}\right)^{r-j} \beta_j(e,c,q^2)$$

$$= \sum_{j=0}^{n} \frac{1-cq^{4j}}{1-c} \beta_j(e,c,q^2)$$

$$\times \sum_{r=j}^{n} \frac{(k/a)_{n-r}(k)_{n+r}}{(q)_{n-r}(aq)_{n+r}} g_r \frac{1-eq^{4r}}{1-a} \frac{(e/c;q^2)_{r-j}(e;q^2)_{r+j}}{(q^2;q^2)_{r-j}(cq^2;q^2)_{r+j}} \left(\frac{c}{e}\right)^{r-j}$$

$$= \sum_{j=0}^{n} \frac{1-cq^{4j}}{1-c} \beta_j(e,c,q^2) \sum_{r=0}^{n-j} \frac{(k/a)_{n-r-j}(k)_{n+r+j}}{(q)_{n-r-j}(aq)_{n+r+j}}$$

$$\times g_{r+j} \frac{1-eq^{4r+4j}}{1-e} \frac{(e/c;q^2)_r(e;q^2)_{r+2j}}{(q^2;q^2)_r(cq^2;q^2)_{r+2j}} \left(\frac{c}{e}\right)^r$$

$$= \sum_{j=0}^{n} \frac{1-cq^{4j}}{1-c} \beta_j(e,c,q^2) \frac{(k/a)_{n-j}(k)_{n+j}(eq^2;q^2)_{2j}}{(q)_{n-j}(aq)_{n+j}(cq^2;q^2)_{2j}} \sum_{r=0}^{n-j} \frac{1-eq^{4r+4j}}{1-eq^{4j}}$$

$$\times \frac{(q^{-(n-j)},kq^{n+j};q)_r(e/c,eq^{4j};q^2)_r}{(aq^{1-(n-j)}/k,aq^{n+j+1};q)_r(cq^{4j+2},q^2;q^2)_r} \left(\frac{qac}{ke}\right)^r g_{r+j},$$

where the second equality follows by (13.4). $\qquad\square$

We are now in a position to prove the remaining chain.

Proof of (12.35). Set $e = a^2$, $c = ak/q$, $g_n = (1+a)q^n/(1+aq^{2n})$ in Theorem 12.1, and use (5.11) to put the inner sum over r at (12.37) in closed form, and simplify the resulting sum. □

If we now follow the method used to prove the other chains to prove (12.35), the following summation formulae are implied.

Corollary 12.6. *For each non-negative integer n, the following summation formulae hold:*

$$\sum_{j=0}^{n} \frac{1 - aq^{4j}}{1 - a} \frac{(a, b, bq, a/b^2, a^2q^{2+2n}/b^2, q^{-2n}; q^2)_j}{(aq/b, aq^2/b, b^2q^2, b^2q^{-2n}/a, aq^{2+2n}, q^2; q^2)_j} \left(\frac{b^2q}{a} \right)^j$$
$$= \frac{(-q, aq/b^2; q)_n (aq^2; q^2)_n}{(-bq, aq/b; q)_n (aq^2/b^2; q^2)_n}. \quad (12.38)$$

$$\sum_{j=0}^{n} \frac{1 - aq^{4j}}{1 - a} \frac{(a, b, bq, aq^2/b^2, a^2q^{2+2n}/b^2, q^{-2n}; q^2)_j}{(aq/b, aq^2/b, b^2, b^2q^{-2n}/a, aq^{2+2n}, q^2; q^2)_j} \left(\frac{b^2}{aq} \right)^j$$
$$= \frac{(-q, aq^2/b^2; q)_n (aq^2; q^2)_n q^{-n}}{(-b, aq/b; q)_n (aq^2/b^2; q^2)_n}. \quad (12.39)$$

Proof. Since (12.35) is true, it follows that the coefficient of $\alpha_j^{(2)}(a, k, q)$ in each of the two expressions for $\beta_n^{(2)}(a, k, q)$ (the one coming from the chain at (12.35) and the one coming from (12.3)) have to be equal (although it is normal to look for $\alpha_n(a, k)$ that lead to $\beta_n(a, k)$ having closed form, there is nothing in the analysis that prevents $\{\alpha_n(a, k)\}$ from being a completely arbitrary sequence). The summation formula at (12.38) follows upon setting these coefficients equal, and then relabelling. The transformation at (12.39) follows upon applying (5.13) to the left side of (12.38) and then replacing a with a/q and b with b/q. □

Exercises

12.1 By making the appropriate choice for the sequence $\{\theta_r\}_{r \geq 0}$ in (12.1), prove the transformation in part (iii) of Exercise 4.4.

12.2 By making the appropriate choice for the sequence $\{\theta_r\}_{r \geq 0}$ in (12.1), and using the q-Pfaff–Saalschütz sum (4.5) or otherwise, prove the Sears $_4\phi_3$ transformation in Exercise 5.15 (with d replaced with e and e replaced with g).

12.3 (i) By employing the $_6\phi_5$ summation formula (5.5) or otherwise, prove the summation formula at (12.6).
(ii) By employing Heine's transformation (4.1) or otherwise, prove the summation formula at (12.7).

12.4 (Bressoud [84, p. 217]) (i) By specializing r in (12.8) and using the appropriate summation formula, prove the transformation

$$\sum_{n=0}^{\infty} \frac{(a\beta, \beta/g; q)_n}{(agq, q; q)_n} \left(\frac{qg}{\beta^2}\right)^n$$
$$= \frac{(aq/\beta, q/\beta; q)_\infty}{(aq, q/\beta^2; q)_\infty} \sum_{n=0}^{\infty} \frac{1 - aq^{2n}}{1 - a} \frac{(a, g^{-1}; q)_n (a\beta; q)_{2n}}{(agq, q; q)_n (aq/\beta; q)_{2n}} \left(\frac{qg}{\beta^2}\right)^n. \quad (12.40)$$

(ii) Specialize β in (12.40) to get, for each positive integer N (note that both series are actually finite in (12.41)), that

$$\sum_{n=0}^{\infty} \frac{(q^{-N}, q^{-N}/ag; q)_n}{(agq, q; q)_n} \left(a^2 g q^{1+2N}\right)^n$$
$$= \frac{(a^2 q; q)_{2N}}{(aq, a^2 q; q)_N} \sum_{n=0}^{\infty} \frac{1 - aq^{2n}}{1 - a} \frac{(a, g^{-1}; q)_n (q^{-N}; q)_{2n}}{(agq, q; q)_n (a^2 q^{1+N}; q)_{2n}} \left(a^2 g q^{1+2N}\right)^n,$$
$$(12.41)$$

and hence show that

$$\sum_{n=0}^{\infty} a^n q^{n^2} \begin{bmatrix} N \\ n \end{bmatrix}_q \frac{(agq; q)_N}{(agq; q)_n (agq; q)_{N-n}}$$
$$= \frac{(a^2 q; q)_{2N} (q; q)_N}{(aq; q)_N} \sum_{n=0}^{\infty} \frac{1 - aq^{2n}}{1 - a} \frac{(a, g^{-1}; q)_n q^{2n^2} a^{2n} g^n}{(agq, q; q)_n (q; q)_{N-2n} (a^2 q; q)_{N+2n}}.$$
$$(12.42)$$

(iii) By specializing a and g in (12.42), prove that the identities

$$\sum_{n=0}^{\infty} q^{n^2} \begin{bmatrix} N \\ n \end{bmatrix}_q = \sum_{n=-\infty}^{\infty} (-1)^n q^{n(5n+1)/2} \begin{bmatrix} 2N \\ N+2n \end{bmatrix}_q, \quad (12.43)$$

$$\sum_{n=0}^{\infty} q^{n^2+n} \begin{bmatrix} N \\ n \end{bmatrix}_q = \frac{1}{1-q^{N+1}} \sum_{n=-\infty}^{\infty} (-1)^n q^{n(5n+3)/2} \begin{bmatrix} 2N+2 \\ N+2n+2 \end{bmatrix}_q, \quad (12.44)$$

hold for each non-negative integer N, and hence derive the
Rogers–Ramanujan identities (10.17), (10.18).
*Note that the polynomial versions of the Rogers–Ramanujan identities
above, which were found by Bressoud, are not the same as those described
in Section 11.2.*

12.5 Fill in the details in the proof of Lemma 12.1.

12.6 By employing Singh's identity (12.12), or otherwise, prove the
summation formula

$$\sum_{n=0}^{\infty} \frac{(1-aq^{2n})(a,a/k;q)_n(k;q)_{2n}}{(1-a)(kq,q;q)_n(a^2q/k;q)_{2n}} \left(\frac{aq}{k}\right)^n = \frac{(aq, qa^2/k^2; q)_\infty}{(qa/k, qa^2/k; q)_\infty}. \quad (12.45)$$

12.7 (Andrews [31]) Mirror the proof of Theorem 12.2 to prove the second
WP-Bailey chain of Andrews in Theorem 12.3 (the principal difference is
that the q-Pfaff–Saalschütz sum (4.5) is used instead of Jackson's $_8\phi_7$
summation formula at (5.11)).

12.8 Prove the claims at (12.22) and (12.23).

12.9 (i) Fill in the details in the proof of (12.24).
(ii) Hence specialize the parameters to show (using (5.11)) that, for all
non-negative integers n,

$$\sum_{r=0}^{n} \frac{(q\sqrt{a}, -q\sqrt{a}, a, b, q^{-n}; q)_r \, q^r}{(\sqrt{a}, -\sqrt{a}, \frac{qa}{b}, q^{2-n}b^2, q; q)_r} = \frac{\left(\frac{q\sqrt{a}}{\sqrt{q}b}, -\frac{q\sqrt{a}}{\sqrt{q}b}, \frac{a}{qb^2}, \frac{1}{qb}; q\right)_n}{\left(\frac{\sqrt{a}}{\sqrt{q}b}, -\frac{\sqrt{a}}{\sqrt{q}b}, \frac{1}{qb^2}, \frac{qa}{b}; q\right)_n}. \quad (12.46)$$

(iii) Apply Bailey's transformation for a $_{10}\phi_9$ series (5.14) to the series in
the summation formula (12.23) to show, for each non-negative integer n,
that

$$\sum_{j=0}^{n} \frac{(a, q\sqrt{a}, -q\sqrt{a}, a\sqrt{\frac{q}{k}}, -a\sqrt{\frac{q}{k}}, \frac{a}{\sqrt{k}}, -\frac{aq}{\sqrt{k}}, \frac{k}{a}, kq^n, q^{-n}; q)_j q^j}{(\sqrt{a}, -\sqrt{a}, \sqrt{kq}, -\sqrt{kq}, q\sqrt{k}, -\sqrt{k}, \frac{a^2q}{k}, \frac{aq^{1-n}}{k}, aq^{n+1}, q; q)_j}$$

$$= \frac{\left(aq, \sqrt{k}, \frac{k^2}{a^2}; q\right)_n}{\left(k, \frac{k}{a}, q\sqrt{k}; q\right)_n}. \quad (12.47)$$

12.10 Complete the proof of Theorem 12.4 by filling in the details that were just indicated.

12.11 (i) By employing one of the chains of Warnaar from Theorem 12.4, derive the WP-Bailey pair $(\alpha_n(a,k),\ \beta_n(a,k))$, where

$$\alpha_{2n}(a,k) = \frac{1-aq^{4n}}{1-a}\frac{(a,a^2/k^2;q^2)_n}{(k^2q^2/a,q^2;q^2)_n}\frac{k^{2n}}{a^{2n}}, \qquad \alpha_{2n+1}(a,k) = 0, \quad (12.48)$$

$$\beta_n(a,k) = \frac{(k^2q/a;q^2)_n}{(aq;q^2)_n}\frac{(a/k,k;q)_n}{(k^2q/a,q;q)_n}\left(\frac{-k}{a}\right)^n.$$

(ii) Hence prove that, for each non-negative integer n,

$$\sum_{r=0}^{n/4}\frac{1-aq^{8r}}{1-a}\frac{(a,a^4/k^4;q^4)_r}{(k^4q^4/a^3,q^4;q^4)_r}\frac{(kq^n,q^{-n};q)_{4r}\,q^{4r}}{(aq^{n+1},aq^{1-n}/k;q)_{4r}}$$

$$= \left(\frac{-k}{a}\right)^n\frac{(k^2q/a;q^2)_n(a/k,aq;q)_n}{(aq;q^2)_n(k/a,k^2q/a;q)_n}$$

$$\times\sum_{r=0}^{n/2}\frac{1-\frac{k^2q^{4r}}{a}}{1-\frac{k^2}{a}}\frac{(kq^n,q^{-n};q)_{2r}}{\left(\frac{k^2q^{n+1}}{a},\frac{kq^{1-n}}{a};q\right)_{2r}}\frac{\left(\frac{a^2}{k^2},\frac{k^2}{a};q^2\right)_r}{\left(\frac{k^4q^2}{a^3},q^2;q^2\right)_r}\frac{\left(\frac{k^4q^2}{a^3};q^4\right)_r}{(aq^2;q^4)_r}\left(\frac{-q^2k^2}{a^2}\right)^r.$$

$$(12.49)$$

12.12 (i) Use the chain of Liu and Ma in Theorem 12.5 to derive the WP-Bailey pair $(\alpha_n(a,k),\ \beta_n(a,k))$, where

$$\alpha_n(a,k) = \frac{(-1)^n\left(1-aq^{2n}\right)\left(a,\frac{k}{a};q\right)_n\left(\frac{a^2q}{k};q^2\right)_n}{(1-a)\left(\frac{a^2q}{k},q;q\right)_n(kq;q^2)_n}, \qquad (12.50)$$

$$\beta_{2n}(a,k) = \frac{\left(k,\frac{k^2}{a^2};q^2\right)_n}{\left(\frac{a^2q^2}{k},q^2;q^2\right)_n}, \qquad \beta_{2n+1}(a,k) = 0.$$

(ii) Show that moving two steps along the chain of Liu and Ma in Theorem 12.5 returns the original WP-Bailey pair.

(iii) By starting with the unit WP-Bailey pair (12.15), moving one step along the second chain of Andrews (see Corollary 12.3) and then one step along the chain of Liu and Ma at (12.33), derive the formula

$$\sum_{j=0}^{\lfloor\frac{n}{2}\rfloor}\frac{\left(1-\frac{kq^{-2n+4j}}{a^2}\right)(q^{-n};q)_{2j}\left(\frac{k^2}{a^2},\frac{kq^{-2n}}{a^2};q^2\right)_j\,q^{2j}}{\left(1-\frac{k}{a^2}\right)\left(\frac{k^2q^{2-n}}{a^2};q\right)_{2j}\left(\frac{q^{2-2n}}{k},q^2;q^2\right)_j}$$

$$= \frac{\left(k, \frac{k}{a}; q\right)_n \left(\frac{a^2 q^2}{k}; q^2\right)_n}{\left(aq, \frac{a^2}{k^2 q}; q\right)_n (k; q^2)_n} \left(\frac{-a}{q^2 k}\right)^n$$

$$\times \sum_{j=0}^{n} \frac{\left(1 - aq^{2j}\right) \left(a, \frac{a}{kq}, q^{-n}, kq^n; q\right)_j (kq^2; q^2)_j}{(1 - a) \left(kq^2, \frac{aq^{1-n}}{k}, aq^{n+1}, q; q\right)_j \left(\frac{a^2}{k}; q^2\right)_j} \left(-\frac{aq}{k}\right)^j . \quad (12.51)$$

12.13 Rewrite the transformations at (12.18) and (12.47) to prove the WP-Bailey pairs in part (i) of Corollary 12.5. Hence derive the identities in part (ii) of Corollary 12.5.

12.14 Find a direct proof of either of the summation formulae in Corollary 12.6.

12.15 (i) Use one of the chains in Theorem 12.6 to derive following WP-Bailey pair:

$$\alpha_n(a, k) = \frac{(1 - \sqrt{a}q^n) \left(\sqrt{a}, \frac{a}{k}; \sqrt{q}\right)_n}{(1 - \sqrt{a}) \left(\frac{k\sqrt{q}}{\sqrt{a}}, \sqrt{q}; \sqrt{q}\right)_n} \left(\frac{k}{a}\right)^n , \quad (12.52)$$

$$\beta_n(a, k) = \frac{\left(k, \frac{a}{k}, -\frac{k}{\sqrt{a}}, -\frac{k\sqrt{q}}{\sqrt{a}}; q\right)_n}{\left(-\sqrt{aq}, -\sqrt{aq}, \frac{k^2 q}{a}, q; q\right)_n} \left(\frac{k\sqrt{q}}{a}\right)^n .$$

(ii) Hence derive the identity

$$\sum_{j=0}^{n} \frac{\left(1 - \frac{k^2 q^{8j}}{a^2}\right) \left(\frac{a^2}{k}, -\frac{k}{a^{3/2}}, \frac{k}{a}, -\frac{kq}{a^{3/2}}; q^2\right)_j \left(q^{-4n}, kq^{4n}; q^4\right)_j}{\left(1 - \frac{k^2}{a^2}\right) \left(-\sqrt{a}q, -\sqrt{a}q^2, \frac{k^2 q^2}{a^3}, q^2; q^2\right)_j \left(\frac{kq^{4-4n}}{a^2}, \frac{k^2 q^{4n+4}}{a^2}; q^4\right)_j} \left(\frac{kq^3}{a^2}\right)^j$$

$$= \frac{\left(\frac{k}{a^2}, -aq^2, -aq^4, \frac{k^2 q^4}{a^2}; q^4\right)_n}{\left(\frac{a^2}{k}, -\frac{k}{a}, -\frac{kq^2}{a}, a^2 q^4; q^4\right)_n} \left(\frac{a^2}{kq^2}\right)^n$$

$$\times \sum_{j=0}^{n} \frac{\left(1 - \sqrt{a}q^{2j}\right) \left(\sqrt{a}, \frac{a^2}{k}; q\right)_j \left(q^{-4n}, kq^{4n}; q^4\right)_j q^{4j}}{(1 - \sqrt{a}) \left(\frac{kq}{a^{3/2}}, q; q\right)_j \left(\frac{a^2 q^{4-4n}}{k}, a^2 q^{4n+4}; q^4\right)_j} . \quad (12.53)$$

12.16 By employing the WP-Bailey chain of Andrews in Theorem 12.2 and relabelling parameters, show that if $(\alpha_n(a, k, q), \beta_n(a, k, q))$ is a WP-Bailey pair, then subject to suitable convergence conditions,

$$\sum_{n=0}^{\infty} \frac{(q\sqrt{k}, -q\sqrt{k}, y, z; q)_n}{(\sqrt{k}, -\sqrt{k}, qk/y, qk/z; q)_n} \left(\frac{qa}{yz}\right)^n \beta_n$$

$$= \frac{(qk, qk/yz, qa/y, qa/z; q)_\infty}{(qk/y, qk/z, qa, qa/yz; q)_\infty} \sum_{n=0}^{\infty} \frac{(y, z; q)_n}{(qa/y, qa/z; q)_n} \left(\frac{qa}{yz}\right)^n \alpha_n. \quad (12.54)$$

12.17 Let $(\alpha_n(a, k), \beta_n(a, k))$ be a WP-Bailey pair with respect to a and k, and let N be an integer. By using the ${}_6\phi_5$ summation at (5.5) or otherwise, prove that if

$$\alpha_n^*(aq^N, k) := \frac{1 - aq^{2n+N}}{1 - aq^N} q^{-nN} \sum_{j=0}^{n} \frac{(q^N; q)_{n-j}(aq^N; q)_{n+j}}{(q; q)_{n-j}(aq; q)_{n+j}} \alpha_j(a, k), \quad (12.55)$$

$$\beta_n^*(aq^N, q) := q^{-nN} \beta_n(a, k),$$

then $(\alpha_n^*(aq^N, q), \beta_n^*(aq^N, q))$ is a WP-Bailey pair with respect to aq^N and k.

Chapter 13

Further Results on Bailey/WP-Bailey Pairs and Chains

In this chapter we consider some further aspects of Bailey- and WP-Bailey pairs and chains. These include various formulae for deriving new pairs from existing pairs, ways of changing the base in a pair, and variations and extensions of the definitions of Bailey- and WP-Bailey pairs.

13.1 From $(\alpha_n(aq^N, q), \beta_n(aq^N, q))$ to $(\alpha_n(a, q), \beta_n(a, q))$

As was seen in Section 11.4, a key step in the proof of the Andrews–Gordon identities (and other identities of the same type) is the derivation of a Bailey pair $(\alpha_n^*(a, q), \beta_n^*(a, q))$ (with respect to a) from a Bailey pair $(\alpha_n'(aq, q), \beta_n'(aq, q))$ (with respect to aq). In this section we prove the formulae employed at that step of the proof, and indeed prove more general formulae.

We first note the elementary identities

$$1 = \frac{1 - aq^{n+r+1}}{1 - aq^{2r+1}} - aq^{2r+1}\frac{1 - q^{n-r}}{1 - aq^{2r+1}}, \tag{13.1}$$

$$1 = q^{-n+r}\frac{1 - aq^{n+r+1}}{1 - aq^{2r+1}} - q^{-n+r}\frac{1 - q^{n-r}}{1 - aq^{2r+1}}, \tag{13.2}$$

which are used in the proof of the next lemma.

Lemma 13.1. *Suppose* $(\alpha_n(a,q), \beta_n(a,q))$ *is a Bailey pair with respect to a. Then so are the pairs* $(\alpha_n^*(a,q), \beta_n^*(a,q))$ *and* $(\alpha_n^\dagger(a,q), \beta_n^\dagger(a,q))$, *where* $\alpha_0^*(a,q) = \beta_0^*(a,q) = \alpha_0^\dagger(a,q) = \beta_0^\dagger(a,q) = 1$, *and for* $n > 0$,

$$\alpha_n^*(a,q) = (1-aq)\left(\frac{\alpha_n(aq,q)}{1-aq^{2n+1}} - aq^{2n-1}\frac{\alpha_{n-1}(aq,q)}{1-aq^{2n-1}}\right), \tag{13.3}$$

$$\beta_n^*(a,q) = \beta_n(aq,q),$$

$$\alpha_n^\dagger(a,q) = (1-aq)\left(q^n\frac{\alpha_n(aq,q)}{1-aq^{2n+1}} - q^{n-1}\frac{\alpha_{n-1}(aq,q)}{1-aq^{2n-1}}\right), \tag{13.4}$$

$$\beta_n^\dagger(a,q) = q^n\beta_n(aq,q).$$

Proof. From the definition of a Bailey pair,

$$\beta_n(aq,q) = \sum_{r=0}^{n}\frac{\alpha_r(aq,q)}{(q;q)_{n-r}(aq^2;q)_{n+r}}$$

$$= \sum_{r=0}^{n-1}\frac{\alpha_r(aq,q)}{(q;q)_{n-r}(aq^2;q)_{n+r}} + \frac{\alpha_n(aq,q)}{(aq^2;q)_{2n}}$$

$$= (1-aq)\sum_{r=0}^{n-1}\frac{\alpha_r(aq,q)}{(q;q)_{n-r}(aq;q)_{n+r+1}} + \frac{\alpha_n(aq,q)}{(aq^2;q)_{2n}}$$

$$= (1-aq)\sum_{r=0}^{n-1}\frac{\alpha_r(aq,q)}{(1-aq^{2r+1})(q;q)_{n-r}(aq;q)_{n+r}}$$

$$\quad - (1-aq)\sum_{r=0}^{n-1}\frac{aq^{2r+1}\alpha_r(aq,q)}{(1-aq^{2r+1})(q;q)_{n-r-1}(aq;q)_{n+r+1}} + \frac{\alpha_n(aq,q)}{(aq^2;q)_{2n}}$$

$$= (1-aq)\sum_{r=0}^{n-1}\frac{\alpha_r(aq,q)}{(1-aq^{2r+1})(q;q)_{n-r}(aq;q)_{n+r}}$$

$$\quad - (1-aq)\sum_{r=1}^{n}\frac{aq^{2r-1}\alpha_{r-1}(aq,q)}{(1-aq^{2r-1})(q;q)_{n-r}(aq;q)_{n+r}} + \frac{\alpha_n(aq,q)}{(aq^2;q)_{2n}}.$$

The next-to-last equality follows from (13.1) and the result now follows for $(\alpha_n^*(a,q), \beta_n^*(a,q))$. The result for $(\alpha_n^\dagger(a,q), \beta_n^\dagger(a,q))$ follows similarly, except we use (13.2) at the next to last step (see Q13.1). $\square$

Note that the identities (13.1) and (13.2) express a Bailey pair with respect to a in terms of a Bailey pair with respect to aq. The more general result gives two expressions for Bailey pairs with respect to a in terms of Bailey pairs with respect to aq^N, where N is a positive integer.

Theorem 13.1. *Let N be a non-negative integer, and suppose $(\alpha_n(a,q)$, $\beta_n(a,q))$ is a Bailey pair with respect to a. Then so are the pairs $(\alpha_n^*(a,q)$, $\beta_n^*(a,q))$ and $(\alpha_n^\dagger(a,q), \beta_n^\dagger(a,q))$, where $\alpha_0^*(a,q) = \beta_0^*(a,q) = \alpha_0^\dagger(a,q) = \beta_0^\dagger(a,q) = 1$, and for $n > 0$,*

$$\alpha_n^*(a,q) = (1 - aq^{2n})(aq;q)_N \tag{13.5}$$
$$\times \sum_{j=0}^{N} (-a)^j q^{2nj - j(j+1)/2} \begin{bmatrix} N \\ j \end{bmatrix} \frac{(aq;q)_{2n-j-1}}{(aq;q)_{2n-j+N}} \alpha_{n-j}(aq^N, q),$$
$$\beta_n^*(a,q) = \beta_n(aq^N, q);$$

$$\alpha_n^\dagger(a,q) = (1 - aq^{2n})(aq;q)_N \tag{13.6}$$
$$\times \sum_{j=0}^{N} \frac{(-1)^j q^{N(n-j) + j(j-1)/2}}{(aq^{2n-j};q)_{N+1}} \begin{bmatrix} N \\ j \end{bmatrix} \alpha_{n-j}(aq^N, q),$$
$$\beta_n^\dagger(a,q) = q^{nN} \beta_n(aq^N, q).$$

Before coming to the proof, we note that results somewhat similar to (13.5) may be found elsewhere, for example [4, Lemma 1.2] and [254, Theorem 3.1]. However, the Bailey pairs given by these authors derive from a composition of (13.5) and (11.3), and the proofs are possibly less transparent than the proof given here. We believe that (13.6) is new. We also note that Lovejoy [185, Theorem 2.3] gave a quite general result going in the opposite direction, namely, a pair of formulae expressing a Bailey pair $(\alpha_n'(aq^N, q), \beta_n'(aq^N, q))$ in terms of a Bailey pair $(\alpha_n(a,q), \beta_n(a,q))$.

Proof of Theorem 13.1. The proof is straightforward mathematical induction. The claim at (13.5) is seen to be trivially true for $N = 0$, and true for $N = 1$ by (13.1). Suppose it is true for $N = 0, 1, \ldots, M$. By (13.1) followed by the induction step (with a replaced with aq), $\beta_n^*(a,q) = \beta_n(aq,q) = \beta_n(aq^{M+1}, q)$, and

$$\alpha_n^*(a,q) = \frac{1 - aq}{1 - aq^{2n+1}} \alpha_n(aq, q) - \frac{aq^{2n-1}(1 - aq)}{1 - aq^{2n-1}} \alpha_{n-1}(aq, q),$$
$$= (aq;q)_{M+1} \sum_{j=0}^{M} (-aq)^j q^{2nj - j(j+1)/2} \begin{bmatrix} M \\ j \end{bmatrix} \frac{(aq^2;q)_{2n-j-1}}{(aq^2;q)_{2n-j+M}} \alpha_{n-j}(aq^{M+1}, q)$$

$$
- (aq;q)_{M+1} aq^{2n-1}
$$

$$
\times \sum_{j=0}^{M} (-aq)^j q^{2(n-1)j - j(j+1)/2} \begin{bmatrix} M \\ j \end{bmatrix} \frac{(aq^2;q)_{2(n-1)-j-1}}{(aq^2;q)_{2(n-1)-j+M}} \alpha_{n-1-j}(aq^{M+1}, q)
$$

$$
= (aq;q)_{M+1} \frac{(aq;q)_{2n}}{(aq;q)_{2n+M+1}} \alpha_n(aq^{M+1}, q)
$$

$$
+ (aq;q)_{M+1} \sum_{j=1}^{M} (-a)^j q^{2nj - j(j+1)/2 + j} \begin{bmatrix} M \\ j \end{bmatrix} \frac{(aq;q)_{2n-j}}{(aq;q)_{2n-j+M+1}} \alpha_{n-j}(aq^{M+1}, q)
$$

$$
- (aq;q)_{M+1} aq^{2n-1}
$$

$$
\times \sum_{j=1}^{M+1} (-aq)^{j-1} q^{2(n-1)(j-1) - j(j-1)/2} \begin{bmatrix} M \\ j-1 \end{bmatrix} \frac{(aq;q)_{2n-j-1}}{(aq;q)_{2n-j+M}} \alpha_{n-j}(aq^{M+1}, q)
$$

$$
= (aq;q)_{M+1} \frac{(aq;q)_{2n}}{(aq;q)_{2n+M+1}} \alpha_n(aq^{M+1}, q)
$$

$$
+ (aq;q)_{M+1} \sum_{j=1}^{M} (-a)^j q^{2nj - j(j+1)/2} \alpha_{n-j}(aq^{M+1}, q)
$$

$$
\times \left\{ q^j \begin{bmatrix} M \\ j \end{bmatrix} \frac{(aq;q)_{2n-j}}{(aq;q)_{2n-j+M+1}} + \begin{bmatrix} M \\ j-1 \end{bmatrix} \frac{(aq;q)_{2n-j-1}}{(aq;q)_{2n-j+M}} \right\} + (-a)^{M+1}
$$

$$
\times (aq;q)_{M+1} q^{2n(M+1) - (M+1)(M+2)/2} \frac{(aq;q)_{2n-M-2}}{(aq;q)_{2n-1}} \alpha_{n-M-1}(aq^{M+1}, q).
$$

The factor inside the braces above may be written as

$$
\frac{(aq;q)_{2n-j-1}}{(aq;q)_{2n-j+M+1}} \left\{ q^j \begin{bmatrix} M \\ j \end{bmatrix} (1 - aq^{2n-j}) + \begin{bmatrix} M \\ j-1 \end{bmatrix} (1 - aq^{2n-j+M+1}) \right\}
$$

$$
= \frac{(aq;q)_{2n-j-1}}{(aq;q)_{2n-j+M+1}}
$$

$$
\left\{ \left(q^j \begin{bmatrix} M \\ j \end{bmatrix} + \begin{bmatrix} M \\ j-1 \end{bmatrix} \right) - aq^{2n} \left(\begin{bmatrix} M \\ j \end{bmatrix} + q^{M+1-j} \begin{bmatrix} M \\ j-1 \end{bmatrix} \right) \right\}
$$

$$
= \frac{(aq;q)_{2n-j-1}}{(aq;q)_{2n-j+M+1}} (1 - aq^{2n}) \begin{bmatrix} M+1 \\ j \end{bmatrix},
$$

where the last identity follows from the well-known recurrence formulae for the q-binomial coefficientss (see (14.9) and (14.10)). Upon substituting this final expression for the expression in braces in the last expression for $\alpha_n^*(a,q)$

above, we see that

$$\alpha_n^*(a,q) = (1 - aq^{2n})(aq;q)_{M+1}$$
$$\times \sum_{j=0}^{M+1} (-a)^j q^{2nj-j(j+1)/2} \begin{bmatrix} M+1 \\ j \end{bmatrix} \frac{(aq;q)_{2n-j-1}}{(aq;q)_{2n-j+M+1}} \alpha_{n-j}(aq^{M+1},q),$$

so that the induction claim has been extended to $N = M + 1$ (after noting that the $j = 0$ and $j = M + 1$ terms match up with the first and last terms above not contained in the sum from $j = 1$ to $j = M$), and the proof of (13.5) is thus complete.

The proof of (13.6) is similar, and is left as an exercise (see Q13.3). $\square$

13.2 A transformation of Warnaar

The relation (12.3) shows that once a $\{\alpha_n(a,k)\}$ has been defined, then the corresponding sequence $\{\beta_n(a,k)\}$ is uniquely determined. A result of Warnaar shows that the converse is also true. More precisely, the following result is true.

Lemma 13.2. *(i) ([256, Lemma 2.1]) For a and k indeterminates the following two equations are equivalent:*

$$\beta_n(a,k) = \sum_{j=0}^{n} \frac{(\frac{k}{a};q)_{n-j}(k;q)_{n+j}}{(q;q)_{n-j}(aq;q)_{n+j}} \alpha_j(a,k), \tag{13.7}$$

$$\alpha_n(a,k) = \frac{1 - aq^{2n}}{1 - a} \sum_{j=0}^{n} \frac{1 - kq^{2j}}{1 - k} \frac{(\frac{a}{k};q)_{n-j}(a;q)_{n+j}}{(q;q)_{n-j}(kq;q)_{n+j}} \left(\frac{k}{a}\right)^{n-j} \beta_j(a,k). \tag{13.8}$$

(ii) ([25, Eq. (2.7)]) The pair of sequences $(\alpha_n(a,q), \beta_n(a,q))$ is a Bailey pair with respect to a if and only if

$$\alpha_n(a,q) = \frac{1 - aq^{2n}}{1 - a} \sum_{j=0}^{n} \frac{(a;q)_{n+j}(-1)^{n-j}q^{(n-j)(n-j-1)/2}}{(q;q)_{n-j}} \beta_j(a,q). \tag{13.9}$$

Proof. For (i), as usual, it is sufficient to show that each relation implies the other. Firstly, assuming (13.8),

$$\sum_{j=0}^{n} \frac{(\frac{k}{a};q)_{n-j}(k;q)_{n+j}}{(q;q)_{n-j}(aq;q)_{n+j}} \alpha_j(a,k)$$

$$= \sum_{j=0}^{n} \frac{(\frac{k}{a};q)_{n-j}(k;q)_{n+j}}{(q;q)_{n-j}(aq;q)_{n+j}}$$

$$\times \frac{1-aq^{2j}}{1-a} \sum_{r=0}^{j} \frac{1-kq^{2r}}{1-k} \frac{(\frac{a}{k};q)_{j-r}(a;q)_{j+r}}{(q;q)_{j-r}(kq;q)_{j+r}} \left(\frac{k}{a}\right)^{j-r} \beta_r(a,k)$$

$$= \sum_{r=0}^{n} \frac{1-kq^{2r}}{1-k}\beta_r(a,k)$$

$$\times \sum_{j=r}^{n} \frac{1-aq^{2j}}{1-a} \frac{(\frac{k}{a};q)_{n-j}(k;q)_{n+j}}{(q;q)_{n-j}(aq;q)_{n+j}} \frac{(\frac{a}{k};q)_{j-r}(a;q)_{j+r}}{(q;q)_{j-r}(kq;q)_{j+r}} \left(\frac{k}{a}\right)^{j-r}$$

$$= \sum_{r=0}^{n} \frac{1-kq^{2r}}{1-k}\beta_r(a,k)$$

$$\times \sum_{j=0}^{n-r} \frac{1-aq^{2j+2r}}{1-a} \frac{(\frac{k}{a};q)_{n-j-r}(k;q)_{n+j+r}}{(q;q)_{n-j-r}(aq;q)_{n+j+r}} \frac{(\frac{a}{k};q)_{j}(a;q)_{j+2r}}{(q;q)_{j}(kq;q)_{j+2r}} \left(\frac{k}{a}\right)^{j}$$

$$= \sum_{r=0}^{n} \frac{1-kq^{2r}}{1-k} \frac{1-aq^{2r}}{1-a} \frac{(\frac{k}{a};q)_{n-r}(k;q)_{n+r}}{(q;q)_{n-r}(aq;q)_{n+r}} \frac{(a;q)_{2r}}{(kq;q)_{2r}}\beta_r(a,k)$$

$$\times \sum_{j=0}^{n-r} \frac{1-aq^{2j+2r}}{1-aq^{2r}} \frac{(aq^{2r}, kq^{n+r}, \frac{a}{k}, q^{-(n-r)};q)_j}{(aq^{1-(n-r)}/k, aq^{n+r+1}, kq^{2r+1}, q;q)_j}q^{j}$$

$$= \sum_{r=0}^{n} \frac{1-kq^{2r}}{1-k} \frac{1-aq^{2r}}{1-a} \frac{(\frac{k}{a};q)_{n-r}(k;q)_{n+r}}{(q;q)_{n-r}(aq;q)_{n+r}} \frac{(a;q)_{2r}}{(kq;q)_{2r}}\beta_r(a,k)\delta_{n-r,0} \text{ (by (5.2))}$$

$$= \frac{1-kq^{2n}}{1-k} \frac{1-aq^{2n}}{1-a} \frac{(k;q)_{2n}}{(aq;q)_{2n}} \frac{(a;q)_{2n}}{(kq;q)_{2n}}\beta_n(a,k) = \beta_n(a,k).$$

The proof of the other direction is similar and is left as an exercise (see Q14.4).

The statement in (ii) follows simply from letting $k \to 0$ in (i).   □

We had previously used Lemma 13.2 to prove the general WP-Bailey chain in Proposition 12.1. However, as we see next, it also provides a mechanism for producing new WP-Bailey pairs from existing pairs (see also Mc Laughlin and Zimmer [201]).

Corollary 13.1. *If $(\alpha_n(a,k), \beta_n(a,k))$ is a WP-Bailey pair, then so is*

$(\alpha'_n(a,k), \beta'_n(a,k))$, *where*

$$\alpha'(a,k) = \frac{(1 - aq^{2n})}{(1-a)} \left(\frac{k}{a}\right)^n \beta_n(k,a), \tag{13.10}$$

$$\beta'(a,k) = \frac{(1-k)}{(1-kq^{2n})} \left(\frac{k}{a}\right)^n \alpha_n(k,a).$$

Proof. It is immediate upon interchanging a and k in (13.8) that the pair $(\alpha'_n(a,k), \beta'_n(a,k))$ satisfy (12.3). $\qquad\square$

If Lemma 13.2 is applied to the WP-Bailey pair at (12.9), nothing essentially different is produced, but applying it to the pair at (12.22) leads to the new pair

$$\alpha_n(a,k) = \frac{(1 - aq^{2n}) \left(\frac{a^2}{k^2 q}; q\right)_n}{(1-a)(q;q)_n} \left(\frac{k}{a}\right)^n, \tag{13.11}$$

$$\beta_n(a,k) = \frac{\left(k, \frac{a}{kq}; q\right)_n \left(\frac{k^2 q}{a}; q\right)_{2n}}{\left(\frac{k^2 q^2}{a}, q; q\right)_n (a;q)_{2n}}.$$

13.3 Bilateral Bailey- and WP-Bailey pairs

In [67, Eq. (4.1)], Berkovich, McCoy and Schilling extended the definition of a Bailey pair by defining a *bilateral Bailey pair with respect to a* to be a pair of sequences $(\alpha_n(a,q), \beta_n(a,q))$ satisfying

$$\beta_n(a,q) = \sum_{j=-\infty}^{n} \frac{\alpha_j(a,q)}{(q;q)_{n-j}(aq;q)_{n+j}}, \quad \forall n \in \mathbb{Z}. \tag{13.12}$$

This topic was further investigated by Jouhet [163], who defined a *WP-bilateral Bailey pair* to be a pair of sequences $(\alpha_n(a,k), \beta_n(a,k)) = (\alpha_n(a,k,q), \beta_n(a,k,q))$ (for consistency reasons, we replace Jouhet's λ with k) satisfying

$$\beta_n(a,k) = \sum_{j=-\infty}^{n} \frac{(k/a;q)_{n-j}(k;q)_{n+j}}{(q;q)_{n-j}(aq;q)_{n+j}} \alpha_j(a,k), \quad \forall n \in \mathbb{Z}. \tag{13.13}$$

Jouhet proceeded to prove bilateral versions of the two WP-Bailey chains of Andrews. The extension of the first chain may be described as follows.

Theorem 13.2. *If* $(\alpha_n(a,k), \beta_n(a,k))_{n\in\mathbb{Z}}$ *satisfy* (13.13), *then so do* $(\alpha'_n(a,k), \beta'_n(a,k))_{n\in\mathbb{Z}}$, *where*

$$\alpha'_n(a,k) = \frac{(\rho_1, \rho_2; q)_n}{(aq/\rho_1, aq/\rho_2; q)_n} \left(\frac{aq}{\rho_1\rho_2}\right)^n \alpha_n(a,c), \tag{13.14}$$

$$\beta'_n(a,k) = \frac{(k\rho_1/a, k\rho_2/a; q)_n}{(aq/\rho_1, aq/\rho_2; q)_n} \times$$

$$\sum_{j=-\infty}^{n} \frac{(1 - cq^{2j})(\rho_1, \rho_2; q)_j (k/c; q)_{n-j} (k; q)_{n+j}}{(1 - c)(k\rho_1/a, k\rho_2/a; q)_j (q; q)_{n-j} (qc; q)_{n+j}} \left(\frac{aq}{\rho_1\rho_2}\right)^j \beta_j(a,c),$$

with $c = k\rho_1\rho_2/aq$.

For each integer n,

$$\sum_{j=0}^{\infty} \frac{\left(\dfrac{q^{-2n}}{a}, \dfrac{\rho_1 q^{-n}}{a}, \dfrac{\rho_2 q^{-n}}{a}, \dfrac{qk}{\rho_1\rho_2}; q\right)_j}{\left(\dfrac{\rho_1\rho_2 q^{-2n}}{ak}, \dfrac{q^{1-n}}{\rho_1}, \dfrac{q^{1-n}}{\rho_2}, q; q\right)_j} \left(\frac{aq}{k}\right)^j \alpha_{n-j}(a,k) = \frac{(aq; q)_{2n}}{(kq; q)_{2n}}$$

$$\times \sum_{j=0}^{\infty} \frac{1 - kq^{2n-2j}}{1 - k} \frac{\left(\dfrac{q^{-2n}}{k}, \dfrac{\rho_1 q^{-n}}{k}, \dfrac{\rho_2 q^{-n}}{k}, \dfrac{qa}{\rho_1\rho_2}; q\right)_j}{\left(\dfrac{\rho_1\rho_2 q^{-2n}}{ak}, \dfrac{q^{1-n}}{\rho_1}, \dfrac{q^{1-n}}{\rho_2}, q; q\right)_j} \left(\frac{k^2 q}{a^2}\right)^j \beta_{n-j}(a,k).$$

$$\tag{13.15}$$

$$\sum_{n=-\infty}^{\infty} \frac{(1 - kq^{2n})(\rho_1, \rho_2; q)_n}{(1-k)(kq/\rho_1, kq/\rho_2; q)_n} \left(\frac{aq}{\rho_1\rho_2}\right)^n \beta_n(a,k) =$$

$$\frac{(kq, kq/\rho_1\rho_2, aq/\rho_1, aq/\rho_2; q)_\infty}{(aq, aq/\rho_1\rho_2, kq/\rho_1, kq/\rho_2; q)_\infty} \sum_{n=-\infty}^{\infty} \frac{(\rho_1, \rho_2; q)_n}{(aq/\rho_1, aq/\rho_2; q)_n} \left(\frac{aq}{\rho_1\rho_2}\right)^n \alpha_n(a,k).$$

$$\tag{13.16}$$

Proof. To get (13.14), the proof of Andrews [31] follows through virtually without change (although since one of the series is now infinite, changing the order of summation is no longer automatically justified, and some absolute convergence conditions are necessary to justify the change in the order of summation). Use (13.13) to substitute for $\beta_j(a,c)$ on the right side of (13.14). Next, change the order of summation. Use Jackson's $_8\phi_7$ summation formula (5.11) to sum the new inner sum (after some elementary

q-product rearrangements) and finally show that the resulting expression is equal to the formula for $\beta_n(a, k)$ at (13.13) with $\alpha_n(a, k)$ replaced with $\alpha'_n(a, k)$. The details are omitted (see the proof of (12.13)).

For (13.15), set the two expressions for $\beta'_n(a, k)$ that are implied by (13.14) equal, and then replace c everywhere with k (so that k is replaced with $kaq/\rho_1\rho_2$). This gives

$$
\sum_{j=-\infty}^{n} \frac{(qk/\rho_1\rho_2; q)_{n-j}(qak/\rho_1\rho_2; q)_{n+j}(\rho_1, \rho_2; q)_j}{(q; q)_{n-j}(aq; q)_{n+j}(aq/\rho_1, aq/\rho_2; q)_j} \left(\frac{aq}{\rho_1\rho_2}\right)^j \alpha_j(a, k)
$$

$$
= \frac{(qk/\rho_1, qk\rho_2; q)_n}{(aq/\rho_1, aq/\rho_2; q)_n} \times
$$

$$
\sum_{j=-\infty}^{n} \frac{(1 - kq^{2j})(\rho_1, \rho_2; q)_j \left(\dfrac{aq}{\rho_1\rho_2}; q\right)_{n-j} \left(\dfrac{kaq}{\rho_1\rho_2}; q\right)_{n+j}}{(1 - k)\left(\dfrac{qk}{\rho_1}, \dfrac{qk}{\rho_2}; q\right)_j (q; q)_{n-j}(qk; q)_{n+j}} \left(\frac{aq}{\rho_1\rho_2}\right)^j \beta_j(a, k).
$$

$$(13.17)$$

Next, replace j with $n - j$ and use the formula at (2.13) a number of times. The result at (13.15) follows after some cancellation.

The identity at (13.16) follows from (13.17), upon letting $n \to \infty$. $\qquad\square$

Before considering some implications of this bilateral WP-Bailey chain, we describe the bilateral extension of the second chain of Andrews.

Theorem 13.3. *If* $(\alpha_n(a, k), \beta_n(a, k))_{n \in \mathbb{Z}}$ *satisfy* (13.13), *then so do* $(\alpha''_n(a, k), \beta''_n(a, k))_{n \in \mathbb{Z}}$, *where*

$$
\alpha''_n(a, k) = \frac{(qa^2/k; q)_{2n}}{(k; q)_{2n}} \left(\frac{k^2}{qa^2}\right)^n \alpha_n\left(a, \frac{qa^2}{k}\right), \qquad (13.18)
$$

$$
\beta''_n(a, k) = \sum_{j=-\infty}^{n} \frac{(k^2/a^2q; q)_{n-j}}{(q; q)_{n-j}} \left(\frac{k^2}{qa^2}\right)^j \beta_j\left(a, \frac{qa^2}{k}\right).
$$

For each integer n,

$$
\sum_{j=0}^{\infty} \frac{(kq^{-2n}/a^2; q)_{2j}}{(q^{1-2n}/k; q)_{2j}} \frac{(q^{-2n}/a, qa/k; q)_j}{(kq^{-2n}/a^2, q; q)_j} \left(\frac{qa}{k}\right)^j \alpha_{n-j}(a, k)
$$

$$
= \frac{(aq; q)_{2n}}{(k; q)_{2n}} \sum_{j=0}^{\infty} \frac{(qa^2/k^2; q)_j}{(q; q)_j} \left(\frac{k^2}{qa^2}\right)^j \beta_{n-j}(a, k). \quad (13.19)
$$

$$\sum_{n=-\infty}^{\infty} \frac{(k;q)_{2n}}{(qa^2/k;q)_{2n}} \left(\frac{qa^2}{k^2}\right)^n \alpha_n(a,k)$$

$$= \frac{(aq,qa^2/k^2;q)_\infty}{(qa/k,qa^2/k;q)_\infty} \sum_{n=-\infty}^{\infty} \left(\frac{qa^2}{k^2}\right)^n \beta_n(a,k). \quad (13.20)$$

Proof. The proof of (13.18) also follows that of Andrews virtually without change. Use (13.13) to substitute for $\beta_j(a, qa^2/k)$ on the right side of (13.18). This time, after changing the order of summation, the q-Pfaff–Saalschütz summation formula (4.5) is used to get the resulting new inner sum in closed form.

Set the two expressions for $\beta_n''(a,k)$ that are implied by (13.18) equal, and replace k with qa^2/k. This gives

$$\sum_{j=-\infty}^{n} \frac{(k;q)_{2j}(qa/k;q)_{n-j}(qa^2/k;q)_{n+j}}{(qa^2/k;q)_{2j}(q;q)_{n-j}(aq;q)_{n+j}} \left(\frac{qa^2}{k^2}\right)^j \alpha_j(a,k)$$

$$= \sum_{j=-\infty}^{n} \frac{(qa^2/k^2;q)_{n-j}}{(q;q)_{n-j}} \left(\frac{qa^2}{k^2}\right)^j \beta_j(a,k). \quad (13.21)$$

The identity at (13.19) follows after replacing j with $n-j$, using (2.13), and simplifying the resulting expressions. The identity at (13.20) follows from letting $n \to \infty$ in (13.21) above. $\qquad \square$

Notice that each of the bilateral chains in Theorems 13.2 and 13.3 revert back to one of the WP-Bailey chains of Andrews, if $\alpha_n(a,k) = 0$ for $n < 0$, so the chief interest will be in finding pairs for which this is not the case. One possibility, which Jouhet [163] examines in detail, is to set $a = q^m$, where m is a non-negative integer, causing the series at (13.13) to terminate below, since

$$\frac{1}{(aq;q)_{n+j}} = \frac{1}{(q^{m+1};q)_{n+j}} = \frac{(q;q)_m}{(q;q)_{m+n+j}} = 0, \forall j < -m-n.$$

We first prove a bilateral version of the unit WP-Bailey pair.

Lemma 13.3. *(Jouhet, [163, Eq. (4.2)]) For each non-negative integer m, the pair $(\alpha_n(a,k,q), \beta_n(a,k,q))$ is a bilateral WP-Bailey pair, where*

$$\alpha_n(a,k,q) = \frac{1-aq^{2n}}{1-a} \frac{(a;q)_{n-m}(a/k;q)_{n+m}}{(q;q)_{n+m}(kq;q)_{n-m}} \frac{1-kq^{-2m}}{1-k} \left(\frac{k}{a}\right)^{n+m}, \quad (13.22)$$

$$\beta_n(a,k,q) = \delta_{n+m,0}.$$

Proof. Substitute the expression for $\alpha_n(a, k, q)$ into (13.13) and use (5.2) — see Q13.5. $\qquad\square$

Jouhet then substituted the unit pair back into the first bilateral chain (13.14) and used the resulting pair to derive the following bilateral transformation.

Proposition 13.1. *(Jouhet, [163, Prop. 4.1]) Let m be a non-negative integer and suppose $|q|$, $|kq/\sigma_1\sigma_2|$, $|aq/\sigma_1\sigma_2| < 1$. Then*

$$\sum_{r=-\infty}^{\infty} \frac{1 - kq^{2r}}{1 - k} \frac{(\sigma_1, \sigma_2, k\rho_1/a, k\rho_2/a, kq^m/c, kq^{-m}; q)_r}{(kq/\sigma_1, kq/\sigma_2, aq/\rho_1, aq/\rho_2, cq^{1-m}, q^{m+1}; q)_r} \left(\frac{aq}{\sigma_1\sigma_2}\right)^r$$

$$= \frac{(kq, kq/\sigma_1\sigma_2, aq/\sigma_1, aq/\sigma_2; q)_\infty}{(aq, aq/\sigma_1\sigma_2, kq/\sigma_1, kq/\sigma_2; q)_\infty} \frac{(a/c, q/k, q/\rho_1, q/\rho_2; q)_m}{(k/c, q/a, aq/k\rho_1, aq/k\rho_2; q)_m}$$

$$\sum_{r=-\infty}^{\infty} \frac{1 - aq^{2r}}{1 - a} \frac{(\sigma_1, \sigma_2, \rho_1, \rho_2, aq^m/c, aq^{-m}; q)_r}{(aq/\sigma_1, aq/\sigma_2, aq/\rho_1, aq/\rho_2, cq^{1-m}, q^{m+1}; q)_r} \left(\frac{kq}{\sigma_1\sigma_2}\right)^r,$$

$$(13.23)$$

where $c = k\rho_1\rho_2/aq$.

Proof. Insert the unit pair (13.22) into (13.14) to get the pair

$$\alpha_n(a, k, q) = \frac{1 - cq^{-2m}}{1 - c} \frac{(1/c, a/c; q)_m}{(q/a, q; q)_m} \left(\frac{c^2 q}{a^2}\right)^m \qquad (13.24)$$

$$\times \frac{1 - aq^{2n}}{1 - a} \frac{(\rho_1, \rho_2, aq^m/c, aq^{-m}; q)_n}{(aq/\rho_1, aq/\rho_2, cq^{1-m}, q^{m+1}; q)_n} \left(\frac{k}{a}\right)^n,$$

$$\beta_n(a, k, q) = \frac{1 - cq^{-2m}}{1 - c} \frac{(1/c, k/c, aq/k\rho_1, aq/k\rho_2; q)_m}{(q/\rho_1, q/\rho_2, q/k, q; q)_m} \left(\frac{c^2 q}{a^2}\right)^m$$

$$\times \frac{(k\rho_1/a, k\rho_2/a, kq^m/c, kq^{-m}; q)_n}{(aq/\rho_1, aq/\rho_2, cq^{1-m}, q^{m+1}; q)_n}.$$

Then insert this pair into (13.16), with the pair of parameters (ρ_1, ρ_2) in (13.16) replaced with a new pair of parameters (σ_1, σ_2). $\qquad\square$

Remarks: 1) We have used a slightly different labelling of parameters from those used by Jouhet.

2) Note that both series terminate below because of the $1/(q^{m+1}; q)_n$ factor.

3) Jouhet also shows that (13.23) is a common extension of Ramanujan's $_1\psi_1$ summation formula (7.1) and Bailey's $_6\psi_6$ summation formula (8.1) (see Q13.6).

Note also that the pairs $(\alpha_n(a,k), \beta_n(a,k))_{n\in\mathbb{Z}}$ defined by (13.22) and (13.24) have $\alpha_n(a,k) = \beta_n(a,k) = 0$ for all $n < -m$ (again because of the $1/(q^{m+1}; q)_n$ factors). Before considering the shifted Bailey pairs of Jouhet (which also have the property that $\alpha_n(a,k) = \beta_n(a,k) = 0$ for all n sufficiently negative) in more detail in the next section, we consider two examples of bilateral WP-Bailey pairs $(\alpha_n(a,k), \beta_n(a,k))$ which are non-zero for all $n \in \mathbb{Z}$.

Lemma 13.4. *The pair of sequences* $(\alpha_n(a,k), \beta_n(a,k))_{n\in\mathbb{Z}}$ *is a bilateral WP-Bailey pair, where,*

$$\alpha_n(a,k) = \frac{1-aq^{2n}}{1-a} \frac{(\sigma_1, \sigma_2; q)_n}{(aq/\sigma_1, aq/\sigma_2; q)_n} \left(\frac{a}{\sigma_1\sigma_2}\right)^n, \qquad (13.25)$$

$$\beta_n(a,k) = \frac{(q/a, aq/k\sigma_1, aq/k\sigma_2, aq/\sigma_1\sigma_2; q)_\infty}{(q/k, q/\sigma_1, q/\sigma_2, a^2q/k\sigma_1\sigma_2; q)_\infty}$$

$$\times \frac{(k\sigma_1/a, k\sigma_2/a; q)_n}{(aq/\sigma_1, aq/\sigma_2; q)_n} \left(\frac{a}{\sigma_1\sigma_2}\right)^n.$$

Proof. Substitute the stated expression for $\alpha_n(a,k)$ into the sum on the left side of (13.13), replace j with $n-j$ and use the formula at (2.13) a number of times. The formula for $\beta_n(a,k)$ follows from applying the $_6\phi_5$ summation formula (5.6) to the resulting series. The details are left as an exercise — see Q13.7.   □

The insertion of this pair into (13.15) leads to a transformation equivalent to the well-known transformation for a non-terminating $_8\phi_7$ series (5.12), while if it is inserted into (13.16), that statement reduces to a tautology, with both sides being summable by Bailey's $_6\psi_6$ summation formula (8.1).

On the other hand, the insertion of this pair into (13.19) and (13.20) leads to the following identities.

Proposition 13.2. *If* $\max\{|q|, |aq/(\sigma_1\sigma_2)|, |a^2q/(k\sigma_1\sigma_2)|\} < 1$, *then*

$$\sum_{j=0}^{\infty} \frac{1-aq^{2j}}{1-a} \frac{(ak; q)_{2j}}{(aq/k; q)_{2j}} \frac{(a, \sigma_1, \sigma_2, q/k; q)_j}{(aq/\sigma_1, aq/\sigma_2, ak, q; q)_j} \left(\frac{aq}{k\sigma_1\sigma_2}\right)^j =$$

$$\frac{(aq, aq/k\sigma_1, aq/k\sigma_2, aq/\sigma_1\sigma_2; q)_\infty}{(aq/k, aq/\sigma_1, aq/\sigma_2, aq/k\sigma_1\sigma_2; q)_\infty} \sum_{j=0}^{\infty} \frac{(\sigma_1, \sigma_2, q/k^2; q)_j}{(aq/k\sigma_1, aq/k\sigma_2, q; q)_j} \left(\frac{aq}{\sigma_1\sigma_2}\right)^j.$$

$$(13.26)$$

If $|a^3q/(k^2\sigma_1\sigma_2)| < 1$, then

$$\sum_{j=-\infty}^{\infty} \frac{1-aq^{2j}}{1-a} \frac{(k;q)_{2j}}{(a^2q/k;q)_{2j}} \frac{(\sigma_1,\sigma_2;q)_j}{(aq/\sigma_1,aq/\sigma_2;q)_j} \left(\frac{a^3q}{k^2\sigma_1\sigma_2}\right)^j$$

$$= \frac{(aq,a^2q/k^2,q/a,aq/k\sigma_1,aq/k\sigma_2,aq/\sigma_1\sigma_2;q)_\infty}{(aq/k,a^2q/k,q/k,q/\sigma_1,q/\sigma_2,a^2q/k\sigma_1\sigma_2;q)_\infty}$$

$$\times \sum_{j=-\infty}^{\infty} \frac{(k\sigma_1/a,k\sigma_2/a;q)_j}{(aq/\sigma_1,aq/\sigma_2;q)_j} \left(\frac{a^3q}{k^2\sigma_1\sigma_2}\right)^j . \quad (13.27)$$

Proof. For (13.26), insert the bilateral WP-Bailey pair at (13.25) in (13.19), use (2.13) and set $n = 0$. Then replace a with $1/a$, k with k/a, σ_1 with σ_1/a and σ_2 with σ_2/a. The identity at (13.27) follows immediately upon inserting this same pair in (13.20) above. The details are left as an exercise (see Q13.8).   □

Remark: The transformation at (13.26) may be regarded as taking a $_{10}\phi_9$ series to a $_3\phi_2$ series, while (13.27) may be regarded as taking a $_8\psi_8$ series to a $_2\psi_2$ series.

A second bilateral WP-Bailey pair with all terms non-zero is given by the next lemma.

Lemma 13.5. *The pair of sequences* $(\alpha_n(a,k), \beta_n(a,k))_{n\in\mathbb{Z}}$ *is a bilateral WP-Bailey pair, where,*

$$\alpha_n(a,k) = \frac{1-aq^{2n}}{1-a} \frac{(k/y,y;q)_n}{(aq/y,aqy/k;q)_n} \frac{(a^2q/k;q^2)_n}{(kq;q^2)_n}(-1)^n, \quad (13.28)$$

$$\beta_{2m}(a,k) = \frac{(q/a,qa/k;q)_\infty(qy/a,kq/ay,aq^2/ky,q^2ay/k^2;q^2)_\infty}{(q/y,qy/k;q)_\infty(q,kq/a^2,q^2/k,q^2a^2/k^2;q^2)_\infty}$$

$$\times \frac{(ky/a,k^2/ay;q^2)_m}{(aq^2/y,q^2ay/k;q^2)_m},$$

$$\beta_{2m+1}(a,k) = -k\frac{(q/a,qa/k;q)_\infty(y/a,k/ay,aq/ky,qay/k^2;q^2)_\infty}{(q/y,qy/k;q)_\infty(q,kq/a^2,q^2/k,q^2a^2/k^2;q^2)_\infty}$$

$$\times \frac{(kyq/a,k^2q/ay;q^2)_m}{(aq/y,qay/k;q^2)_{m+1}}. \quad (13.29)$$

Proof. The proof is similar to the proof of Lemma 13.4, except that the q-analogue of Watson's $_3F_2$ summation formula (5.19) is used to sum the resulting series. Once again, the details are left as an exercise — see Q13.9.   □

The insertion of this second pair into (13.15) and (13.16) produces the following transformations.

Proposition 13.3. *(i) If $|q|, |aq/k| < 1$, then*

$$\frac{(q/y, qy/k; q)_\infty (q, kq/a^2, q^2/k, a^2q^2/k^2; q^2)_\infty}{(q/a, qa/k; q)_\infty}$$

$$\times \sum_{j=0}^\infty \frac{1 - q^{2j}/a}{1 - 1/a} \frac{(1/a, \rho_1/a, \rho_2/a, qk/\rho_1\rho_2, k/(ay), y/a; q)_j (q/k; q^2)_j}{(q/\rho_1, q/\rho_2, \rho_1\rho_2/ak, qy/k, q/y, q; q)_j (kq/a^2; q^2)_j} \left(\frac{-aq}{k}\right)^j$$

$$= (qy/a, kq/ay, aq^2/ky, q^2ay/k^2; q^2)_\infty$$

$$\times \sum_{j=0}^\infty \frac{1 - q^{4j}/k}{1 - 1/k} \frac{(1/k, \rho_1/k, \rho_2/k, qa/\rho_1\rho_2; q)_{2j} (y/a, k/ay; q^2)_j q^{2j}}{(q/\rho_1, q/\rho_2, \rho_1\rho_2/ak, q; q)_{2j} (aq^2/ky, q^2ay/k^2; q^2)_j}$$

$$- (y/a, k/ay, aq/ky, qay/k^2; q^2)_\infty \times$$

$$\sum_{j=0}^\infty \frac{1 - q^{4j+2}/k}{1 - 1/k} \frac{(1/k, \rho_1/k, \rho_2/k, qa/\rho_1\rho_2; q)_{2j+1} (qy/a, qk/ay; q^2)_j q^{2j+1}}{(q/\rho_1, q/\rho_2, \rho_1\rho_2/ak, q; q)_{2j+1} (aq/ky, qay/k^2; q^2)_{j+1}}.$$

$$(13.30)$$

(ii) If $|aq/\rho_1\rho_2| < 1$, then

$$\frac{(kq, kq/\rho_1\rho_2, aq/\rho_1, aq/\rho_2, q/y, qy/k; q)_\infty (q, kq/a^2, q^2/k, a^2q^2/k^2; q^2)_\infty}{(aq, aq/\rho_1\rho_2, kq/\rho_1, kq/\rho_2, q/a, qa/k; q)_\infty}$$

$$\times \sum_{j=-\infty}^\infty \frac{1 - aq^{2j}}{1 - a} \frac{(\rho_1, \rho_2, k/y, y; q)_j (a^2q/k; q^2)_j}{(aq/\rho_1, aq/\rho_2, aqy/k, aq/y; q)_j (kq; q^2)_j} \left(\frac{-aq}{\rho_1\rho_2}\right)^j$$

$$= (qy/a, kq/ay, aq^2/ky, q^2ay/k^2; q^2)_\infty$$

$$\times \sum_{j=-\infty}^\infty \frac{1 - kq^{4j}}{1 - k} \frac{(\rho_1, \rho_2; q)_{2j} (ky/a, k^2/ay; q^2)_j}{(kq/\rho_1, kq/\rho_2; q)_{2j} (aq^2/y, q^2ay/k; q^2)_j} \left(\frac{aq}{\rho_1\rho_2}\right)^{2j}$$

$$- k(y/a, k/ay, aq/ky, qay/k^2; q^2)_\infty \times$$

$$\sum_{j=-\infty}^\infty \frac{1 - kq^{4j+2}}{1 - k} \frac{(\rho_1, \rho_2; q)_{2j+1} (kyq/a, k^2q/ay; q^2)_j}{(kq/\rho_1, kq/\rho_2; q)_{2j+1} (aq/y, qay/k; q^2)_{j+1}} \left(\frac{aq}{\rho_1\rho_2}\right)^{2j+1}.$$

$$(13.31)$$

Proof. This follows the same lines as other proofs above, and is left as an exercise — see Q13.10. $\qquad\square$

Remarks: 1) The transformation at (13.30) is equivalent to that at [200, Eq. (2.3)].

2) The identity at (13.31) may be recast as a transformation of an $_8\psi_8$ series with base q to a combination of two $_8\psi_8$ series with base q^2.

This insertion of this second pair into the transformations (13.19) and (13.20) in Theorem 13.3 likewise leads to a unilateral- and a bilateral transformation

Proposition 13.4. *(i) If* $|aq/k| < 1$, *then*

$$\frac{(q/y, qy/k; q)_\infty (q, kq/a^2, q^2/k, a^2q^2/k^2; q^2)_\infty}{(q/a, qa/k; q)_\infty}$$

$$\times \sum_{j=0}^{\infty} \frac{1 - q^{2j}/a}{1 - 1/a} \frac{(1/a, qa/k, k/ay, y/a; q)_j (k/a^2; q^2)_j}{(k/a^2, qy/k, q/y, q; q)_j (q^2/k; q^2)_j} \left(\frac{-aq}{k}\right)^j$$

$$= (qy/a, kq/ay, aq^2/ky, q^2ay/k^2; q^2)_\infty$$

$$\times \sum_{j=0}^{\infty} \frac{(y/a, k/ay; q^2)_j}{(q^2a/ky, q^2ay/k^2; q^2)_j} \frac{(qa^2/k^2; q)_{2j}}{(q; q)_{2j}} q^{2j}$$

$$- (y/a, k/ay, aq/ky, qay/k^2; q^2)_\infty \times$$

$$\sum_{j=0}^{\infty} \frac{(qy/a, qk/ay; q^2)_j}{(qa/ky, qay/k^2; q^2)_{j+1}} \frac{(qa^2/k^2; q)_{2j+1}}{(q; q)_{2j+1}} q^{2j+1}. \quad (13.32)$$

(ii) If $|a^2q/k^2| < 1$, *then*

$$\frac{(a^2q/k, q/y, qy/k; q)_\infty (q, kq/a^2, q^2/k, a^2q^2/k^2; q^2)_\infty}{(aq, a^2q/k^2, q/a; q)_\infty}$$

$$\times \sum_{j=-\infty}^{\infty} \frac{1 - aq^{2j}}{1 - a} \frac{(k/y, y; q)_j (a^2q/k; q^2)_j}{(aqy/k, aq/y; q)_j (kq; q^2)_j} \frac{(k; q)_{2j}}{(qa^2/k; q)_{2j}} \left(\frac{-a^2q}{k^2}\right)^j$$

$$= (qy/a, kq/ay, aq^2/ky, q^2ay/k^2; q^2)_\infty \sum_{j=-\infty}^{\infty} \frac{(ky/a, k^2/ay; q^2)_j}{(aq^2/y, q^2ay/k; q^2)_j} \left(\frac{a^2q}{k^2}\right)^{2j}$$

$$- k(y/a, k/ay, aq/ky, qay/k^2; q^2)_\infty \sum_{j=-\infty}^{\infty} \frac{(kyq/a, k^2q/ay; q^2)_j}{(aq/y, qay/k; q^2)_{j+1}} \left(\frac{a^2q}{k^2}\right)^{2j+1}.$$

$$(13.33)$$

Proof. The proofs are once more straightforward and are again left as exercises — see Q13.11. □

13.4 Shifted Bailey pairs

Notice that if k is set equal to 0 in (13.14), then a bilateral version of the Bailey chain at (11.3) is produced.

Theorem 13.4. *(Bilateral Bailey lemma, Jouhet [163, Theorem 1.2]) If $(\alpha_n(a,q), \beta_n(a,q))$ satisfy (13.12), then so do $(\alpha'_n(a,q), \beta'_n(a,q))$, where*

$$\alpha'_n(a,q) = \frac{(\rho_1, \rho_2; q)_n}{(aq/\rho_1, aq/\rho_2; q)_n} \left(\frac{aq}{\rho_1\rho_2}\right)^n \alpha_n(a,q), \tag{13.34}$$

$$\beta'_n(a,q) = \frac{1}{(aq/\rho_1, aq/\rho_2; q)_n} \sum_{j=-\infty}^{n} \frac{(\rho_1, \rho_2; q)_j (aq/\rho_1\rho_2; q)_{n-j}}{(q;q)_{n-j}} \left(\frac{aq}{\rho_1\rho_2}\right)^j \beta_j(a,q),$$

subject to convergence conditions on the sequences $\alpha_n(a,q)$ and $\beta_n(a,q)$, which make the relevant infinite series absolutely convergent.

Jouhet [163] considered a special type of bilateral Bailey pair by setting $a = q^m$ (m a non-negative integer) so that the series for $\beta_n(a,q)$ and $\beta'_n(a,q)$ terminate below, thus avoiding problems related to switching the order of summation. He termed a pair of sequences $(\alpha_n(q^m,q), \beta_n(q^m,q))$ satisfying (13.12) (with $a = q^m$) a *shifted Bailey pair*. In [163] he proved that a particular pair of sequences was a shifted Bailey pair and then used two particular cases of (13.34) to prove a number of multi-sum identities. The two instances of (13.34) are

$$\alpha'_n(a,q) = q^{n^2} a^n \alpha_n(a,q), \qquad (\rho_1, \rho_2 \to \infty) \tag{13.35}$$

$$\beta'_n(a,q) = \sum_{j=-\infty}^{n} \frac{q^{j^2} a^j}{(q;q)_{n-j}} \beta_j(a,q),$$

$$\alpha'_n(a,q) = q^{n^2/2} a^{n/2} \alpha_n(a,q), \qquad (\rho_1 \to \infty, \rho_2 = -\sqrt{aq}) \tag{13.36}$$

$$\beta'_n(a,q) = \sum_{j=-\infty}^{n} \frac{q^{j^2/2} a^{j/2}}{(q;q)_{n-j}} \frac{(-\sqrt{aq}; q)_j}{(-\sqrt{aq}; q)_n} \beta_j(a,q).$$

The particular shifted Bailey pair Jouhet used is the pair described in the following lemma.

Lemma 13.6. *For each non-negative integer m, $(\alpha_n(q^m,q), \beta_n(q^m,q))$ is a shifted Bailey pair, where*

$$\alpha_n(q^m, q) = (-1)^n q^{n(n-1)/2}, \tag{13.37}$$

$$\beta_n(q^m, q) = (q;q)_m (-1)^n q^{n(n-1)/2} \begin{bmatrix} m+n \\ m+2n \end{bmatrix}_q.$$

Proof. By definition,

$$
\beta_n(q^m, q) = \sum_{j=-\infty}^{n} \frac{(-1)^j q^{j(j-1)/2}}{(q;q)_{n-j}(q^{m+1};q)_{n+j}} = \sum_{j=0}^{\infty} \frac{(-1)^{n-j} q^{(n-j)(n-j-1)/2}}{(q;q)_j(q^{m+1};q)_{2n-j}}
$$

$$
= \frac{(-1)^n q^{n(n-1)/2}}{(q^{m+1};q)_{2n}} \sum_{j=0}^{\infty} \frac{(-1)^j q^{j(j+1)/2-jn}}{(q;q)_j(q^{m+1+2n};q)_{-j}}
$$

$$
= \frac{(-1)^n q^{n(n-1)/2}}{(q^{m+1};q)_{2n}} \sum_{j=0}^{\infty} \frac{(q^{-m-2n};q)_j}{(q;q)_j} q^{j(1+m+n)} \ (\text{by } (2.5))
$$

$$
= \frac{(-1)^n q^{n(n-1)/2}}{(q^{m+1};q)_{2n}} \frac{(q^{1-n};q)_\infty}{(q^{1+m+n};q)_\infty} (\text{by the } q\text{-binomial theorem } (3.1))
$$

$$
= \frac{(q;q)_m (-1)^n q^{n(n-1)/2}}{(q;q)_{2n+m}} \frac{(q;q)_{m+n}}{(q;q)_{-n}}
$$

$$
= (q;q)_m (-1)^n q^{n(n-1)/2} \begin{bmatrix} m+n \\ m+2n \end{bmatrix}_q .
$$

Note that in the first series for $\beta_n(q^m, q)$, the term $1/(q^{m+1};q)_{n+j} = 0$ if $m + n + j < 0$ and thus, since $j \le n$, $\beta_n(q^m, q) = 0$ if $m + 2n < 0$. Hence it is necessary to consider only $\beta_n(q^m, q) = 0$ for which $m + 2n \ge 0$. In this situation, $1 + m + n > 0$, $|q^{1+m+n}| < 1$, and the q-binomial series above converges. □

The multi-sum identities that Jouhet derived included the two in the next theorem.

Theorem 13.5. *(Jouhet, [163, Theorem 2.3]) Let k be a positive integer and let m be a non-negative integer. Then*

$$
\sum_{-\lfloor m/2 \rfloor \le n_k \le n_{k-1} \le \cdots \le n_1} \frac{q^{n_1^2 + \cdots + n_k^2 + m(n_1 + \cdots + n_k)}}{(q;q)_{n_1-n_2} \cdots (q;q)_{n_{k-1}-n_k}} (-1)^{n_k} q^{n_k(n_k-1)/2}
$$

$$
\times \begin{bmatrix} m+n_k \\ m+2n_k \end{bmatrix}_q = \frac{(q^{k(m+1)}, q^{k(1-m)+1}, q^{2k+1}; q^{2k+1})_\infty}{(q;q)_\infty}; \quad (13.38)
$$

$$
\sum_{-\lfloor m/2 \rfloor \le n_k \le n_{k-1} \le \cdots \le n_1} \frac{q^{n_1^2/2 + n_2^2 + \cdots + n_k^2 + m(n_1/2 + n_2 + \cdots + n_k)} (-q^{(m+1)/2}; q)_{n_1}}{(q;q)_{n_1-n_2} \cdots (q;q)_{n_{k-1}-n_k}}
$$

$$
\times (-1)^{n_k} q^{n_k(n_k-1)/2} \begin{bmatrix} m+n_k \\ m+2n_k \end{bmatrix}_q
$$

$$= \frac{(-q^{(m+1)/2};q)_\infty}{(q;q)_\infty}(q^{(k-1/2)(m+1)},q^{k(1-m)+(m+1)/2},q^{2k};q^{2k})_\infty. \quad (13.39)$$

Proof. The proofs are similar to the proofs of some of the multi-sum identities in Chapter 11. For (13.38), apply the chain at (13.35) k times to the pair at (13.37) to get the shifted Bailey pair

$$\alpha_n^{(k)}(q^m,q) = q^{kn^2}q^{kmn}\alpha_n(q^m,q), \quad (13.40)$$

$$\beta_n^{(k)}(q^m,q) = \sum_{n_k\leq n_{k-1}\leq\cdots\leq n_1\leq n} \frac{q^{n_1^2+\cdots+n_k^2+m(n_1+\cdots+n_k)}\beta_{n_k}(q^m,q)}{(q;q)_{n-n_1}(q;q)_{n_1-n_2}\cdots(q;q)_{n_{k-1}-n_k}}.$$

The result at (13.38) follows upon letting $n \to \infty$ (after an appeal to Tannery's theorem [247] to justify the interchange of limit and summation) in the relation

$$\beta_n^{(k)}(q^m,q) = \sum_{j=-\infty}^{n} \frac{\alpha_j^{(k)}(q^m,q)}{(q;q)_{n-j}(q^{m+1};q)_{n+j}}, \quad (13.41)$$

and using the Jacobi triple product identity (6.1) to sum the right side.

The proof of the identity at (13.39) is similar, except the chain at (13.35) is applied $k - 1$ times, followed by one application of the chain at (13.36). The details are left as an exercise — see Q13.12. $\square$

13.5   Change of base in Bailey pairs

As noted previously, the ability to derive a Bailey pair $(\alpha_n^*(a,q),\beta_n^*(a,q))$ (with respect to a) from a Bailey pair $(\alpha_n'(aq,q),\beta_n'(aq,q))$ (with respect to aq) is an important step in the derivation of many multi-sum identities. Another useful set of formulae were given in [85] by Bressoud, Ismail and Stanton. These formulae relate of pairs of the form $(\alpha_n^*(a^m,q^m),\beta_n^*(a^m,q^m))$ with a pair $(\alpha_n(a,q),\beta_n(a,q))$, where $m \in \{2,1/2,3,1/3\}$, and these in turn lead to further collections of multi-sum identities. We may label these transformations *Bailey chains*, since they provide mechanisms for generating new Bailey pairs from existing Bailey pairs, although the authors also call them versions of the Bailey lemma. A similar collection of chains were stated in [68] by Berkovich and Warnaar, where formulae relating pairs of the form $(\alpha_n^*(a^p,q^m),\beta_n^*(a^p,q^m))$ with a pair $(\alpha_n(a,q),\beta_n(a,q))$, for

$$(p,m) \in \{(1,1),(2,2),(3,3),(1,1/2),(1/2,1/4),(1,1/3),(1,2),(2,4),(1,3)\}.$$

We first consider one of the chains in [85], although our proof is somewhat different. In the original proof in [85], the authors employ a quadratic transformation due to Singh [235], which is not used in the present proof. However, Singh's quadratic transformation is somewhat important, as it makes an appearance in various contexts in the literature, and this seems an appropriate place to give a proof, given the aforementioned part it plays in the original proof in [85] (in fact, it appears as a step in the proof of a chain of Berkovich and Warnaar [68] below).

13.5.1 Singh's quadratic transformation

Before proving the first of these new chains, it is necessary to first prove a quadratic transformation first discovered by Singh [235], with later proofs given by Askey and Wilson [53], and Jain [159]. The proof given here is based in part on that of Guo and Zeng [136], who gave a general method in their paper for proving basic hypergeometric identities. We modify their proof to give one that is a little more direct, in that it avoids the generalities. We will prove Singh's transformation in the following form.

Lemma 13.7. *For each non-negative integer* n,

$$\sum_{j=0}^{n} \frac{(a^2, b^2, c, q^{-n}; q)_j q^j}{(abq^{1/2}, -abq^{1/2}, -cq^{-n}, q; q)_j} = \sum_{j=0}^{n} \frac{(a^2, b^2, c^2, q^{-2n}; q^2)_j q^{2j}}{(a^2 b^2 q, -cq^{-n}, -cq^{1-n}, q^2; q^2)_j}.$$
(13.42)

Proof. Define

$$f_n(a, b, c, q, j) := \frac{(a^2, b^2, c, q^{-n}; q)_j q^j}{(abq^{1/2}, -abq^{1/2}, -cq^{-n}, q; q)_j},$$

$$g_n(a, b, c, q, j) := \frac{(a^2, b^2, c^2, q^{-2n}; q^2)_j q^{2j}}{(a^2 b^2 q, -cq^{-n}, -cq^{1-n}, q^2; q^2)_j},$$

$$F_n(a, b, c, q) := \sum_{j=0}^{n} f_n(a, b, c, q, j),$$

$$G_n(a, b, c, q) := \sum_{j=0}^{n} g_n(a, b, c, q, j).$$

Then the claim is that $F_n(a, b, c, q) = G_n(a, b, c, q)$ for all non-negative integers n. It is easily seen to be true for $n = 0, 1$, and it will be shown that $F_n(a, b, c, q)$ and $G_n(a, b, c, q)$ satisfy the same recurrence relation, giving the result for all $n \geq 0$. Firstly, elementary algebra gives that

$$f_n(a, b, c, q, j) - f_{n-1}(a, b, c, q, j)$$
$$= -\frac{(1 - a^2)(1 - b^2)(1 - c^2)q^{1-n}}{(1 - a^2 b^2 q)(1 + cq^{-n})(1 + cq^{1-n})} f_{n-1}(aq^{1/2}, bq^{1/2}, cq, q, j - 1).$$

Secondly, upon summing the above relation from $j = 0$ to $j = n$ (and noting that $f_{n-1}(a, b, c, q, n) = 0$), it follows that

$$F_n(a, b, c, q) - F_{n-1}(a, b, c, q)$$
$$= -\frac{(1 - a^2)(1 - b^2)(1 - c^2)q^{1-n}}{(1 - a^2 b^2 q)(1 + cq^{-n})(1 + cq^{1-n})} F_{n-1}(aq^{1/2}, bq^{1/2}, cq, q). \quad (13.43)$$

It will be shown that $G_n(a, b, c, q)$ satisfies the same recurrence, giving the result. If the same steps are taken with the sequence $\{g_n(a, b, c, q, j)\}$, we initially get the recurrence

$$G_n(a, b, c, q) - G_{n-1}(a, b, c, q)$$
$$= -\frac{(1 - a^2)(1 - b^2)(1 - c^2)q^{1-n}}{(1 - a^2 b^2 q)(1 + cq^{-n})(1 + cq^{1-n})} \frac{(1 + cq^n)q^{1-n}}{1 + cq^{2-n}}$$
$$\sum_{j=0}^{n-1} \frac{(a^2 q^2, b^2 q^2, c^2 q^2, q^{2-2n}; q^2)_j q^{2j}}{(a^2 b^2 q^3, -cq^{3-n}, -cq^{4-n}, q^2; q^2)_j}.$$

Finally, apply the Sears $_4\phi_3$ transformation (5.23) to the series on the right above, and after a little simplification, it follows that (13.43) holds with $F_i(-, -, -, -)$ replaced with $G_i(-, -, -, -)$, thus giving that (13.42) holds.
$\square$

A particular case of the next transformation is also needed. This transformation, due to Jain and Verma [161, Eq. (1.3)], is also a quadratic transformation, in that it relates a $_{10}\phi_9$ with base q to a $_5\phi_4$ series with base q^2. One step of the proof is reminiscent of the method used to prove Bailey- or WP-Bailey chains, namely switching the order of summation in a finite double sum, and then shifting the summation index on the new inner sum so that the summation index starts at zero, and then applying some known summation formula to get the inner sum in closed form.

Proposition 13.5. *For each non-negative integer N there holds*

$$\sum_{k=0}^{N} \frac{1 - aq^{2k}}{1 - a} \frac{(a, b, x, -x, y, -y, -q^{-N}, q^{-N}; q)_k}{\left(\frac{aq}{b}, \frac{aq}{x}, -\frac{aq}{x}, \frac{aq}{y}, -\frac{aq}{y}, -aq^{N+1}, aq^{N+1}, q; q\right)_k} \left(-\frac{a^3 q^{3+2N}}{bx^2 y^2}\right)^k$$

$$= \frac{\left(a^2q^2, \frac{a^2q^2}{x^2y^2}; q^2\right)_N}{\left(\frac{a^2q^2}{x^2}, \frac{a^2q^2}{y^2}; q^2\right)_N} \sum_{n=0}^{N} \frac{\left(x^2, y^2, -\frac{aq}{b}, -\frac{aq^2}{b}, q^{-2N}; q^2\right)_n q^{2n}}{\left(-aq, -aq^2, \frac{a^2q^2}{b^2}, \frac{x^2y^2q^{-2N}}{a^2}, q^2; q^2\right)_n}. \quad (13.44)$$

Proof. Set $c = -q^{-n}$ in the $_6\phi_5$ summation formula at (5.5) and rearrange the resulting product side to get

$$\sum_{k=0}^{n} \frac{1-aq^{2k}}{1-a} \frac{(a,b,-q^{-n},q^{-n}; q)_k}{\left(\frac{aq}{b}, -aq^{n+1}, aq^{n+1}, q; q\right)_k} \left(\frac{-aq^{2n+1}}{b}\right)^k$$

$$= \frac{\left(-\frac{aq}{b}, -\frac{aq^2}{b}, a^2q^2; q^2\right)_n}{\left(-aq, -aq^2, \frac{a^2q^2}{b^2}; q^2\right)_n}.$$

Multiply both sides of the above identity by

$$\frac{\left(x^2, y^2, q^{-2N}; q^2\right)_n q^{2n}}{\left(a^2q^2, \frac{x^2y^2q^{-2N}}{a^2}, q^2; q^2\right)_n}$$

and sum from $n = 0$ to $n = N$, noting that the resulting sum on the right side is the sum on the right side of (13.44). On the resulting left side, switch the order of summation, use (2.26) on the term $(q^{-2n}; q^2)_k$, write $(a^2q^{2n+2}; q^2)_k = (a^2q^2; q^2)_{n+k}/(a^2q^2; q^2)$, simplify and shift the index of summation on the inner sum to get

$$\sum_{k=0}^{N} \frac{1-aq^{2k}}{1-a} \frac{(a,b;q)_k(x^2,y^2,q^{-2N};q^2)_k}{\left(\frac{aq}{b},q;q\right)_k (x^2y^2q^{-2N}/a^2;q^2)_k(a^2q^2;q^2)_{2k}} \left(\frac{aq}{b}\right)^k q^{k^2+k}$$

$$\times \sum_{n=0}^{N-k} \frac{(x^2q^{2k}, y^2q^{2k}, q^{2(n-k)}; q^2)_n q^{2n}}{(a^2q^{2+4k}, x^2y^2q^{-2(N-k)}/a^2, q^2; q^2)_n}.$$

Apply the q-Pfaff–Saalschütz sum (4.5) to the inner sum, perform some final elementary q-product manipulations, and (13.44) follows. $\qquad\square$

To get the particular case of this transformation needed to prove the Bailey chain in the next theorem, first multiply both sides of (13.44) by $\left(a^2q^2/x^2; q^2\right)_N$ and then let $x \to aq^N$ in the resulting identity. On the right side, the $(y^2; q^2)_n$ now cancels a similar term in the denominator. On the left side, $\left(a^2q^2/x^2; q^2\right)_N \to \left(q^{2-2N}; q^2\right)_N = 0$, so that the series on the left side vanishes, apart from the $k = N$ term, where the $\left(a^2q^2/x^2; q^2\right)_N$ in the denominator cancels the vanishing term multiplying the series. If the resulting equality is simplified and manipulated, we arrive at the identity:

$$\sum_{n=0}^{N} \frac{\left(-\frac{aq}{b}, -\frac{aq^2}{b}, a^2q^{2N}, q^{-2N}; q^2\right)_n q^{2n}}{\left(-aq, -aq^2, \frac{a^2q^2}{b^2}, q^2; q^2\right)_n} = \frac{1+a}{1+aq^{2N}} \frac{(b, -q; q)_N}{(aq/b, -a; q)_N} \left(-\frac{aq}{b}\right)^N.$$

To put it a slightly more transparent and usable form (that given at [85, Eq. (2.1)]), replace a with $-C$ and b with Cq/D to get

$$\sum_{n=0}^{N} \frac{(D, Dq, C^2q^{2N}, q^{-2N}; q^2)_n q^{2n}}{(Cq, Cq^2, D^2, q^2; q^2)_n} = \frac{1-C}{1-Cq^{2N}} \frac{(Cq/D, -q; q)_N}{(-D, C; q)_N} D^N.$$

$$(13.45)$$

As the authors in [85] state, there is a companion identity to (13.45), which may be proved using Singh's transformation (13.42) in combination with the q-Pfaff–Saalschütz sum (4.5):

$$\sum_{n=0}^{N} \frac{(D, Dq, C^2q^{2N}, q^{-2N}; q^2)_n q^{2n}}{(C, Cq, D^2q^2, q^2; q^2)_n} = \frac{(C/D, -q; q)_N}{(-Dq, C; q)_N} D^N. \qquad (13.46)$$

The details are left as an exercise (see Q13.13). This transformation is also used below in the proof of a Bailey chain of Berkovich and Warnaar ([68, Lemma 5.6]).

Singh's transformation is used in the proof of the chain in the next theorem.

Theorem 13.6. *(Bressoud, Ismail and Stanton [85, Theorem 2.1]) Suppose $(\alpha_n(a, q), \beta_n(a, q))$ is a Bailey pair with respect to a. Then so is the pair $(\alpha'_n(a, q), \beta'_n(a, q))$, where*

$$\alpha'_n(a, q) = \frac{(-B; q)_n}{(-aq/B; q)_n} B^{-n} q^{-n(n-1)/2} \alpha_n(a^2, q^2), \qquad (13.47)$$

$$\beta'_n(a, q) = \sum_{k=0}^{n} \frac{(-aq; q)_{2k}(B^2; q^2)_k(q^{-k}/B, Bq^{k+1}; q)_{n-k}}{(-aq/B, B; q)_n(q^2; q^2)_{n-k}B^k q^{k(k-1)/2}} \beta_k(a^2, q^2). \quad (13.48)$$

Proof. Write $(Bq^{k+1}; q)_{n-k} = (Bq; q)_n/(Bq; q)_k$ and

$$\left(\frac{q^{-k}}{B}; q\right)_{n-k} = \left(\frac{q^{-k}}{B}; q\right)_k \left(\frac{1}{B}; q\right)_{n-2k}$$

$$= (Bq; q)_k(-Bq)^{-k}q^{-k(k-1)/2}\left(\frac{1}{B}; q\right)_{n-2k},$$

use (10.5) to replace $\beta_k(a^2, q^2)$ in (13.48), switch the order of summation, re-index the summation variable on the new inner sum, and apply the usual elementary q-product transformations to get that the sum on the right side of (13.48) equals

$$\frac{(Bq; q)_n}{(-aq/B, B; q)_n} \sum_{j=0}^{n} \frac{(-aq; q)_{2j}(B^2; q^2)_j(1/B; q)_{n-2j}}{(a^2q^2; q^2)_{2j}(q^2; q^2)_{n-j}B^{2j}}(-1)^j q^{-j^2}\alpha_j(a^2, q^2)$$

$$\times \sum_{k=0}^{n-j} \frac{(-aq^{1+2j}, -aq^{2+2j}, B^2q^{2j}, q^{-2(n-j)}; q^2)_k q^{2k}}{(Bq^{1-(n-2j)}, Bq^{2-(n-2j)}, a^2q^{2+4j}, q^2; q^2)_k}.$$

Use (13.45) to get the inner sum in closed form, and after some further elementary q-product manipulations of the type stated in Chapter 2, the new single sum can be shown to be equal to the right side of (10.5), with $\alpha_r(a, q)$ replaced with $\alpha'_r(a, q)$, giving the result. □

The authors consider special cases of (13.47), for example ($B \to \infty$)

$$\alpha'_n(a, q) = \alpha_n(a^2, q^2), \qquad\qquad\qquad (13.49)$$

$$\beta'_n(a, q) = \sum_{k=0}^{n} \frac{(-aq; q)_{2k}}{(q^2; q^2)_{n-k}}q^{n-k}\beta_k(a^2, q^2),$$

and then combine these with various cases of Corollary 11.1 and particular Bailey pairs to produce multi-sum identities, in ways similar those described for multi-sum identities encountered earlier.

An example of the chains found by Berkovich and Warnaar is contained in the next theorem.

Theorem 13.7. *Suppose* $(\alpha_n(a, q), \beta_n(a, q))$ *is a Bailey pair with respect to* a. *Then so is the pair* $(\alpha'_n(a, q), \beta'_n(a, q))$, *where*

$$\alpha'_{2n}(a, q) = q^{2n}\frac{1+a}{1+aq^{4n}}\alpha_n(a^2, q^4), \quad \alpha'_{2n+1}(a, q) = 0, \qquad (13.50)$$

$$\beta'_n(a, q) = \frac{(-q; q^2)_n}{(q^2, aq; q^2)_n} \sum_{k=0}^{\lfloor n/2 \rfloor} \frac{(-a, q^{-2n}; q^2)_{2k}q^{4k}}{(-q^{1-2n}; q^2)_{2k}}\beta_k(a^2, q^4).$$

Proof. The proof follows the usual steps, the only substantial difference being the summation formula used to get the inner sum in closed form, after changing the order of summation. Here the formula at (13.46) is used. The details are once again left as an exercise (see Q13.14, see also Q2.9, which may assist with the final, somewhat tedious, q-product manipulations). □

As described elsewhere, all of these chains may be used to produce numerous basic hypergeometric identities, polynomial versions of Slater-type identities, and a vast collection of multi-sum identities, either by iterating a single chain, or by alternating pairs of chains. A list of Bailey chains is given in Appendix IV.

13.6   The Bailey–Rogers–Ramanujan group

Many of the Bailey chains found in the papers mentioned above have the transformation relating $\alpha_n(a,q)$ and $\alpha'_n(a,q)$ of the form

$$\alpha'_n(a,q) = q^{An^2} a^{Cn} \alpha_n(a^D, q^B), \qquad (13.51)$$

or may be specialized to have this form, for rational numbers A, B, C and D. The transformations considered in this section are those listed below (the labels are those used by Stanton in [245], some of which were used previously in [85]).

$$\alpha'_n(a,q) = a^n q^{n^2} \alpha_n(a,q), \qquad (S1)$$

$$\beta'_n(a,q) = \sum_{j=0}^{n} \frac{a^j q^{j^2}}{(q;q)_{n-j}} \beta_j(a,q).$$

$$\alpha'_n(a,q) = a^{n/2} q^{n^2/2} \alpha_n(a,q), \qquad (S2)$$

$$\beta'_n(a,q) = \frac{1}{(-\sqrt{aq};q)_n} \sum_{j=0}^{n} \frac{(-\sqrt{aq};q)_j a^{j/2} q^{j^2/2}}{(q;q)_{n-j}} \beta_j(a,q).$$

$$\alpha'_n(a,q) = \alpha_n(a^2, q^2), \qquad (D1)$$

$$\beta'_n(a,q) = \sum_{k=0}^{n} \frac{(-aq;q)_{2k}}{(q^2;q^2)_{n-k}} q^{n-k} \beta_k(a^2, q^2).$$

$$\alpha'_n(a,q) = a^{-n} q^{-n^2} \alpha_n(a^2, q^2), \qquad (D2)$$

$$\beta'_n(a,q) = \sum_{k=0}^{n} \frac{(-aq;q)_{2k}}{(q^2;q^2)_{n-k}} q^{k^2+k-2kn-n} (-1)^{n-k} a^{-n} \beta_k(a^2, q^2).$$

$$\alpha_n'(a,q) = a^{-n/2}q^{-n^2/2}\alpha_n(a^2,q^2),$$ (D3)

$$\beta_n'(a,q) = \sum_{k=0}^{n} \frac{(-aq;q)_{2k}(q^{-1/2-k}a^{-1/2},q^{3/2+k}a^{1/2};q)_{n-k}}{(aq^{2k+1},q^2;q^2)_{n-k}}$$
$$\times q^{-k^2/2}a^{-k/2}\beta_k(a^2,q^2).$$

$$\alpha_n'(a^4,q^4) = \alpha_n(a^2,q^2),$$ (E1)

$$\beta_n'(a^4,q^4) = \sum_{k=0}^{n} \frac{(-1)^{n-k}q^{2(n-k)^2}}{(-q^2a^2;q^2)_{2n}(q^4;q^4)_{n-k}}\beta_k(a^2,q^2).$$

$$\alpha_n'(a^4,q^4) = a^{2n}q^{2n^2}\alpha_n(a^2,q^2),$$ (E2)

$$\beta_n'(a^4,q^4) = \sum_{k=0}^{n} \frac{a^{2k}q^{2k^2}}{(-q^2a^2;q^2)_{2n}(q^4;q^4)_{n-k}}\beta_k(a^2,q^2).$$

$$\alpha_n'(a^4,q^4) = a^n q^{n^2}\alpha_n(a^2,q^2),$$ (E3)

$$\beta_n'(a^4,q^4) = \sum_{k=0}^{n} \frac{(aq;q^2)_{2n-k}(-aq;q^2)_k a^k q^{k^2}}{(-q^2a^2;q^2)_{2n}(q^4;q^4)_{n-k}(a^2q^2;q^4)_n}\beta_k(a^2,q^2).$$

$$\alpha_n'(a^3,q^3) = a^n q^{n^2}\alpha_n(a,q),$$ (T1)

$$\beta_n'(a^3,q^3) = \frac{1}{(a^3q^3;q^3)_{2n}} \sum_{k=0}^{n} \frac{(aq;q)_{3n-k}a^k q^{k^2}}{(q^3;q^3)_{n-k}}\beta_k(a,q).$$

$$\alpha_n'(a,q) = a^{-n}q^{-n^2}\alpha_n(a^3,q^3),$$ (T2)

$$\beta_n'(a,q) = \frac{1}{(aq;q)_{2n}} \sum_{k=0}^{n} \frac{(aq^{2n+1};q^{-1})_{3k}(a^3q^3;q^3)_{2(n-k)}}{(q^3;q^3)_k}$$
$$\times (-1)^k a^{-n}q^{3k(k-1)/2-n^2}\beta_{n-k}(a^3,q^3).$$

Note that (S1) and (S2) are, respectively, (11.4) and (11.10), that (D1)–(D3) are the cases $B \to \infty$, $B \to 0$, and $B^2 = aq$, respectively, of the chain in Theorem 13.6, that (E1)–(E3) are the cases $B \to \infty$, $B \to 0$, and $B = a$, respectively, of the chain (IV.3) in Appendix IV, and that (T1) and (T2) are the chains (IV.4) and (IV.5) in Appendix IV.

If a is specialized to be 1 in (13.51), then this transformation has the form

$$\alpha_n'(1,q) = q^{An^2}\alpha_n(1,q^B). \tag{13.52}$$

In [245], Stanton encodes the transformation as a 2×2 matrix

$$\begin{bmatrix} 1 & A \\ 0 & B \end{bmatrix}. \tag{13.53}$$

It is easy to see that composition of transformations corresponds to multiplication matrices, in that if $\alpha_n^{(1)}(1,q) = q^{An^2}\alpha_n(1,q^B)$ and $\alpha_n^{(2)}(1,q) = q^{Cn^2}\alpha_n(1,q^D)$, then the composition is

$$\alpha_n{}''(1,q) := q^{Cn^2}\alpha_n^{(1)}(1,q^D) = q^{Cn^2+ADn^2}\alpha_n(1,q^{BD}),$$

while the corresponding multiplication of matrices is

$$\begin{bmatrix} 1 & A \\ 0 & B \end{bmatrix} \begin{bmatrix} 1 & C \\ 0 & D \end{bmatrix} = \begin{bmatrix} 1 & C+AD \\ 0 & BD \end{bmatrix}.$$

Stanton [245] goes on to identify each chain above (with a set equal to 1) with its corresponding matrix:

$$(S1) = \begin{bmatrix} 1 & 1 \\ 0 & 1 \end{bmatrix},\ (S2) = \begin{bmatrix} 1 & 1/2 \\ 0 & 1 \end{bmatrix},\ (D1) = \begin{bmatrix} 1 & 0 \\ 0 & 2 \end{bmatrix},\ (D2) = \begin{bmatrix} 1 & -1 \\ 0 & 2 \end{bmatrix},$$

$$(D3) = \begin{bmatrix} 1 & -1/2 \\ 0 & 2 \end{bmatrix},\ (E1) = \begin{bmatrix} 1 & 0 \\ 0 & 1/2 \end{bmatrix},\ (E2) = \begin{bmatrix} 1 & 1/2 \\ 0 & 1/2 \end{bmatrix},$$

$$(E3) = \begin{bmatrix} 1 & 1/4 \\ 0 & 1/2 \end{bmatrix},\ (T1) = \begin{bmatrix} 1 & 1/3 \\ 0 & 1/3 \end{bmatrix},\ (T2) = \begin{bmatrix} 1 & -1 \\ 0 & 3 \end{bmatrix}.$$

Definition 1. (Stanton [245]) The Bailey–Rogers–Ramanujan group is the subgroup of 2×2 upper triangular rational matrices generated by

$$\{(S1), (S2), (D1), (D2), (D3), (E1), (E2), (E3), (T1), (T2)\}. \tag{13.54}$$

It is clear that $Ei = Di^{-1}$, $i = 1,2,3$, $T2 = T1^{-1}$, $S1 = S2^2$ and $S2 = D1E2$, but it is convenient to include all ten matrices in the list of generators.

The key point of all this is that any element of the Bailey–Rogers–Ramanujan group corresponds to a finite identity of Rogers–Ramanujan type as exhibited at (13.57), and if the limit as $n \to \infty$ in (13.57) exists, then a single- or multi-sum identity of Rogers–Ramanujan identity results. To explain Stanton's result in Theorem 13.8, it is necessary to introduce some further notation. Let

$$w = w_1 w_2 \ldots w_k w_{k+1}$$

be an element of the Bailey–Rogers–Ramanujan group, with each w_i a generator from the set at (13.54). Let the corresponding Bailey pairs be denoted

$$(\alpha_n^{(0)}, \beta_n^{(0)}), \ (\alpha_n^{(1)}, \beta_n^{(1)}), \ \ldots, \ (\alpha_n^{(k)}, \beta_n^{(k)}), \ (\alpha_n^{(k+1)}, \beta_n^{(k+1)}).$$

Let the relation between $\beta_n^{(i)}$ and $\beta_n^{(i+1)}$ corresponding to w_{i+1} be represented as

$$\beta_n^{(i+1)} = \sum_{s_i=0}^{n} M_{n,s_i}^{(i)} \beta_{s_i}^{(i)}, \ 0 \le i \le k, \tag{13.55}$$

where each $M^{(i)}$ is an infinite lower triangular matrix. As an example of this notation (continuing to follow Stanton [245]), if $w_1 = (S1)$, then

$$M_{n,k}^{(0)} = \frac{q^{k^2}}{(q;q)_{n-k}}.$$

In what follows, we will always take

$$(\alpha_n^{(0)}, \beta_n^{(0)}) = ((-1)^n q^{n(n-1)/2}(1+q^n), \delta_{n,0}), \tag{13.56}$$

the unit Bailey pair with respect to $a = 1$ (of course $\alpha_0^{(0)} = 1$).

Theorem 13.8. *(Stanton [245]) If $w = w_1 w_2 \ldots w_k w_{k+1}$ is an element of the Bailey–Rogers–Ramanujan group, the corresponding finite Rogers–Ramanujan identity is given by*

$$\beta_n^{(k+1)} = \sum_{n \ge s_k \ge s_{k-1} \ge \cdots \ge s_1 \ge 0} M_{n,s_k}^{(k)} M_{s_k,s_{k-1}}^{(k-1)} \cdots M_{s_2,s_1}^{(1)} M_{s_1,0}^{(0)} \tag{13.57}$$

$$= \sum_{j=0}^{n} \frac{\alpha_j^{(k+1)}}{(q;q)_{n-j}(q;q)_{n+j}}.$$

Proof. The first equality follows from iterating (13.55), starting with (13.56). The second equality simply expresses the fact that $(\alpha_n^{(k+1)}, \beta_n^{(k+1)})$ is a Bailey pair with respect to $a = 1$. $\qquad\square$

If

$$w = \begin{bmatrix} 1 & A \\ 0 & B \end{bmatrix},$$

then (recalling that $\alpha_n^{(0)} = (-1)^n q^{n(n-1)/2}(1+q^n)$, from (13.56)) the corresponding transformation is given by $\alpha_0^{(k+1)}(1,q) = 1$, and for $n > 0$,

$$\alpha_n^{(k+1)} = q^{An^2}(-1)^n q^{Bn(n-1)/2}(1+q^{Bn}) = q^{(A+B/2)n^2}(-1)^n(q^{-Bn/2}+q^{Bn/2}),$$

so that if $A + B/2 > 0$, the right side of (13.57) converges as $n \to \infty$, by the Jacobi triple product identity (6.1), to

$$\frac{(q^A, q^{A+B}, q^{2A+B}; q^{2A+B})_\infty}{(q;q)_\infty^2}. \tag{13.58}$$

This now makes the proof of many of the Slater identities [237] automatic, as Stanton goes on to illustrate with several examples.

The left side of (13.57) converges termwise if

$$\lim_{n\to\infty} M_{n,s_k}^{(k)} = M_{\infty,s_k}^{(k)}$$

exists, and $\lim_{s_k\to\infty} M_{\infty,s_k}^{(k)} = 0$.[1] From the series representations above for $\beta_n'(1,q)$, it can be seen that this occurs if w_{k+1} is any one of (S1), (S2), (E2), (E3), or (T1). If w_{k+1} is (D1) or (E1), it can likewise be seen that if k is replaced with $n - k$, then

$$\lim_{n\to\infty} M_{n-s_k,s_k}^{(k)} = M_{\infty,s_k}^{(k)} \text{ exists, provided } \lim_{n\to\infty} M_{n-s_k,s_{k-1}}^{(k-1)} = M_{\infty,s_{k-1}}^{(k-1)}$$

exists, which will happen if w_k is any one of (S1), (S2), (E2), (E3), or (T1), otherwise repeat the previous step and continue. In the case where w_{k+1} is either (D2), (D3), or (T2), then provided $A + B/2 > 0$ as mentioned above, the limit as $n \to \infty$ of the left side of (13.57) still exists, even though the term-wise limit does not exist.

As an example of how a Slater-type identity may be proved more-or-less automatically, consider the product

$$(E2)(S1) = \begin{bmatrix} 1 & 3/2 \\ 0 & 1/2 \end{bmatrix},$$

---

[1]Stanton [245] omitted this second requirement, as it happens automatically for (S1), (S2), (E2), (E3) and (T1), but it is necessary to account for the fact that the term-wise limit does not exist for (D3).

so $A = 3/2$, $B = 1/2$ and $2A + B = 7/2$. With $w_1 = (E2)$ and $\beta_n^{(0)}(1, q) = \delta_{n,0}$,

$$\beta_n^{(1)}(1, q^4) = \frac{1}{(-q^2; q^2)_{2n}(q^4; q^4)_n}.$$

Next, replacing q with $q^{1/4}$ and substituting $\beta_n(1, q)$ in $w_2 = (S1)$ gives

$$\beta_n^{(2)}(1, q) = \sum_{j=0}^{n} \frac{q^{j^2}}{(q; q)_{n-j}(-q^{1/2}; q^{1/2})_{2j}(q; q)_j}. \qquad (13.59)$$

Finally, replacing q with q^2 in both (13.59) and (13.58) (after substituting for A and B), and then letting $n \to \infty$ in (13.59) leads (after cancelling a factor of $(q^2; q^2)_\infty$ on each side) to the Slater identity ([237, (33)])

$$\sum_{j=0}^{\infty} \frac{q^{2j^2}}{(-q; q)_{2j}(q^2; q^2)_j} = \frac{(q^3, q^4, q^7; q^7)_\infty}{(q^2; q^2)_\infty}.$$

Stanton [245] listed a total of twenty six single-sum identities, both old and new, and two examples are give in the exercises — see Q13.15 and Q13.16.

Now that the machinery has been set up, multi-sum identities follow equally easily from Stanton's result (13.57). For example, the group element

$$(S1)^{k+1} = \begin{bmatrix} 1 & k+1 \\ 0 & 1 \end{bmatrix}$$

leads to the Andrews–Gordon identity

$$\sum_{s_k \geq s_{k-1} \geq \dots s_1 \geq 0} \frac{q^{s_k^2 + \dots + s_1^2}}{(q; q)_{s_k - s_{k-1}}(q; q)_{s_{k-1} - s_{k-2}} \cdots (q; q)_{s_2 - s_1}(q; q)_{s_1}}$$
$$= \frac{(q^{k+1}, q^{k+2}, q^{2k+3}; q^{2k+3})_\infty}{(q; q)_\infty}. \qquad (13.60)$$

Stanton gave seven example of multi-sum identities in [245], two of which are included as exercises — see Q13.17 and Q13.18.

It is also straightforward to produce double-sum identities, and Stanton [245] give six examples of such identities in which the infinite products have modulus q^{11}. For example, the group element

$$(S1)(T1)(S1) = \begin{bmatrix} 1 & 5/3 \\ 0 & 1/3 \end{bmatrix},$$

and leads to the identity

$$\sum_{s_2 \geq s_1 \geq 0} \frac{q^{3s_2^2 + s_1^2}(q;q)_{3s_2 - s_1}}{(q^3;q^3)_{2s_2}(q^3;q^3)_{s_2 - s_1}(q;q)_{s_1}} = \frac{(q^5, q^6, q^{11}; q^{11})_\infty}{(q^3;q^3)_\infty}. \tag{13.61}$$

It is equally easy to give double-sum identities to many other moduli — see Q13.19.

13.7   Conjugate Bailey pairs and conjugate WP-Bailey pairs

Recall that a conjugate Bailey pair $(\delta_n, \gamma_n) = (\delta_n(a,q), \gamma_n(a,q))$ with respect to a is defined to be a pair of sequences $\{\delta_n\}$, $\{\gamma_n\}$ related by

$$\gamma_n = \sum_{r=n}^{\infty} \delta_r U_{r-n} V_{r+n}, \tag{13.62}$$

with $U_n = 1/(q;q)_n$, $V_n = 1/(aq;q)_n$, and that a conjugate WP-Bailey pair $(\delta_n, \gamma_n) = (\delta_n(a,k,q), \gamma_n(a,k,q))$ with respect to a and k is a pair of sequences $\{\delta_n\}$, $\{\gamma_n\}$ related by (13.62), with $U_n = (k/a;q)_n/(q;q)_n$, $V_n = (k;q)_n/(aq;q)_n$. If the context is clear, (δ_n, γ_n) will be used to denote a conjugate pair, with the dependence on a, k and q being suppressed. Recall also (10.3), that if (α_n, β_n) is a (WP-) Bailey pair, and (δ_n, γ_n) is a conjugate (WP-) Bailey pair, then

$$\sum_{n=0}^{\infty} \alpha_n \gamma_n = \sum_{n=0}^{\infty} \beta_n \delta_n. \tag{13.63}$$

This section is mostly about conjugate Bailey pairs, but before getting to those, we remark that each of the ten WP-Bailey chain encountered in Chapter 12 implies a conjugate WP-Bailey pair for each positive integer n. We state and prove this only for WP-Bailey chains of the form stated in Proposition 13.6. This covers eight of the 10 chains encountered in Chapter 12, and the two remaining two cases (the fourth chain of Warnaar at (12.32) and the chain of Liu and Ma at (12.33)) may be proved similarly with slightly more work, involving the use of indicator functions to make δ_r identically 0 for r either in certain intervals or certain congruence classes modulo 2.

Proposition 13.6. *Suppose a WP-Bailey chain is defined by the pair of transformations*

$$\alpha_n'(a, k, q) = G_n(a, c, q)\alpha_n(a^m, c, q^p), \tag{13.64}$$

$$\beta_n'(a, k, q) = \sum_{j=0}^{n} D_{n,j}(a, c, q)\beta_j(a^m, c, q^p),$$

where $c = f(k)$ for some invertible function f, and m and p are non-zero rationals (if $a = 1$, we take m to be 1). Let N be a positive integer. Then $(\delta_r(a, k, q), \gamma_r(a, k, q))$ is a conjugate WP-Bailey pair with respect to a and k, where

$$\delta_r(a, k, q) = D_{N,r}(a^{1/m}, k, q^{1/p}) \text{ if } 0 \leq r \leq N, \quad \delta_r = 0 \text{ if } r > N, \tag{13.65}$$

$$\gamma_r(a, k, q) = \frac{(f^{-1}(k)/a^{1/m}; q^{1/p})_{N-r}(f^{-1}(k); q^{1/p})_{N+r}}{(q^{1/p}; q^{1/p})_{N-r}(a^{1/m}q^{1/p}; q^{1/q})_{N+r}} G_r(a^{1/m}, k, q^{1/p}).$$

Proof. By the definition of a WP-Bailey pair at (12.3), and (13.64),

$$\beta_N'(a, k, q) = \sum_{r=0}^{N} \frac{(k/a; q)_{N-r}(k; q)_{N+r}}{(q; q)_{N-r}(aq; q)_{N+r}} \alpha_r'(a, k, q)$$

$$= \sum_{r=0}^{N} \frac{(f^{-1}(c)/a; q)_{N-r}(f^{-1}(c); q)_{N+r}}{(q; q)_{N-r}(aq; q)_{N+r}} G_r(a, c, q)\alpha_r(a^m, c, q^p)$$

$$= \sum_{j=0}^{N} D_{N,j}(a, c, q) \sum_{r=0}^{j} \frac{(c/a^m; q^p)_{j-r}(c; q^p)_{r+j}}{(q^p; q^p)_{j-r}(a^m q^p; q^p)_{r+j}} \alpha_r(a^m, c, q^p)$$

$$= \sum_{r=0}^{N} \alpha_r(a^m, c, q^p) \sum_{j=r}^{N} D_{N,j}(a, c, q) \frac{(c/a^m; q^p)_{j-r}(c; q^p)_{r+j}}{(q^p; q^p)_{j-r}(a^m q^p; q^p)_{r+j}}.$$

The result now follows from the definition of a conjugate WP-Bailey pair with respect to a and k, after replacing a with $a^{1/m}$, c with k and q with $q^{1/p}$, and then comparing coefficients of $\alpha_r(a, k, q)$ in the second and fourth expressions for $\beta_N'(a, k, q)$ above.   □

Remark: As with δ_r, note that $\gamma_r = 0$ if $r > N$ also, because of the $(q^{1/p}; q^{1/p})_{N-r}$ factor in the denominator of the expression defining γ_r above.

As an illustration, the first WP-Bailey chain of Andrews (12.13) leads to conjugate WP-Bailey pair (δ_r, γ_r) (for each positive integer N), where

$$\delta_r(a, k, q) = \frac{(kq/\rho_1, kq/\rho_2; q)_N}{(aq/\rho_1, aq/\rho_2; q)_N} \frac{(aq/\rho_1\rho_2; q)_{N-r}(akq/\rho_1\rho_2; q)_{N+r}}{(q; q)_{N-r}(kq; q)_{N+r}} \tag{13.66}$$

$$\times \frac{1 - kq^{2r}}{1 - k} \frac{(\rho_1, \rho_2; q)_r}{(kq/\rho_1, kq/\rho_2; q)_r} \left(\frac{aq}{\rho_1 \rho_2} \right)^r,$$

$$\gamma_r(a, k, q) = \frac{(kq/\rho_1\rho_2; q)_{N-r}(akq/\rho_1\rho_2; q)_{N+r}}{(q; q)_{N-r}(aq; q)_{N+r}} \frac{(\rho_1, \rho_2; q)_r}{(aq/\rho_1, aq/\rho_2; q)_r} \left(\frac{aq}{\rho_1 \rho_2} \right)^r.$$

It is straightforward to derive the corresponding result for conjugate Bailey pairs, either by setting $k = c = f(k) = 0$ in Proposition 13.6, or directly from the definitions of Bailey pairs and conjugate Bailey pairs.

Corollary 13.2. *Suppose a Bailey chain is defined by the pair of transformations*

$$\alpha'_n(a, q) = G_n(a, q)\alpha_n(a^m, q^p), \tag{13.67}$$

$$\beta'_n(a, q) = \sum_{j=0}^{n} D_{n,j}(a, q)\beta_j(a^m, q^p),$$

where m and p are non-zero rationals (if $a = 1$, we take m to be 1). Let N be a positive integer. Then (δ_r, γ_r) is a conjugate Bailey pair with respect to a, where

$$\delta_r(a, q) = D_{N,r}(a^{1/m}, q^{1/p}) \text{ if } 0 \leq r \leq N, \quad \delta_r = 0 \text{ if } r > N, \tag{13.68}$$

$$\gamma_r(a, q) = \frac{1}{(q^{1/p}; q^{1/p})_{N-r}(a^{1/m}q^{1/p}; q^{1/q})_{N+r}} G_r(a^{1/m}, q^{1/p}).$$

As an example, if result in Corollary 13.2 is applied to the Bailey chain of Andrews stated at (IV.1) in Appendix IV, then it produces the conjugate Bailey pair (which holds for each positive integer N):

$$\delta_r(a, q) = \frac{(\rho_1, \rho_2; q)_r}{(aq/\rho_1, aq/\rho_2; q)_N} \frac{(aq/\rho_1\rho_2; q)_{N-r}}{(q; q)_{N-r}} \left(\frac{aq}{\rho_1 \rho_2} \right)^r, \tag{13.69}$$

$$\gamma_r(a, q) = \frac{1}{(q; q)_{N-r}(aq; q)_{N+r}} \frac{(\rho_1, \rho_2; q)_r}{(aq/\rho_1, aq/\rho_2; q)_r} \left(\frac{aq}{\rho_1 \rho_2} \right)^r.$$

Notice that $\delta_r = \gamma_r = 0$ for $r > N$ (recall $1/(q; q)_j = 0$ for $j < 0$), and that letting $N \to \infty$ leads, up to a multiplicative factor, to conjugate Bailey pair stated by Bailey [61, pp. 2–3], and which is stated at (10.4) and (10.7). The insertion of this conjugate pair and the Bailey pair at (10.16) in (10.3) leads to Watson's [257] transformation for a terminating $_8\phi_7$ series (5.10).

13.7.1 A conjugate Bailey pair of Schilling and Warnaar

It appears that the first variations of the conjugate Bailey pair used by Bailey and Slater at (10.4) and (10.7) are to be found in the papers of Bressoud [84] and Singh [234], and, as has been noted above, their transformations were essentially WP-Bailey chains. However, there have been a number of recent papers in which new conjugate Bailey pairs have been stated. Schilling and Warnaar [222] generalized a special case of the conjugate pair (10.4), (10.7) to an infinite family of conjugate pairs, and in [223] the same authors slightly modified the definition of a conjugate Bailey pair and gave a different infinite hierarchy of conjugate Bailey pairs. However the applications of these pairs go beyond the scope of the present volume, so we instead state and prove a result of the authors from a third paper [224], where the authors state the conjugate Bailey pair in Theorem 13.9 below (we have modified the notation in their statement of the theorem slightly). As a side benefit, the method used to prove a key lemma (Lemma 13.8 below) used in the proof of the theorem may be modified to prove a generalization of the Jacobi triple product due to Warnaar ([255, Theorem 2.5]) — see Q13.20.

Theorem 13.9. *(Schilling and Warnaar [224]) For $n \in \mathbb{Z}_+$ and $j \in \mathbb{Z}$, the pair (δ_r, γ_r) is a conjugate Bailey pair relative to $a = q^n$, where*

$$
\delta_r = \begin{bmatrix} 2r + n \\ r - j \end{bmatrix} - \begin{bmatrix} 2r + n \\ r - j - 1 \end{bmatrix} \tag{13.70}
$$

$$
\gamma_r = \frac{1}{(q, q, aq; q)_\infty} \sum_{i=1}^{\infty} (-1)^i q^{i(i-1)/2} \left\{ q^{i(r+j+n+1)} - q^{i(r-j)} \right\}.
$$

We continue with the proof given by Schilling and Warnaar [224], which depends on the following lemma.

Lemma 13.8. *(Schilling and Warnaar [224]) For a and b indeterminates,*

$$
\sum_{r=0}^{\infty} \frac{(ab; q)_{2r}}{(ab, q; q)_r} \left\{ \frac{1}{(aq; q)_{r-1}(bq; q)_r} - \frac{1}{(aq; q)_r (bq; q)_{r-1}} \right\}
$$

$$
= \frac{1}{(aq, bq, q; q)_\infty} \sum_{i=1}^{\infty} (-1)^i q^{i(i-1)/2} (a^i - b^i). \tag{13.71}
$$

Proof. After combining the terms inside the curly braces to get the single term $(b - a)q^r/(aq, bq; q)_r$, writing $(ab; q)_{2r} = (ab; q)_r (abq^r; q)_r$ and multiplying across by $(aq, bq; q)_\infty/(b - a)$, (13.71) may be rewritten as

$$\sum_{r=0}^{\infty} \frac{(abq^r; q)_r (aq^r, bq^r; q)_\infty q^r}{(q; q)_r} = \frac{1}{(q; q)_\infty} \sum_{i=1}^{\infty} (-1)^{i+1} q^{i(i-1)/2} \frac{a^i - b^i}{a - b}. \quad (13.72)$$

Next, after applying (3.5) and (3.4) to the products in the numerator of the left side of (13.72), this left side may be rewritten as

$$\sum_{r=0}^{\infty} \sum_{i=0}^{\infty} \sum_{j=0}^{\infty} \sum_{k=0}^{r} (-1)^{i+j+k} a^{i+k} b^{j+k} \frac{q^{i(i+1)/2 + j(j+1)/2 + k(k-1)/2 + r(i+j+k+1)}}{(q; q)_i (q; q)_j (q; q)_k (q; q)_{r-k}}.$$

$$(13.73)$$

After shifting the summation variables i, j and r by the replacements $i \to i - k$, $j \to j - k$ and $r \to r + k$, the summation ranges become $i \geq k \geq 0$, $j \geq k \geq 0$ (or $0 \leq k \leq \min\{i, j\}$) and $r \geq 0$, so that (13.73) may be written as

$$\sum_{i=0}^{\infty} \sum_{j=0}^{\infty} (-1)^{i+j} a^i b^j q^{i(i+1)/2 + j(j+1)/2}$$

$$\times \sum_{k=0}^{\min\{i,j\}} \frac{(-1)^k q^{k(k-1)/2}}{(q; q)_{i-k} (q; q)_{j-k} (q; q)_k} \sum_{r=0}^{\infty} \frac{q^{r(i+j-k+1)}}{(q; q)_r}. \quad (13.74)$$

By (3.3) the sum over r equals $1/(q^{i+j-k+1}; q)_\infty = (q; q)_{i+j-k}/(q; q)_\infty$, and then

$$\sum_{k=0}^{\min\{i,j\}} \frac{(-1)^k q^{k(k-1)/2} (q; q)_{i+j-k}}{(q; q)_{i-k} (q; q)_{j-k} (q; q)_k}$$

$$= \frac{(q; q)_{i+j}}{(q; q)_i (q; q)_j} \sum_{k=0}^{\min\{i,j\}} \frac{(q^{-i}, q^{-j}; q)_k}{(q^{-i-j}, q; q)_k} \quad \text{(by (2.12))}$$

$$= \frac{(q; q)_{i+j}}{(q; q)_i (q; q)_j} \frac{(q^{-l}; q)_l}{(q^{-i-j}; q)_l} \quad \text{(by (3.13), where } l = \min\{i, j\})$$

$$= q^{ij} \quad \text{(after using (2.26))}.$$

Thus (13.74) may be expressed as

$$\frac{1}{(q; q)_\infty} \sum_{i=0}^{\infty} \sum_{j=0}^{\infty} (-1)^{i+j} a^i b^j q^{(i+j)(i+j+1)/2}$$

$$= \frac{1}{(q;q)_\infty} \sum_{j=0}^{\infty} \sum_{i=j+1}^{\infty} (-1)^{i+1} a^{i-j-1} b^j q^{i(i-1)/2} \text{ (after } i \to i - j - 1)$$

$$= \sum_{i=1}^{\infty} (-1)^{i+1} a^{i-1} q^{i(i-1)/2} \sum_{j=0}^{i-1} (b/a)^j$$

$$= \sum_{i=1}^{\infty} (-1)^{i+1} a^{i-1} q^{i(i-1)/2} \frac{1 - (b/a)^i}{1 - b/a}$$

$$= \frac{1}{(q;q)_\infty} \sum_{i=1}^{\infty} (-1)^{i+1} q^{i(i-1)/2} \frac{a^i - b^i}{a - b}.$$

$\square$

We now use the formula in Lemma 13.8 to prove the theorem.

Proof of Theorem 13.9. After substituting for δ_r in the defining relation for a conjugate Bailey pair ((10.2) with $U_k = 1/(q;q)_k$, $V_k = 1/(aq;q)_k$ and $a = q^n$), shifting the summation index in the sum containing the δ_k's so that it starts at 0, and multiplying both sides by $-1/(q;q)_n$, it follows that what needs to be shown is that

$$\sum_{k=0}^{\infty} \frac{1}{(q;q)_k(q;q)_{k+2r+n}} \left\{ \begin{bmatrix} 2k+2r+n \\ r+k-j-1 \end{bmatrix} - \begin{bmatrix} 2k+2r+n \\ r+k-j \end{bmatrix} \right\}$$

$$= \frac{1}{(q;q)_\infty^3} \sum_{i=1}^{\infty} (-1)^i q^{i(i-1)/2} \left\{ q^{i(r-j)} - q^{i(r+j+n+1)} \right\}. \quad (13.75)$$

After some elementary manipulations, it may be seen that the left side of (13.75) equals

$$\frac{1}{(q;q)_{r-j}(q;q)_{r+n+j+1}} \sum_{k=0}^{\infty} \frac{(q^{2r+n+1};q)_{2k}}{(q;q)_k(q^{2r+n+1};q)_k}$$

$$\times \left\{ \frac{1}{(q^{r-j+1};q)_{k-1}(q^{r+n+j+2};q)_k} - \frac{1}{(q^{r-j+1};q)_k(q^{r+n+j+2};q)_{k-1}} \right\}$$

$$= \frac{1}{(q;q)_{r-j}(q;q)_{r+n+j+1}} \frac{1}{(q^{r-j+1}, q^{r+n+j+2}, q; q)_\infty}$$

$$\times \sum_{i=1}^{\infty} (-1)^i q^{i(i-1)/2} \left\{ (q^{r-j})^i - (q^{r+j+n+1})^i \right\}$$

$$= \frac{1}{(q;q)_\infty^3} \sum_{i=1}^{\infty} (-1)^i q^{i(i-1)/2} \left\{ q^{i(r-j)} - q^{i(r+j+n+1)} \right\},$$

where the first equality follows from Lemma 13.8, with $a = q^{r-j}$ and $b = q^{r+j+n+1}$. ∎

13.7.2　The conjugate Bailey pair of Rowell

Another quite general conjugate Bailey pair was stated by Rowell in [221], some special cases of which were found previously by Andrews and Warnaar [49].

Theorem 13.10. *(Rowell [221]) The pair (δ_r, γ_r) is a conjugate Bailey pair relative to f, where*

$$\delta_n = \frac{(efq^2/abc, efq/a; q)_\infty}{(efq^2/ab, efq^2/ac; q)_\infty} \frac{(a,b,c;q)_n}{(eq;q)_n} \left(\frac{efq^2}{abc}\right)^n \tag{13.76}$$

$$\gamma_n = \frac{(efq/a, a; q)_n}{(fq, fq/a; q)_n} \left(\frac{-1}{a}\right)^n q^{-n(n-1)/2} \times$$

$$\sum_{j=n}^\infty \frac{(1 - efq^{2j+1}/a)(efq^{n+1}/a, fq/a, b, c; q)_j (eq/a; q)_{j-n}}{(efq^2/ab, efq^2/ac, fq^{n+1}, eq; q)_j (q; q)_{j-n}} \left(\frac{-ef}{bc}\right)^j q^{j(j+3)/2}.$$

Proof.

$$\sum_{j\geq n} \frac{\delta_j}{(q;q)_{j-n}(fq;q)_{n+j}}$$

$$= \frac{(efq^2/abc, efq/a; q)_\infty}{(efq^2/ab, efq^2/ac; q)_\infty} \sum_{j\geq n} \frac{(a,b,c;q)_j}{(eq;q)_j(q;q)_{j-n}(fq;q)_{n+j}} \left(\frac{efq^2}{abc}\right)^j$$

$$= \frac{(efq^2/abc, efq/a; q)_\infty}{(efq^2/ab, efq^2/ac; q)_\infty} \frac{(a,b,c;q)_n}{(eq;q)_n(fq;q)_{2n}} \left(\frac{efq^2}{abc}\right)^n$$

$$\times \sum_{j\geq 0} \frac{(aq^n, bq^n, cq^n; q)_j}{(eq^{n+1}; q)_j(q;q)_j(fq^{2n+1};q)_j} \left(\frac{efq^2}{abc}\right)^j$$

$$= \frac{(efq^2/abc, efq/a; q)_\infty}{(efq^2/ab, efq^2/ac; q)_\infty} \frac{(a,b,c;q)_n}{(eq;q)_n(fq;q)_{2n}} \left(\frac{efq^2}{abc}\right)^n \quad \text{(by (5.27))}$$

$$\times \frac{(efq^{n+2}/ab, efq^{n+2}/ac; q)_\infty}{(efq^2/abc, efq^{2n+2}/a; q)_\infty} \sum_{k=0}^\infty \frac{1 - efq^{2n+2k+1}/a}{1 - efq^{2n+1}/a}$$

$$\times \frac{(efq^{2n+1}/a, fq^{n+1}/a, eq/a, bq^n, cq^n; q)_k}{(eq^{n+1}, fq^{2n+1}, efq^{n+2}/ab, efq^{n+2}/ac, q; q)_k} \left(\frac{-efq^{n+2}}{bc}\right)^k q^{k(k-1)/2}$$

$$= \frac{(a,b,c;q)_n(efq/a;q)_{2n+1}}{(eq, efq^2/ab, efq^2/ac; q)_n(fq;q)_{2n}} \left(\frac{efq^2}{abc}\right)^n \sum_{k=0}^\infty \frac{1 - efq^{2n+2k+1}/a}{1 - efq^{2n+1}/a}$$

$$\times \frac{(efq^{2n+1}/a, fq^{n+1}/a, eq/a, bq^n, cq^n; q)_k}{(eq^{n+1}, fq^{2n+1}, efq^{n+2}/ab, efq^{n+2}/ac, q; q)_k} \left(\frac{-efq^{n+2}}{bc}\right)^k q^{k(k-1)/2}$$

$$= \frac{(efq/a, a; q)_n}{(fq/a, fq; q)_n} \left(\frac{-q^2}{a}\right)^n \sum_{k=0}^{\infty} (1 - efq^{2n+2k+1}/a) \frac{(eq/a; q)_k}{(q; q)_k}$$

$$\times \frac{(efq^{n+1}/a, fq/a, b, c; q)_{k+n}}{(eq, fq^{n+1}, efq/ab, efq/ac; q)_{k+n}} \left(\frac{-ef}{bc}\right)^{k+n} (q^{n+2})^k q^{k(k-1)/2}$$

$$= \gamma_n,$$

where the final equality follows by shifting the summation index $(k \to k-n)$, and simplifying. $\square$

Note that the conjugate Bailey pair employed by Bailey and Slater, and stated in Theorem 10.1, is a special case of the conjugate Bailey pair of Rowell. To see this, let $e = aq$ in the conjugate Bailey pair of Rowell above, so that the series in the expression for γ_n vanishes, except for the $j = n$ term. Simplify the resulting expressions for δ_n and γ_n, replace b with ρ_1, c with ρ_2 and f with a, and finally, after substituting in (13.62), cancel a factor of $1 - a$ and move the infinite product so that it is associated with γ_n.

Rowell tabled many other special cases of the pair (δ_r, γ_r) in Theorem 13.10, and then used some of these to prove (or reprove), amongst other results, several identities of Ramanujan for false- and partial theta functions (proved by Andrews and Warnaar in [49]), a number of identities for the generating function for triangular numbers, several identities for double-sums related to indefinite binary quadratic forms, and various identities for integer partitions. To prove many of these identities, Rowell employed the symmetric- and asymmetric Bailey transforms found by Andrews and Warnaar, and which are described in the next section.

Many of these special cases of conjugate Bailey pairs may be classified further as special cases of the general pairs in the next two corollaries, which arise after setting $e = a$ in (13.76), one effect of this substitution being to cancel four of the q-products in the series in the expression for γ_n. The next two corollaries were stated by Rowell ([221, Corollaries 3.1, 3.2]).

Corollary 13.3. *The pair* $(\delta_r, \gamma_r) = (\delta_r(1, q^2), \gamma_r(1, q^2))$ *is a conjugate*

Bailey pair, where

$$\delta_n(1, q^2) = \frac{(q^4/bc, q^2; q^2)_\infty}{(q^4/b, q^4/c; q)_\infty} \frac{(a, b, c; q^2)_n}{(aq^2; q^2)_n} \left(\frac{q^4}{bc}\right)^n \qquad (13.77)$$

$$\gamma_n(1, q^2) = \frac{(a; q^2)_n}{(q^2/a; q^2)_n} \left(\frac{-1}{a}\right)^n q^{-n(n-1)}$$

$$\times \sum_{j=n}^{\infty} \frac{(1 - q^{4j+2})(q^2/a, b, c; q^2)_j}{(q^4/b, q^4/c, aq^2; q^2)_j} \left(\frac{-a}{bc}\right)^j q^{j(j+3)}.$$

Proof. In (13.76), replace q with q^2, set $f = 1$ and $e = a$. $\qquad \square$

Corollary 13.4. *The pair* $(\delta_r, \gamma_r) = (\delta_r(q^2, q^2), \gamma_r(q^2, q^2))$ *is a conjugate Bailey pair, where*

$$\delta_n(q^2, q^2) = \frac{(q^6/bc, q^4; q^2)_\infty}{(q^6/b, q^6/c; q)_\infty} \frac{(a, b, c; q^2)_n}{(aq^2; q^2)_n} \left(\frac{q^6}{bc}\right)^n \qquad (13.78)$$

$$\gamma_n(q^2, q^2) = \frac{(a; q^2)_n}{(q^4/a; q^2)_n} \left(\frac{-1}{a}\right)^n q^{-n(n-1)}$$

$$\times \sum_{j=n}^{\infty} \frac{(1 - q^{4j+4})(q^4/a, b, c; q^2)_j}{(q^6/b, q^6/c, aq^2; q^2)_j} \left(\frac{-a}{bc}\right)^j q^{j(j+5)}.$$

Proof. In (13.76), replace q with q^2, set $f = q^2$ and $e = a$. $\qquad \square$

Generally, with the values for γ_n and δ_n at (13.76), substituting particular sequences in the identity $\sum_{n=0}^{\infty} \beta_n \delta_n = \sum_{n=0}^{\infty} \alpha_n \gamma_n$ will give an expression on the right side that is a double sum. However, there are circumstances in which the double sum may be converted to a single sum. Specifically, if $\gamma_n = f_n \sum_{j \geq n} g_j$, then

$$\sum_{n=0}^{\infty} \alpha_n \gamma_n = \sum_{n=0}^{\infty} \alpha_n f_n \sum_{j \geq n} g_j = \sum_{j=0}^{\infty} g_j \sum_{n=0}^{j} \alpha_n f_n,$$

and it may be the case that the inner sum may be put in closed form, thus converting the double sum to a single sum. See Q13.22 for one example.

13.8 Symmetric- and asymmetric Bailey pairs and transforms

In [49], while proving a number of false theta series identities due to Ramanujan, Andrews and Warnaar introduced two bilateral versions of the Bailey transform.

Lemma 13.9. *(Symmetric bilateral Bailey transform, [49, Lemma 2.1]). If*

$$\beta_n = \sum_{r=-n}^{n} \alpha_r U_{n-r} V_{n+r} \quad and \quad \gamma_n = \sum_{r=|n|}^{\infty} \delta_r U_{r-n} V_{r+n}, \tag{13.79}$$

then

$$\sum_{n=-\infty}^{\infty} \alpha_n \gamma_n = \sum_{n=0}^{\infty} \beta_n \delta_n \tag{13.80}$$

subject to conditions on the four sequences α_n, β_n, γ_n and δ_n which make all the relevant infinite series absolutely convergent.

Proof.

$$\sum_{n=0}^{\infty} \beta_n \delta_n = \sum_{n=0}^{\infty} \delta_n \sum_{r=-n}^{n} \alpha_r U_{n-r} V_{n+r}$$

$$= \sum_{r=-\infty}^{\infty} \alpha_r \sum_{n=|r|}^{\infty} \delta_n U_{n-r} V_{n+r}$$

$$= \sum_{r=-\infty}^{\infty} \alpha_r \gamma_r.$$

$\square$

Lemma 13.10. *(Asymmetric bilateral Bailey transform, [49, Lemma 2.2]) Let $m = \max\{n, -n-1\}$. If*

$$\beta_n = \sum_{r=-n-1}^{n} \alpha_r U_{n-r} V_{n+r+1} \quad and \; \gamma_n = \sum_{r=m}^{\infty} \delta_r U_{r-n} V_{r+n+1}, \tag{13.81}$$

then

$$\sum_{n=-\infty}^{\infty} \alpha_n \gamma_n = \sum_{n=0}^{\infty} \beta_n \delta_n \tag{13.82}$$

subject to conditions on the four sequences α_n, β_n, γ_n and δ_n which make all the relevant infinite series absolutely convergent.

Proof. The proof is similar to the proof of Lemma 13.9. $\square$

We will follow Andrews and Warnaar [49] and refer to a pair (α_n, β_n) satisfying either (13.79) or (13.79) as a *Bailey pair*, and to a (γ_n, δ_n) satisfying

either (13.79) or (13.79) as a *conjugate Bailey pair*, in each case suppress-
ing the labels "bilateral" and "(a)symmetric". In order to prove the false
theta series identities, Andrews and Warnaar [49] first derived a number of
conjugate Bailey pairs.

Theorem 13.11. *(Andrews and Warnaar [49]) Let $U_n = V_n = 1/(q^2; q^2)_n$
in the symmetric bilateral Bailey transform. Then (γ_n, δ_n) is a conjugate
Bailey pair, if*

$$\delta_n = \frac{(q^2; q^2)_{2n}}{(-q; q)_{2n+1}} q^n, \qquad \gamma_n = q^{-n^2} \sum_{j \geq |n|} q^{j^2+j}, \qquad (13.83)$$

or

$$\delta_n = (q; q)_{2n} q^n, \qquad \gamma_n = q^{-2n^2} \sum_{j \geq 2|n|} q^{j(j+1)/2}. \qquad (13.84)$$

Proof. We will give Rowell's proof using his conjugate Bailey pair, rather
than the proof of Andrews and Warnaar. By definition of γ_n in the symmetric Bailey transform, $\gamma_n = \gamma_{-n}$, so it is necessary to consider $n > 0$. In
this case the relation between γ_n and δ_n in Theorem 13.11 is the same as
that between γ_n and δ_n in Corollary 13.3. Thus it is just necessary to show
that the conjugate pairs at (13.83) and (13.84) are special cases of Rowell's
pair at (13.77). For (13.83), set $a = -q$, $b = q$ and $c = q^2$ in (13.77), divide
both δ_n and γ_n by $1 - q^2$, and simplify. The pair at (13.84) follows after
setting $b = q$ and $c = q^2$ in (13.77), letting $a \to 0$, dividing both δ_n and γ_n
by $1 - q$ and finally using the fact that

$$\sum_{j \geq |n|} (1 + q^{2j+1}) q^{2j^2+j} = \sum_{j \geq |n|} q^{(2j)(2j+1)/2} + q^{(2j+1)(2j+2)/2} = \sum_{j \geq 2|n|} q^{j(j+1)/2}.$$

$\square$

Theorem 13.12. *(Andrews and Warnaar [49]) Let $U_n = V_n = 1/(q^2; q^2)_n$
in the asymmetric bilateral Bailey transform. Then (γ_n, δ_n) is a conjugate
Bailey pair, if*

$$\delta_n = \frac{(q^2; q^2)_{2n+1}}{(-q; q)_{2n+2}} q^n, \qquad \gamma_n = q^{-n^2-n} \sum_{j \geq m} q^{j^2+2j}, \qquad (13.85)$$

or

$$\delta_n = (q; q)_{2n+1} q^n, \qquad \gamma_n = q^{-2n^2-2n} \sum_{j \geq 2m} q^{j(j+3)/2}, \qquad (13.86)$$

where $m = \max\{n, -n - 1\}$.

Proof. It can be seen that, with $U_n = V_n = 1/(q^2; q^2)_n$ in the asymmetric bilateral Bailey transform,

$$\gamma_n = \sum_{j=m}^{\infty} \frac{\delta_j}{(q^2; q^2)_{j-n}(q^2; q^2)_{j+n+1}} = \sum_{j=m}^{\infty} \frac{\delta_j}{(q^2; q^2)_{j-m}(q^2; q^2)_{j+m+1}},$$

irrespective of whether $m = n \, (n \geq 0)$ or $m = -n - 1 \, (n < 0)$.

Now consider the case $a = -q^2$, $b = q^2$ and $c = q^3$ in (13.78). After simplifying, this gives (for δ_n and γ_n in (13.78))

$$\delta_j = \frac{(1+q^2)(q^2; q^2)_{2j+1} q^j}{(-q; q)_{2j+2}}, \qquad \gamma_m = (1 - q^4) q^{-m^2 - m} \sum_{j \geq m} q^{j^2 + 2j}.$$

After substituting in the defining relation connecting the conjugate Bailey pair $(\delta_r(q^2, q^2), \gamma_r(q^2, q^2))$ and dividing through by $1 - q^4$, this gives

$$q^{-m^2 - m} \sum_{j \geq m} q^{j^2 + 2j} = \sum_{j=m}^{\infty} \frac{1}{(q^2; q^2)_{j-m}(q^2; q^2)_{j+m+1}} \frac{(q^2; q^2)_{2j+1} q^j}{(-q; q)_{2j+2}}$$

$$= \sum_{j=m}^{\infty} \frac{1}{(q^2; q^2)_{j-n}(q^2; q^2)_{j+n+1}} \frac{(q^2; q^2)_{2j+1} q^j}{(-q; q)_{2j+2}}.$$

Since $-m^2 - m = -n^2 - n$, irrespective of whether $m = n \, (n \geq 0)$ or $m = -n - 1 \, (n < 0)$, this concludes the proof of (13.85).

The proof of (13.86) is similar, except we take $b = q^2$, $c = q^3$ and $a \to 0$ in (13.78), at the end using the fact that

$$\sum_{j \geq m} (1 + q^{2j+2}) q^{2j^2 + 3j} = \sum_{j \geq m} q^{(2j)(2j+3)/2} + q^{(2j+1)(2j+4)/2} = \sum_{j \geq 2m} q^{j(j+3)/2}.$$

$\square$

By combining the symmetric Bailey transform with each of the conjugate Bailey pairs in the two theorems above, Andrews and Warnaar derived fours more specific transformations for bilateral Bailey pairs. One example is the following.

Theorem 13.13. *(Andrews and Warnaar [49, Theorem 5]) If (α_n, β_n) is a bilateral symmetric Bailey pair, related by*

$$\beta_n = \sum_{r=-n}^{n} \frac{\alpha_r}{(q^2; q^2)_{n-r}(q^2; q^2)_{n+r}}, \tag{13.87}$$

then

$$\sum_{n=0}^{\infty} \frac{(q^2;q^2)_{2n}q^n\beta_n}{(-q;q)_{2n+1}} = \sum_{j=0}^{\infty} q^{j(j+1)} \sum_{n=-j}^{j} q^{-n^2}\alpha_n. \tag{13.88}$$

Proof. With (γ_n,δ_n) as at (13.83),

$$\sum_{n=0}^{\infty} \frac{(q^2;q^2)_{2n}q^n\beta_n}{(-q;q)_{2n+1}} = \sum_{n=0}^{\infty} \beta_n\delta_n = \sum_{n=-\infty}^{\infty} \alpha_n\gamma_n$$

$$= \sum_{n=-\infty}^{\infty} \alpha_n q^{-n^2} \sum_{j\geq|n|} q^{j^2+j} = \sum_{j=0}^{\infty} q^{j^2+j} \sum_{n=-j}^{j} q^{-n^2}\alpha_n.$$

$\square$

Next, Andrews and Warnaar [49] point out that if $(\alpha'_n(1,q),\beta'_n(1,q)) = (\alpha'_n,\beta'_n)$ is a symmetric bilateral Bailey pair, then

$$\beta'_n = \sum_{j=-n}^{n} \frac{\alpha'_j}{(q;q)_{n-j}(q;q)_{n+j}} = \frac{\alpha'_0}{(q;q)_n(q;q)_n} + \sum_{j=1}^{n} \frac{\alpha'_{-j}+\alpha'_j}{(q;q)_{n-j}(q;q)_{n+j}},$$

so that $(\alpha_n,\beta_n) = (\alpha_n(1,q),\beta_n(1,q))$ is a standard Bailey pair, where $\alpha_0 = \alpha'_0$, $\alpha_n = \alpha'_{-n} + \alpha'_n$ for $n \geq 1$, and $\beta_n = \beta'_n, n \geq 0$. Similarly, the process may be reversed to derive a symmetric bilateral Bailey pair from a standard Bailey pair.

Andrews and Warnaar then convert some particular Bailey pairs to bilateral pairs and use them to prove the false theta series identities of Ramanujan which were mentioned at the start of the section.

Theorem 13.14. *(Andrews and Warnaar [49, Theorem 6]) If $|q| < 1$, then*

$$\sum_{n=0}^{\infty} \frac{(-1)^n q^{n(n+1)}(q;q^2)_n}{(-q;q)_{2n+1}} = \sum_{n=0}^{\infty} (-1)^n q^{n(n+1)/2}. \tag{13.89}$$

Proof. In the Bailey pair at (V.2.6), replace q with q^2, set $e = -q^{3/2}$ and let $d \to 0$ to derive the standard Bailey pair

$$\alpha'_{2r}(1,q^2) = q^{(2r)(2r+1)/2} + q^{(-2r)(-2r+1)/2},$$

$$\alpha'_{2r+1}(1,q^2) = -q^{(2r+1)(2r+2)/2} - q^{(-2r-1)(-2r)/2},$$

$$\beta'_n(1,q^2) = \frac{(-1)^n q^{n^2}(q;q^2)_n}{(q^2;q^2)_{2n}},$$

and hence the bilateral symmetric pair

$$\alpha_n(1, q^2) = (-1)^n q^{n(n+1)/2}, \qquad \beta_n(1, q^2) = \frac{(-1)^n q^{n^2} (q; q^2)_n}{(q^2; q^2)_{2n}}.$$

Hence, by (13.88),

$$\sum_{n=0}^{\infty} \frac{(-1)^n q^{n(n+1)} (q; q^2)_n}{(-q; q)_{2n+1}} = \sum_{n=0}^{\infty} \frac{(q^2; q^2)_{2n} q^n \beta_n}{(-q; q)_{2n+1}}$$

$$= \sum_{j=0}^{\infty} q^{j(j+1)} \sum_{n=-j}^{j} q^{-n^2} \alpha_n$$

$$= \sum_{j=0}^{\infty} q^{j(j+1)} \sum_{n=-j}^{j} (-1)^n q^{-n(n-1)/2}.$$

The result now follows since the n and $-n+1$ terms in the inner sum cancel, leaving only the $n = -j$ term.   □

The authors in [49] indicate many other applications of their conjugate Bailey pairs and their bilateral transforms, but we restrict ourselves to one additional example — see Q13.21.

13.9   Miscellaneous identities

In this section a number of identities relating the pair $(\alpha_n(a, k), \beta_n(a, k))$ and the pairs $(\alpha_n(a^{-1}, k^{-1}), \beta_n(a^{-1}, k^{-1}))$, $(\alpha_n(-a, -k), \beta_n(-a, -k))$ and $(\alpha_n(a^2, k^2), \beta_n(a^2, k^2))$ are described (see [193, 202] for further details). The corresponding identities for Bailey pairs are also described. Some of the identities will also have relevance to the chapter on Lambert series.

Two of the principal results are contained in the following theorems.

Theorem 13.15. *If $(\alpha_n(a, k), \beta_n(a, k))$ is a WP-Bailey pair, then subject to suitable convergence conditions,*

$$\sum_{n=1}^{\infty} \frac{(q\sqrt{k}, -q\sqrt{k}, z; q)_n (q; q)_{n-1}}{\left(\sqrt{k}, -\sqrt{k}, qk, \frac{qk}{z}; q\right)_n} \left(\frac{qa}{z}\right)^n \beta_n(a, k)$$

$$- \sum_{n=1}^{\infty} \frac{\left(q\sqrt{\frac{1}{k}}, -q\sqrt{\frac{1}{k}}, \frac{1}{z}; q\right)_n (q; q)_{n-1}}{\left(\sqrt{\frac{1}{k}}, -\sqrt{\frac{1}{k}}, \frac{q}{k}, \frac{qz}{k}; q\right)_n} \left(\frac{qz}{a}\right)^n \beta_n\left(\frac{1}{a}, \frac{1}{k}\right) -$$

$$\sum_{n=1}^{\infty} \frac{(z;q)_n (q;q)_{n-1}}{(qa, \frac{qa}{z}; q)_n} \left(\frac{qa}{z}\right)^n \alpha_n(a, k) + \sum_{n=1}^{\infty} \frac{(\frac{1}{z}; q)_n (q;q)_{n-1}}{(\frac{q}{a}, \frac{qz}{a}; q)_n} \left(\frac{qz}{a}\right)^n \alpha_n \left(\frac{1}{a}, \frac{1}{k}\right)$$

$$= \frac{(a-k)\left(1-\frac{1}{z}\right)\left(1-\frac{ak}{z}\right)}{(1-a)(1-k)\left(1-\frac{a}{z}\right)\left(1-\frac{k}{z}\right)} + \frac{z\left(z, \frac{q}{z}, \frac{k}{a}, \frac{qa}{k}, \frac{ak}{z}, \frac{qz}{ak}, q, q; q\right)_{\infty}}{k\left(\frac{z}{k}, \frac{qk}{z}, \frac{z}{a}, \frac{qa}{z}, a, \frac{q}{a}, k, \frac{q}{k}; q\right)_{\infty}}. \quad (13.90)$$

Theorem 13.16. *If $(\alpha_n(a, k, q), \beta_n(a, k, q))$ is a WP-Bailey pair, then subject to suitable convergence conditions,*

$$\sum_{n=1}^{\infty} \frac{(1-kq^{2n})(z;q)_n (q;q)_{n-1}}{(1-k)(qk, qk/z; q)_n} \left(\frac{qa}{z}\right)^n \beta_n(a, k, q)$$

$$+ \sum_{n=1}^{\infty} \frac{(1+kq^{2n})(z;q)_n (q;q)_{n-1}}{(1+k)(-qk, -qk/z; q)_n} \left(\frac{-qa}{z}\right)^n \beta_n(-a, -k, q)$$

$$- 2 \sum_{n=1}^{\infty} \frac{(1-k^2 q^{4n})(z^2; q^2)_n (q^2; q^2)_{n-1}}{(1-k^2)(q^2 k^2, q^2 k^2/z^2; q^2)_n} \left(\frac{q^2 a^2}{z^2}\right)^n \beta_n(a^2, k^2, q^2)$$

$$= \sum_{n=1}^{\infty} \frac{(z;q)_n (q;q)_{n-1}}{(qa, qa/z; q)_n} \left(\frac{qa}{z}\right)^n \alpha_n(a, k, q)$$

$$+ \sum_{n=1}^{\infty} \frac{(z;q)_n (q;q)_{n-1}}{(-qa, -qa/z; q)_n} \left(\frac{-qa}{z}\right)^n \alpha_n(-a, -k, q)$$

$$- 2 \sum_{n=1}^{\infty} \frac{(z^2; q^2)_n (q^2; q^2)_{n-1}}{(q^2 a^2, q^2 a^2/z^2; q^2)_n} \left(\frac{q^2 a^2}{z^2}\right)^n \alpha_n(a^2, k^2, q^2). \quad (13.91)$$

These identities require several preliminary lemmas.

Lemma 13.11. *If $(\alpha_n, \beta_n) = (\alpha_n(a, k), \beta_n(a, k))$ is a WP-Bailey pair, then subject to suitable convergence conditions,*

$$\sum_{n=1}^{\infty} \frac{(q\sqrt{k}, -q\sqrt{k}, z; q)_n (q;q)_{n-1}}{(\sqrt{k}, -\sqrt{k}, qk, qk/z; q)_n} \left(\frac{qa}{z}\right)^n \beta_n - \sum_{n=1}^{\infty} \frac{(z;q)_n (q;q)_{n-1}}{(qa, qa/z; q)_n} \left(\frac{qa}{z}\right)^n \alpha_n$$

$$= \sum_{n=1}^{\infty} \frac{(q\sqrt{k}, -q\sqrt{k}, k, z, k/a; q)_n}{(\sqrt{k}, -\sqrt{k}, qk, qk/z, qa; q)_n (1-q^n)} \left(\frac{qa}{z}\right)^n. \quad (13.92)$$

Proof. Rewrite (12.54) as

$$\sum_{n=1}^{\infty} \frac{(q\sqrt{k}, -q\sqrt{k}, z; q)_n (yq; q)_{n-1}}{(\sqrt{k}, -\sqrt{k}, qk/y, qk/z; q)_n} \left(\frac{qa}{yz}\right)^n \beta_n$$

$$- \frac{(qk, qk/yz, qa/y, qa/z; q)_\infty}{(qk/y, qk/z, qa, qa/yz; q)_\infty} \sum_{n=1}^\infty \frac{(z; q)_n (yq; q)_{n-1}}{(qa/y, qa/z; q)_n} \left(\frac{qa}{yz}\right)^n \alpha_n$$

$$= \frac{1}{1-y} \left(\frac{(qk, qk/yz, qa/y, qa/z; q)_\infty}{(qk/y, qk/z, qa, qa/yz; q)_\infty} - 1 \right).$$

From (5.6) it can be seen that

$$\frac{(qk, qk/yz, qa/y, qa/z; q)_\infty}{(qa, qa/yz, qk/y, qk/z; q)_\infty} = \sum_{n=0}^\infty \frac{(q\sqrt{k}, -q\sqrt{k}, k, y, z, k/a; q)_n}{(\sqrt{k}, -\sqrt{k}, qk/y, qk/z, qa, q; q)_n} \left(\frac{qa}{yz}\right)^n,$$

$$(13.93)$$

and the result now follows upon letting $y \to 1$.   □

Let $f(a, k, z, q)$ denote the series on the right side of (13.92):

$$f(a, k, z, q) := \sum_{n=1}^\infty \frac{(q\sqrt{k}, -q\sqrt{k}, k, z, k/a; q)_n}{(\sqrt{k}, -\sqrt{k}, qk, qk/z, qa; q)_n (1-q^n)} \left(\frac{qa}{z}\right)^n. \qquad (13.94)$$

Lemma 13.12. *If $|qa|, |qk| < |z|$ and if none of the denominators vanish, then*

$$f(a, k, z, q) = -f(k, a, z, q). \qquad (13.95)$$

Proof. Write

$$\frac{1}{1-y} \left(\frac{(qk, qk/yz, qa/y, qa/z; q)_\infty}{(qk/y, qk/z, qa, qa/yz; q)_\infty} - 1 \right)$$

$$= \frac{1}{1-y} \left(1 - \frac{(qk/y, qk/z, qa, qa/yz; q)_\infty}{(qk, qk/yz, qa/y, qa/z; q)_\infty} \right) \frac{(qk, qk/yz, qa/y, qa/z; q)_\infty}{(qk/y, qk/z, qa, qa/yz; q)_\infty},$$

use (5.6) on the infinite product inside the brackets, and again let $y \to 1$.   □

Lemma 13.13. *If $|qa| < |z|$ and if none of the denominators vanish, then*

$$f(a, k, z, q) = \sum_{n=1}^\infty \frac{kq^n}{1-kq^n} + \sum_{n=1}^\infty \frac{q^n a/z}{1-q^n a/z} - \sum_{n=1}^\infty \frac{aq^n}{1-aq^n} - \sum_{n=1}^\infty \frac{q^n k/z}{1-q^n k/z}.$$

$$(13.96)$$

Proof. Define

$$G(y) := \frac{(qk, qk/yz, qa/y, qa/z; q)_\infty}{(qk/y, qk/z, qa, qa/yz; q)_\infty}$$

so that

$$f(a,k,z,q) = \lim_{y \to 1} \frac{1}{1-y} \left(\frac{(qk, qk/yz, qa/y, qa/z; q)_\infty}{(qk/y, qk/z, qa, qa/yz; q)_\infty} - 1 \right)$$

$$= \lim_{y \to 1} \frac{G(y) - G(1)}{1-y} = -G'(1).$$

That $-G'(1)$ equals the right side of (13.96) follows by logarithmic differentiation.   □

Lemma 13.14. *If $|qa| < |z|$ and if none of the denominators vanish, then*

$$f(a,k,z,q) + f(-a,-k,z,q) = 2f(a^2, k^2, z^2, q^2). \tag{13.97}$$

Proof. This follows immediately from Lemma 13.13, after employing the elementary identity

$$\frac{x}{1-x} + \frac{(-x)}{1-(-x)} = \frac{2x^2}{1-x^2}.$$

                                                                                      □

    Theorem 13.16 is now easily proved.

Proof of Theorem 13.16. This follows from (13.94) and Lemma 13.14.   □

Lemma 13.15. *If $|qa| < |z| < |a/q|$ and if none of the denominators vanish, then*

$$f(a,k,z,q) - f\left(\frac{1}{a}, \frac{1}{k}, \frac{1}{z}, q\right) = \frac{(a-k)(1-1/z)(1-ak/z)}{(1-a)(1-k)(1-a/z)(1-k/z)}$$

$$+ \frac{z}{k} \frac{(z, q/z, k/a, qa/k, ak/z, qz/ak, q, q; q)_\infty}{(z/k, qk/z, z/a, qa/z, a, q/a, k, q/k; q)_\infty}. \tag{13.98}$$

Proof. This follows upon using (13.96) to express the left side of (13.98) in terms of Lambert series, and using (17.7) (as indicated there, the Lambert series combine to give a special case of Bailey's $_6\psi_6$ summation formula (8.1)) and some elementary algebra.   □

    Theorem 13.15 now follows.

Proof of Theorem 13.15. Take the identity at (13.92) and replace a with $1/a$, k with $1/k$ and z with $1/z$. Then subtract the resulting identity from the original identity, use (13.98) to replace two of the sums on the right, and (13.90) follows.   □

The following identities for standard Bailey pairs follow upon letting $k \to 0$ in Lemma 13.11 and Lemmas 13.12 and 13.13.

Corollary 13.5. *If (α_n, β_n) is a Bailey pair with respect to a, then subject to suitable convergence conditions,*

$$\sum_{n=1}^{\infty} (z;q)_n (q;q)_{n-1} \left(\frac{qa}{z}\right)^n \beta_n - \sum_{n=1}^{\infty} \frac{(z;q)_n (q;q)_{n-1}}{(qa, qa/z; q)_n} \left(\frac{qa}{z}\right)^n \alpha_n = f_1(a, z, q),$$

(13.99)

where

$$f_1(a, z, q) = -\sum_{n=1}^{\infty} \frac{(q\sqrt{a}, -q\sqrt{a}, a, z; q)_n q^{n(n+1)/2}}{(\sqrt{a}, -\sqrt{a}, qa, qa/z; q)_n (1 - q^n)} \left(\frac{-a}{z}\right)^n$$

$$= \sum_{n=1}^{\infty} \frac{(z;q)_n}{(qa;q)_n (1 - q^n)} \left(\frac{qa}{z}\right)^n$$

$$= \sum_{n=1}^{\infty} \frac{aq^n/z}{1 - aq^n/z} - \sum_{n=1}^{\infty} \frac{aq^n}{1 - aq^n}.$$

Upon letting $z \to \infty$ in the corollary above, the following identities are obtained.

Corollary 13.6. *If (α_n, β_n) is a Bailey pair with respect to a, then subject to suitable convergence conditions,*

$$\sum_{n=1}^{\infty} (q;q)_{n-1} (-a)^n q^{n(n+1)/2} \beta_n - \sum_{n=1}^{\infty} \frac{(q;q)_{n-1} (-a)^n q^{n(n+1)/2}}{(qa;q)_n} \alpha_n = f_2(a, q),$$

(13.100)

where

$$f_2(a, q) = -\sum_{n=1}^{\infty} \frac{(1 - aq^{2n}) q^{n^2} a^n}{(1 - aq^n)(1 - q^n)}$$

(13.101)

$$= \sum_{n=1}^{\infty} \frac{q^{n(n+1)/2} (-a)^n}{(qa;q)_n (1 - q^n)}$$

$$= -\sum_{n=1}^{\infty} \frac{aq^n}{1 - aq^n}.$$

Another useful special case (partly because of the form of the Lambert series in the third expression for $f_3(a, q)$ below) is the following.

Corollary 13.7. *If* (α_n, β_n) *is a Bailey pair with respect to* a, *then subject to suitable convergence conditions,*

$$\sum_{n=1}^{\infty} (q^2; q^2)_{n-1} (-qa)^n \beta_n - \sum_{n=1}^{\infty} \frac{(q^2; q^2)_{n-1} (-qa)^n}{(q^2 a^2; q^2)_n} \alpha_n = f_3(a, q), \quad (13.102)$$

where

$$f_3(a, q) = -\sum_{n=1}^{\infty} \frac{(q\sqrt{a}, -q\sqrt{a}, a; q)_n (-q; q)_{n-1} q^{n(n+1)/2} a^n}{(\sqrt{a}, -\sqrt{a}; q)_n (q^2 a^2; q^2)_n (1 - q^n)}$$

$$= \sum_{n=1}^{\infty} \frac{(-q; q)_{n-1} (-qa)^n}{(qa; q)_n (1 - q^n)}$$

$$= -\sum_{n=1}^{\infty} \frac{aq^n}{1 - a^2 q^{2n}}.$$

Proof. Let $z \to -1$ in (13.99) and simplify. □

For later use when discussing Lambert series, and for ease of notation, define

$$L_a(q) := \sum_{n=1}^{\infty} \frac{aq^n}{1 - aq^n}.$$

Several alternative representations for $L_a(q)$ are to be found in Corollary 13.6. Two additional representations are given next.

Corollary 13.8.

$$L_a(q) = \frac{-1}{(aq; q)_\infty} \sum_{n=1}^{\infty} \frac{n(-a)^n q^{n(n+1)/2}}{(q; q)_n}, \quad (13.103)$$

$$= (aq; q)_\infty \sum_{n=1}^{\infty} \frac{na^n q^n}{(q; q)_n}.$$

Proof. Let $k \to 0$ and $z \to \infty$ in (12.54), and rearrange to get

$$\sum_{n=1}^{\infty} (yq; q)_{n-1} \left(\frac{-a}{y} \right)^n q^{n(n+1)/2} \beta_n$$

$$- \frac{(qa/y; q)_\infty}{(qa; q)_\infty} \sum_{n=1}^{\infty} \frac{(yq; q)_{n-1}}{(qa/y; q)_n} \left(\frac{-a}{y} \right)^n q^{n(n+1)/2} \alpha_n$$

$$= \frac{1}{1 - y} \left(\frac{(qa/y; q)_\infty}{(qa; q)_\infty} - 1 \right),$$

where (α_n, β_n) is a Bailey pair with respect to a. The result now follows, from Corollary 13.6, upon using the special cases of the q-binomial theorem (3.3) and (3.4) to expand in the infinite products inside the parentheses below as infinite series and then computing the limits:

$$L_a(q) = -\lim_{y \to 1} \frac{1}{1-y} \frac{((qa/y; q)_\infty - (qa; q)_\infty)}{(qa; q)_\infty}$$

$$= -\lim_{y \to 1} \frac{(qa/y; q)_\infty}{1-y} \left(\frac{1}{(qa; q)_\infty} - \frac{1}{(qa/y; q)_\infty} \right).$$

$\square$

Exercises

13.1 Prove the relation at (13.4).

13.2 (i) Invert the relations at (13.3) and (13.4), to show that if $(\alpha_n(a, q), \beta_n(a, q))$ is a Bailey pair with respect to a, then $(\alpha'_n(aq, q), \beta'_n(aq, q))$ and $(\alpha^*_n(aq, q), \beta^*_n(aq, q))$ are Bailey pairs with respect to aq, where

$$\alpha'_n(aq, q) = \frac{1 - aq^{2n+1}}{1 - aq} \sum_{j=0}^{n} a^{n-j} q^{n^2 - j^2} \alpha_j(a, q), \qquad (13.104)$$

$$\beta'_n(aq, q) = \beta_n(a, q).$$

$$\alpha^*_n(aq, q) = \frac{1 - aq^{2n+1}}{(1 - aq)q^n} \sum_{j=0}^{n} \alpha_j(a, q), \qquad (13.105)$$

$$\beta^*_n(aq, q) = q^{-n} \beta_n(a, q).$$

(ii) By starting with the "trivial" Bailey pair (with respect to $a = 1$) defined by $\alpha_n(1, q) = \delta_{n,0}$, use the relations above to derive the following Bailey pairs (which were not stated by Slater [236, 237]), with respect to $a = q$,

$$\alpha'_n(q, q) = \frac{1 - q^{2n+1}}{1 - q} q^{n^2}, \qquad (13.106)$$

$$\beta'_n(q, q) = \frac{1}{(q, q; q)_n}.$$

$$\alpha^*_n(q, q) = \frac{1 - q^{2n+1}}{(1 - q)q^n}, \qquad (13.107)$$

$$\beta_n^*(q,q) = \frac{1}{q^n(q,q;q)_n}.$$

13.3 By mimicking the proof of (13.5) or otherwise, prove the relation at (13.6).

13.4 Complete the proof of Corollary 13.2.

13.5 Prove the claim in Lemma 13.3.

13.6 Specialize the parameters in (13.23) to derive Ramanujan's $_1\psi_1$ summation formula (7.1) and Bailey's $_6\psi_6$ summation formula (8.1).

13.7 Fill in the details in the proof of Lemma 13.4.

13.8 Fill in the details in the proof of Lemma 13.5.

13.9 Fill in the details in the proof of Proposition 13.2.

13.10 Prove the transformations in Proposition 13.3.

13.11 Prove the transformations in Proposition 13.4.

13.12 Prove the multi-sum identity at (13.39).

13.13 By starting with the identity at (13.45) and using Singh's transformation (13.42) in combination with the q-Pfaff–Saalschütz sum (4.5), or otherwise, prove the summation formula at (13.46).

13.14 By employing (13.46) or otherwise, prove the Bailey chain of Berkovich and Warnaar at (13.50).

13.15 (Stanton [245]) By considering the product (E3)(S1), derive the Slater identity ([237, (53)])

$$\sum_{j=0}^{\infty} \frac{q^{4j^2}(q;q^2)_{2j}}{(q^4;q^4)_{2j}} = \frac{(q^5,q^7,q^{12};q^{12})_\infty}{(q^4;q^4)_\infty}. \qquad (13.108)$$

13.16 (Stanton [245]) By considering the product (T1)(D3), derive the identity

$$\lim_{n\to\infty} \sum_{j=0}^{n} \frac{(q^{-3-6j},q^{6j+9};q^6)_{n-j}(q^4,q^8;q^{12})_j q^{-3j^2}}{(q^{12j+6},q^{12};q^{12})_{n-j}(q^6;q^6)_{2j}} = \frac{(q,q^5;q^6)_\infty}{(q^6;q^6)_\infty}. \qquad (13.109)$$

13.17 (Stanton [245]) Compute the matrix $(D1)(S1)^{k-1}$ ($k \geq 2$) and hence derive the Bressoud identity

$$\sum_{s_{k-1}\geq s_{k-2}\geq\cdots s_1\geq 0} \frac{q^{s_{k-1}^2+\cdots+s_1^2+s_1}}{(q;q)_{s_{k-1}-s_{k-2}}(q;q)_{s_{k-2}-s_{k-3}}\cdots(q;q)_{s_2-s_1}(q^2;q^2)_{s_1}} = \frac{(q^{k-1},q^{k+1},q^{2k};q^{2k})_\infty}{(q;q)_\infty}. \quad (13.110)$$

13.18 (Stanton [245]) Compute the matrix $(D1)^k(S1)^i$ $(k+i \geq 2)$ and hence derive the identity

$$\sum_{s_{k+i-1}\geq s_{k+i-2}\geq\cdots s_1\geq s_0=0} \frac{q^{s_{k+i-1}^2+\cdots+s_k^2}}{\prod_{j=k}^{k+i-2}(q;q)_{s_{j+1}-s_j}}$$

$$\times \prod_{j=0}^{k-1} \frac{q^{2^j(s_{k-j}-s_{k-j-1})}(-q^{2^j};q^{2^j})_{2s_{k-j-1}}}{(q^{2^{j+1}};q^{2^{j+1}})_{s_{k-j}-s_{k-j-1}}} = \frac{(q^i,q^{2k+i},q^{2k+2i};q^{2k+2i})_\infty}{(q;q)_\infty}.$$

$$(13.111)$$

13.19 (Stanton [245]) Compute the matrix $(T1)(T1)(S1)$ and hence derive the double-sum identity

$$\sum_{s_2\geq s_1\geq 0} \frac{q^{9s_2^2+3s_1^2}(q^3;q^3)_{3s_2-s_1}(q;q)_{3s_1}}{(q^9;q^9)_{2s_2}(q^9;q^9)_{s_2-s_1}(q^3;q^3)_{2s_1}(q^3;q^3)_{s_1}} = \frac{(q^{13},q^{14},q^{27};q^{27})_\infty}{(q^9;q^9)_\infty}.$$

$$(13.112)$$

13.20 (Warnaar [255, Theorem 1.5]) Mimic the proof of Lemma 13.8 to prove that

$$1 + \sum_{n=1}^\infty (-1)^n q^{n(n-1)/2}(a^n + b^n) = (a,b,q;q)_\infty \sum_{n=0}^\infty \frac{(ab/q;q)_{2n}q^n}{(a,b,ab,q;q)_n}. \quad (13.113)$$

Remark: Note that this identity is an extension of the Jacobi triple product identity (6.1), in that the latter identity is the special case $b = q/a$ of (13.113).

13.21 (Andrews and Warnaar [49, Theorem 7]) (i) By starting with the standard Bailey pair at (11.25), derive the symmetric bilateral Bailey pair

$$\alpha_n(1,q^2) = z^n q^{n^2} \qquad \beta_n(1,q^2) = \frac{(-qz,-q/z;q^2)_n}{(q^2;q^2)_{2n}}, \quad (13.114)$$

and hence derive the identity

$$\sum_{n=0}^{\infty} \frac{(-qz, -q/z; q^2)_n q^n}{(-q; q)_{2n+1}} = \sum_{j=0}^{\infty} \frac{z^{-j} - z^{j+1}}{1 - z} q^{j(j+1)}. \tag{13.115}$$

(ii) Hence, by specializing z to have the values 1 and q^2, derive the identities

$$\sum_{n=0}^{\infty} \frac{(-q; q^2)_n q^n}{(-q^2; q^2)_n (1 + q^{2n+1})} = \sum_{j=0}^{\infty} (2j + 1) q^{j(j+1)}. \tag{13.116}$$

$$(1 - q) \sum_{n=0}^{\infty} \frac{(-q; q^2)_n q^n}{(-q^2; q^2)_{n+1}} = 1. \tag{13.117}$$

13.22 By using the Bailey pair at (V.1.1), the conjugate pair at (13.76), the summation formula at (5.21) and finally some relabelling of parameters, prove the identity

$$\sum_{r=0}^{\infty} \frac{1 - aq^{2r}}{1 - a} \frac{(a, b, c, d, e; q)_r}{(aq/b, aq/c, aq/d, aq/e, q; q)_r} \left(\frac{-a^2}{bcde} \right)^r q^{r(r+3)/2}$$

$$= \frac{(aq, aq/bc; q)_\infty}{(aq/b, aq/c; q)_\infty} \sum_{r=0}^{\infty} \frac{(b, c, aq/de; q)_r}{(aq/d, aq/e, q; q)_r} \left(\frac{aq}{bc} \right)^r. \tag{13.118}$$

Remark: This identity is equivalent to the case $d \to \infty$ of the $_8\phi_7$ transformation at (5.12).

13.23 (i) By starting with the Bailey chain of Bressoud, Ismail and Stanton at (13.47), derive the conjugate Bailey pair $(\delta_r(a, q), \gamma_r(a, q))$ (which holds for each positive integer N), where

$$\delta_r(a^2, q^2) = \frac{(Bq; q)_N}{(B, -aq/B; q)_N} \frac{(-aq; q)_{2r} (B^2; q^2)_r (1/B; q)_{N-2r} (-1)^r}{(q^2; q^2)_{N-r} B^{2r} q^{r^2}}, \tag{13.119}$$

$$\gamma_r(a^2, q^2) = \frac{(-B; q)_r B^{-r} q^{-r(r-1)/2}}{(-aq/B; q)_r (q; q)_{N-r} (aq; q)_{N+r}}. \tag{13.120}$$

(ii) Hence employ the Bailey pair at (10.16) to derive the transformation

(which holds for each non-negative integer N)

$$\sum_{r=0}^{N}\frac{1-a^2q^4}{1-a^2}\frac{(a^2,b,c;q^2)_r(-B,q^{-N};q)_r}{(a^2q^2/b,a^2q^2/c,q^2;q^2)_r(-aq/B,aq^{N+1};q)_r}\left(\frac{a^2q^{2+N}}{bcB}\right)^r$$

$$=\frac{(Bq,aq,1/B;q)_N}{(B,-aq/B,-q;q)_N}\sum_{r=0}^{N}\frac{(a^2q^2/bc,B^2,-aq,-aq^2,q^{-2N};q^2)_rq^{2r}}{(a^2q^2/b,a^2q^2/c,Bq^{1-N},Bq^{2-N},q^2;q^2)_r}.$$
$$\tag{13.121}$$

Remark: The identity (13.121) may be regarded as a transformation between a $_{12}\phi_{11}$ series with base q and $_5\phi_4$ series with base q^2.

13.24 (i) (Andrews [22]) Use the definition at (10.5) to prove that if $(\alpha_n(a,q),\beta_n(a,q))$ is a Bailey pair relative to a, then $(\alpha_n^*(a,q),\beta_n^*(a,q))$ (called *the dual* of $(\alpha_n(a,q),\beta_n(a,q))$) is also a Bailey pair relative to a, where

$$\alpha_n^*(a,q)=a^nq^{n^2}\alpha_n(1/a,1/q),$$
$$\beta_n^*(a,q)=a^{-n}q^{-n^2-n}\beta_n(1/a,1/q).$$

(ii) Show from the definition at (12.3) that if $(\alpha_n(a,k,q),\beta_n(a,k,q))$ is a WP-Bailey pair, then so is $(\alpha_n^*(a,k,q),\beta_n^*(a,k,q))$ (also called *the dual* of $(\alpha_n(a,k,q),\beta_n(a,k,q))$), where

$$\alpha_n^*(a,k,q)=\alpha_n(1/a,1/k,1/q),$$
$$\beta_n^*(a,k,q)=\left(\frac{k}{aq}\right)^{2n}\beta_n(1/a,1/k,1/q).$$

13.25 Show that if $(\alpha_n(a,q),\beta_n(a,q))$ is a Bailey pair with respect to a, then subject to suitable convergence conditions,

$$\sum_{n=1}^{\infty}(q,q;q)_{n-1}(qa)^n\beta_n(a,q)-\sum_{n=1}^{\infty}\frac{(q,q;q)_{n-1}}{(qa,qa;q)_n}(qa)^n\alpha_n(a,q)=f_4(a,q),$$

where $f_4(a,q)=\displaystyle\sum_{n=1}^{\infty}\frac{(q;q)_{n-1}}{(qa;q)_n(1-q^n)}(qa)^n$

$$=-\sum_{n=1}^{\infty}\frac{(q\sqrt{a},-q\sqrt{a},a;q)_n(q;q)_{n-1}}{(\sqrt{a},-\sqrt{a},qa,qa;q)_n(1-q^n)}(-a)^n\,q^{n(n+1)/2}$$

$$=\sum_{n=1}^{\infty}\frac{q^na}{(1-q^na)^2}=\sum_{n=1}^{\infty}\frac{na^nq^n}{1-q^n}.$$

Chapter 14

Gaussian Polynomials

The Gaussian polynomials, or q-binomial coefficientss, are defined by

$$g(n,r;q) := \begin{bmatrix} n \\ r \end{bmatrix} := \begin{bmatrix} n \\ r \end{bmatrix}_q := \begin{cases} \dfrac{(q;q)_n}{(q;q)_r(q;q)_{n-r}}, & 0 \le r \le n, \\ \\ 0, & \text{otherwise.} \end{cases} \qquad (14.1)$$

We are adopting the notation $g(n,r;q)$ of Andrews [30, p. 33]. It is clear from (14.1) that

$$g(n,r;q) = \begin{bmatrix} n \\ r \end{bmatrix} = \begin{bmatrix} n \\ n-r \end{bmatrix} = g(n,n-r;q). \qquad (14.2)$$

If the expression for $g(n,r;q)$ is simplified, in the case $0 \le r \le n$, it follows that

$$g(n,r;q) = \frac{(1-q^{n-r+1})(1-q^{n-r+2})\dots(1-q^{n-1})(1-q^n)}{(1-q)(1-q^2)\dots(1-q^{r-1})(1-q^r)}. \qquad (14.3)$$

It is not clear that these polynomials, possibly first studied by Gauss, actually are polynomials, but this will follow easily from the recurrence relations in Theorem 14.1. If we examine any particular Gaussian polynomial, for example

$$\begin{bmatrix} 7 \\ 3 \end{bmatrix} = 1 + q + 2q^2 + 3q^3 + 4q^4 + 4q^5 + 5q^6 + 4q^7 + 4q^8 + 3q^9 + 2q^{10} + q^{11} + q^{12},$$

a number of interesting properties are suggested. Firstly, all of the coefficients in each of the Gaussian polynomials are positive (which follows from the partition interpretation of $g(n+r, r; q)$ discussed below). Secondly, $g(n, r; q)$ is a *reciprocal polynomial*, in that if its coefficients are labelled by

$$g(n, r; q) =: \sum_{i=0}^{r(n-r)} a_i q^i,\qquad(14.4)$$

then

$$a_i = a_{r(n-r)-i}, \quad \text{for } i = 0, 1, 2, \dots, r(n-r).\qquad(14.5)$$

This follows easily from (14.11), and is left as an exercise (see Q14.2) (that $g(n, r; q)$ has degree $n(n-r)$ follows easily from (14.3)).

The sequence a_i defined by (14.4) is also *unimodal*, in that

$$a_0 \leq a_1 \leq a_2 \leq \cdots \leq a_{m-1} \leq a_m \geq a_{m+1} \geq a_{m+2} \geq \dots a_{r(n-r)-1} \geq a_{r(n-r)},$$

where $m = \lfloor r(n-r)/2 \rfloor$.

The q-binomial coefficients has a number of interpretations. If

$$\begin{bmatrix} n + r \\ r \end{bmatrix} = \sum_i a_{n,r}(i) q^i,\qquad(14.6)$$

then it is well known that $a_{n,r}(i)$ equals the number of partitions of the integer i into at most r parts, each $\leq n$. A proof of this statement may be found, for example, in the book Andrews [30, Theorem 3.1, page 33]), and we will reproduce Andrews proof in Theorem 15.7. If q is a power of some prime p, then it is also well known that $g(n, r; q)$ counts the number of r-dimensional subspaces of the vector space of dimension n over the finite field $GF(q)$ of q elements.

It has already been seen that the q-binomial coefficientss appear in finite forms of the q-binomial theorem at (3.3) and (3.4). It is not difficult to see that letting $q \to 1$ in these identities results in special cases of the (ordinary) binomial theorem, and likewise that letting $q \to 1$ in the q-binomial coefficients results in the corresponding regular binomial coefficient (see (14.12) below). Since there are a multitude of identities involving the standard binomial coefficient, it is of some interest to derive basic hypergeometric identities involving the Gaussian polynomials that reduce to known identities involving the corresponding regular binomial coefficients upon letting $q \to 1$. Two examples of such identities may be found in Q3.7, and some additional examples are given below.

The q-binomial coefficientss appear in the finite form of the Jacobi triple product identity stated at (6.2). They also appear in polynomial version

of identities of Rogers–Ramanujan–Slater type. See, for example, the polynomial versions of the celebrated Rogers–Ramanujan identities at (11.15)
and (11.37). Note that the theory of Bailey pairs leads similarly to finite or
polynomial version of all of the identities on the Slater list [236, 237]. Other
methods may lead to other polynomial versions of these identities that involve q-binomial coefficientss — see in particular the paper of Sills [233],
where the method of q-difference equations described by Andrews in [24]
was further developed to produce finite version all the identities on the
Slater list. The finite versions of the Rogers–Ramanujan identities found in
Sills paper [233] are

$$\sum_{j \geq 0} q^{j(j+1)} \begin{bmatrix} n-j \\ j \end{bmatrix}_q = \sum_{j=-\infty}^{\infty} (-1)^j q^{j(5j+3)} \begin{bmatrix} n+1 \\ \lfloor (n+5j+3)/2 \rfloor \end{bmatrix}_q, \qquad (14.7)$$

$$\sum_{j \geq 0} q^{j^2} \begin{bmatrix} n-j \\ j \end{bmatrix}_q = \sum_{j=-\infty}^{\infty} (-1)^j q^{j(5j+1)} \begin{bmatrix} n+1 \\ \lfloor (n+5j+1)/2 \rfloor \end{bmatrix}_q. \qquad (14.8)$$

Note that all sums are indeed finite, that the left sides of (14.7) and (14.8)
were first stated by MacMahon [188], that the right sides were first given by
Schur [226], and that alternative right sides were found by Andrews [27].

Theorem 14.1. *For each positive integer n and each non-negative integer
r with $0 \leq r \leq n$,*

$$\begin{bmatrix} n \\ r \end{bmatrix} = q^r \begin{bmatrix} n-1 \\ r \end{bmatrix} + \begin{bmatrix} n-1 \\ r-1 \end{bmatrix}; \qquad (14.9)$$

$$\begin{bmatrix} n \\ r \end{bmatrix} = \begin{bmatrix} n-1 \\ r \end{bmatrix} + q^{n-r} \begin{bmatrix} n-1 \\ r-1 \end{bmatrix}, \qquad (14.10)$$

$$\begin{bmatrix} n \\ r \end{bmatrix}_q = q^{r(n-r)} \begin{bmatrix} n \\ r \end{bmatrix}_{1/q}, \qquad (14.11)$$

$$\lim_{q \to 1} \begin{bmatrix} n \\ r \end{bmatrix}_q = \binom{n}{r}. \qquad (14.12)$$

Proof. Each of the above statements follows in a straightforward manner
from the definitions and elementary algebra, and so are left as exercises (see
Q14.1). Note for the proof of (14.12) that

$$\lim_{q \to 1} \frac{1 - q^j}{1 - q} = j,$$

by L'Hospital's theorem. $\square$

Some elegant identities involving the Gaussian polynomials may be found in the text and problems of Chapter 3, and some additional properties are developed in the current chapter.

14.1 q-Trinomial Coefficients

The trinomial coefficients $\begin{pmatrix} m \\ A \end{pmatrix}_2$ are defined by the expansion

$$(1 + x + x^{-1})^m = \sum_{A=-m}^{m} \begin{pmatrix} m \\ A \end{pmatrix}_2 x^A. \tag{14.13}$$

By applying the binomial theorem twice to the two expressions

$$(1 + x + x^{-1})^m = (x^{-1} + (1 + x))^m = (-1 + (x(1 + x^{-1})^2))^m, \tag{14.14}$$

it is easy to derive two explicit representations for the trinomial coefficients:

$$\begin{pmatrix} m \\ A \end{pmatrix}_2 = \sum_{j \geq 0} \frac{m!}{j!(j+A)!(m-2j-A)!} = \sum_{j \geq 0} (-1)^j \begin{pmatrix} m \\ j \end{pmatrix} \begin{pmatrix} 2m - 2j \\ m - j - A \end{pmatrix}. \tag{14.15}$$

Andrews and Baxter [35, p. 299, Eqs. (2.7)–(2.12)] introduced several q-analogs of the trinomial coefficients, each of which reverts back to one of the explicit forms for the standard trinomial coefficient at (14.15), as $q \to 1$.

$$\begin{pmatrix} m; B; q \\ A \end{pmatrix}_2 := \sum_{j \geq 0} \frac{q^{j(j+B)}(q;q)_m}{(q;q)_j(q;q)_{j+A}(q;q)_{m-2j-A}}; \tag{14.16}$$

$$T_0(m, A, q) := \sum_{j=0}^{m} (-1)^j \begin{bmatrix} m \\ j \end{bmatrix}_{q^2} \begin{bmatrix} 2m - 2j \\ m - A - j \end{bmatrix}; \tag{14.17}$$

$$T_1(m, A, q) := \sum_{j=0}^{m} (-q)^j \begin{bmatrix} m \\ j \end{bmatrix}_{q^2} \begin{bmatrix} 2m - 2j \\ m - A - j \end{bmatrix}; \tag{14.18}$$

$$t_0(m, A, q) := \sum_{j=0}^{m} (-1)^j q^{j^2} \begin{bmatrix} m \\ j \end{bmatrix}_{q^2} \begin{bmatrix} 2m - 2j \\ m - A - j \end{bmatrix}; \tag{14.19}$$

$$t_1(m, A, q) := \sum_{j=0}^{m} (-1)^j q^{j^2-j} \begin{bmatrix} m \\ j \end{bmatrix}_{q^2} \begin{bmatrix} 2m - 2j \\ m - A - j \end{bmatrix}; \tag{14.20}$$

$$\tau_0(m, A, q) := \sum_{j=0}^{m} (-1)^j q^{mj - j(j-1)/2} \begin{bmatrix} m \\ j \end{bmatrix} \begin{bmatrix} 2m - 2j \\ m - A - j \end{bmatrix}. \tag{14.21}$$

As Warnaar states in [253], since these q-trinomial coefficients were defined by Andrews and Baxter [35], they have found numerous applications in combinatorics, integer partitions, and statistical mechanics. These include q-trinomial proofs of a famous partition theorem of Schur and of a (then) conjecture of Capparelli on certain restricted partition functions (Andrews [29]). Another application was a q-trinomial analogue of Bailey's lemma, stated by Andrews and Berkovich [36]. Several other applications by other authors are stated in Warnaar's paper [253], in which Warnaar also states and proves several additional applications.

For space reasons, here we are restricted to recalling some of the elementary properties of some of the q-trinomial coefficients (14.16)–(14.21).

Observe first that each of (14.17)–(14.21) is symmetric in A and $-A$. For (14.16), it is straightforward to show (see Q14.4) that

$$\binom{m; B; q}{-A}_2 = q^{A(A+B)} \binom{m; B + 2A; q}{A}_2. \tag{14.22}$$

It is an easy consequence of the definition at (14.14) that

$$\binom{L}{A}_2 = \binom{L-1}{A-1}_2 + \binom{L-1}{A}_2 + \binom{L-1}{A+1}_2. \tag{14.23}$$

There are numerous recurrences for the q-analogues. Some of the more elementary to be found in [35] are the following (see Q14.5).

$$T_1(m, A, q) - q^{m-A} T_0(m, A, q) - T_1(m, A+1, q)$$
$$+ q^{m+A+1} T_0(m, A+1, q) = 0, \tag{14.24}$$

$$\binom{m; B-1; q}{A}_2 = \binom{m; B; q}{A}_2 + q^B (1 - q^m) \binom{m-1; B+1; q}{A+1}_2, \tag{14.25}$$

$$\binom{m; B; q}{A}_2 = q^{m-A} \binom{m; B-2; q}{A}_2 + (1 - q^m) \binom{m-1; B; q}{A}_2. \tag{14.26}$$

Andrews and Baxter [35] also state a number of relations between the various q-trinomial coefficients, the simplest being (see Q14.6)

$$T_i(m, A, q^{-1}) = q^{A^2 - m^2} t_i(m, A, q), \qquad i = 0, 1. \tag{14.27}$$

The authors in [35] also consider the limit as $m \to \infty$ for the q-trinomial coefficients, using special cases of the q-binomial theorem (3.1) or Heine's

q-Gauss summation formula (4.8).

$$\lim_{m\to\infty} \binom{m; A; q}{A} = \frac{1}{(q;q)_\infty}; \tag{14.28}$$

$$\lim_{m\to\infty} \binom{m; A-1; q}{A} = \frac{1+q^A}{(q;q)_\infty}; \tag{14.29}$$

$$\lim_{m\to\infty} t_0(m, A, q) = \frac{1}{(q^2; q^2)_\infty}; \tag{14.30}$$

$$\lim_{m\to\infty} T_1(m, A, q) = \frac{(-q^2; q^2)_\infty}{(q^2; q^2)_\infty}. \tag{14.31}$$

Further relations for the q-trinomial coefficients may be found in [35] or in the paper of Sills [233, Section 1.2].

Exercises

14.1 Prove each of the statements in Theorem 14.1.

14.2 If the integers a_i are defined as at (14.4), prove the assertion at (14.5).

14.3 By using the q-Pfaff–Saalschütz sum (4.5) (the method Carlitz [91] used to prove (14.32)) or otherwise, prove the identity

$$\sum_{k=0}^{\min(a,b)} \begin{bmatrix} x+y+k \\ k \end{bmatrix} \begin{bmatrix} x+a-b \\ a-k \end{bmatrix} \begin{bmatrix} y+b-a \\ b-k \end{bmatrix} q^{(a-k)(b-k)} = \begin{bmatrix} x+a \\ a \end{bmatrix} \begin{bmatrix} y+b \\ b \end{bmatrix}, \tag{14.32}$$

which was proved by Gould [133] and is also found in Stanley [244].

14.4 Prove the identity at (14.22).

14.5 Prove the recurrence relations at (14.23) and (14.24)–(14.26).

14.6 Prove the transformations at (14.27).

14.7 Prove the limits at (14.28)–(14.31).

14.8 Prove that for positive integers $p \le m \le n$,

$$\begin{bmatrix} n \\ m \end{bmatrix} \begin{bmatrix} m \\ p \end{bmatrix} = \begin{bmatrix} n \\ p \end{bmatrix} \begin{bmatrix} n-p \\ m-p \end{bmatrix}. \tag{14.33}$$

14.9 Prove that for integers $0 \le m \le n$,

$$\sum_{k=0}^{n} \begin{bmatrix} n \\ k \end{bmatrix} \begin{bmatrix} k \\ m \end{bmatrix} (-1)^k = \begin{cases} 0 & n-m \text{ odd,} \\ (-1)^m \begin{bmatrix} n \\ m \end{bmatrix} (q; q^2)_{(n-m)/2} & n-m \text{ even.} \end{cases} \tag{14.34}$$

14.10 (Comtet [107]) (i) Prove the orthogonality relations

$$\sum_{j=i}^{n} (-1)^{n-j} q^{\binom{n-j}{2}} \begin{bmatrix} n \\ j \end{bmatrix} \begin{bmatrix} j \\ i \end{bmatrix} = \sum_{j=i}^{n} (-1)^{j-i} q^{\binom{j-i}{2}} \begin{bmatrix} n \\ j \end{bmatrix} \begin{bmatrix} j \\ i \end{bmatrix} = \delta_{n,i}. \tag{14.35}$$

(ii) Hence derive the inverse relations

$$f_n = \sum_{k=0}^{n} \begin{bmatrix} n \\ k \end{bmatrix} g_k \iff g_n = \sum_{k=0}^{n} (-1)^{n-k} q^{\binom{n-k}{2}} \begin{bmatrix} n \\ k \end{bmatrix} f_k. \tag{14.36}$$

14.11 By employing Q14.10 (ii), or otherwise, prove each of the following statements for each non-negative integer n.

$$\sum_{k=0}^{n} (-1)^k q^{k(k+1)/2-kn} \begin{bmatrix} n \\ k \end{bmatrix} (z; q)_k = z^n; \tag{14.37}$$

$$\sum_{k=0}^{\lfloor n/2 \rfloor} q^{k(2k+1)-2kn} \begin{bmatrix} n \\ 2k \end{bmatrix} (q; q^2)_k = q^{-n(n-1)/2}. \tag{14.38}$$

14.12 Prove that

$$\begin{pmatrix} m; A; q \\ A \end{pmatrix}_2 = \tau_0(m, A, q). \tag{14.39}$$

Chapter 15

Bijective Proofs of Basic Hypergeometric Identities

In this chapter, some combinatorial methods of proving basic hypergeometric identities will be examined. The basic idea is the following. Suppose $A(n)$ counts the number of (restricted) partitions (see below for examples various restricted partition counting functions) of a certain type, and $B(n)$ counts the number of (restricted) partitions of a second type. Suppose it is known that

$$\sum_{n=0}^{\infty} A(n)q^n = A(q), \qquad \sum_{n=0}^{\infty} B(n)q^n = B(q), \qquad (15.1)$$

where $A(q)$ and $B(q)$ are some basic hypergeometric series or products. Then

$$A(n) = B(n), \forall\, n \in \mathbb{N} \iff A(q) = B(q). \qquad (15.2)$$

Since this book is primarily concerned with describing methods for proving basic hypergeometric identities, we will primarily concerned with the problem of showing $A(q) = B(q)$, by proving $A(n) = B(n), \forall\, n \in \mathbb{N}$. However, we will occasionally go in the opposite direction, by interpreting a basic hypergeometric identity in terms of the equality to two restricted partition counting functions. There is an extensive literature containing bijective proofs of basic hypergeometric identities, but in the present chapter it is possible to present only some general principles and techniques. For a comprehensive survey of partition bijections, see Pak's wonderful paper [208].

We next describe some basic concepts. A *partition* of a positive integer n is an expression of n as a sum of positive integers, where the order does not matter. By convention, these sums are written in non-increasing order from left to right. For example, the partitions of 4 are 4, $3 + 1$, $2 + 2$, $2 + 1 + 1$ and $1 + 1 + 1 + 1$. The summands of a partition are termed the *parts* of the partition, so that, for example, the parts of the partition $2 + 1 + 1$ are 2, 1 and 1. The ordinary or unrestricted partition function $p(n)$ is the *number* of partitions of n, so that $p(4) = 5$.

There is a simple generating function for the sequence $p(n)$ (where we define $p(0) := 1$):

$$\sum_{n=0}^{\infty} p(n)q^n = \frac{1}{(q;q)_\infty}. \qquad (15.3)$$

Informally, this follows from expanding each factor in the infinite product on the right side of (15.3) as a geometric series.

$$\begin{aligned}
\frac{1}{(q;q)_\infty} &= \frac{1}{\prod_{k=1}^{\infty}(1 - q^k)} \\
&= \prod_{k=1}^{\infty}(1 + q^{1.k} + q^{2.k} + q^{3.k} + \cdots + + q^{j.k} + \dots) \\
&= \sum_{n_1=0}^{\infty}\sum_{n_2=0}^{\infty}\sum_{n_3=0}^{\infty}\cdots\sum_{n_k=0}^{\infty}\dots q^{n_1.1 + n_2.2 + n_3.3 + \cdots + n_k.k + \cdots}. \qquad (15.4)
\end{aligned}$$

For each positive integer n, each occurrence of the term q^n will be of the form $q^{n_1.1 + n_2.2 + n_3.3 + \cdots + n_k.k}$, for some non-negative integers $n_1, n_2, \dots, n_k$, with $k \le n$. This exponent may be considered as a partition of n consisting of n_1 parts of size 1, n_2 parts of size 2, $\dots$, n_k parts of size k. Each occurrence of q^n arises in this way from a partition of n, and each partition of n occurs as an exponent in some term q^n. Hence, after collecting all powers of q^n together, the coefficient of q^n in the expanded multi-series at (15.4) is $p(n)$. This argument may be made rigorous by using the fact that $p(n)$ is the coefficient of q^n in

$$\frac{1}{(q;q)_n} = \frac{1}{\prod_{k=1}^{n}(1 - q^k)},$$

since a partition of n cannot have any part that exceeds n (see for example, Andrews [30, Theorem 1.1, pp. 3–5]).

By similar arguments, one can easily prove the following general results.

Theorem 15.1. *Let S be any set (finite or infinite) of positive integers. Let $p(S, m, n)$ denote the number of partitions of n consisting of exactly m parts (counting repetitions) from S, and let $Q(S, m, n)$ denote the number* *of partitions of n consisting of exactly m distinct parts from S. Then*

$$\sum_{n=0}^{\infty}\sum_{m=0}^{\infty} p(S,m,n)z^m q^n = \frac{1}{\prod_{k\in S}(1 - zq^k)}, \tag{15.5}$$

$$\sum_{n=0}^{\infty}\sum_{m=0}^{\infty} Q(S,m,n)z^m q^n = \prod_{k\in S}(1 + zq^k). \tag{15.6}$$

Proof. See Q15.1.   □

Remark: If $S = \mathbb{N}$, we write $p(m,n)$ for $p(S,m,n)$, and $Q(m,n)$ for $Q(S,m,n)$.

A partition may be represented graphically by a *Ferrers graph*, where a part of size k is represented by a row of k dots, and the rows are left-justified and arranged in non-increasing order from top to bottom. For example, the Ferrers graph of the partition $6 + 5 + 4 + 2 + 2 + 1$ is shown in Figure 15.1.

A Ferrers graph of any partition contains a largest square, say k^2 for some integer k, of dots, this square of dots being termed the *Durfee square* of the partition. In the example in Figure 15.1, the Durfee square contains nine dots and is represented by the square in the figure.

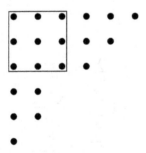

Figure 15.1: Ferrers graph of the partition $6 + 5 + 4 + 2 + 2 + 1$, with 3×3 Durfee square

More generally, for a fixed integer $r \geq 0$, a Ferrers graph contains a largest $k \times (k + r)$ *Durfee rectangle*. For example, if $r = 4$, the largest Durfee rectangle in the partition shown in Figure 15.2 is the 3×7 rectangle shown.

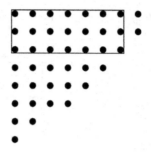

Figure 15.2: The partition $8+8+7+6+5+4+2+1$ with Durfee rectangle of size 3×7 (or $k \times (k+r)$, with $r=4$)

15.1   Some combinatorial proofs without bijections

We are now in a position to give elementary combinatorial proofs of a number of basic hypergeometric identities arising from alternative representations for the generating functions for $p(m,n)$ and $Q(m,n)$ (see the remark after Theorem 15.1). In contrast to other proofs later in the chapter, the proofs in this section do not rely on finding bijections, but simply on grouping the partitions in various (largest part, smallest part, size of Durfee square or rectangle etc.).

Theorem 15.2. *If $|q| < 1$, then*

$$\frac{1}{(zq;q)_\infty} = 1 + \sum_{k=1}^{\infty} \frac{zq^k}{(zq;q)_k}, \tag{15.7}$$

$$= 1 + \sum_{k=1}^{\infty} \frac{zq^k}{(zq^k;q)_\infty}, \tag{15.8}$$

$$= 1 + \sum_{k=1}^{\infty} \frac{z^k q^{k^2}}{(zq,q;q)_k}, \tag{15.9}$$

$$= 1 + \sum_{k=1}^{\infty} \frac{z^k q^{k(k+r)}}{(zq;q)_{k+r}(q;q)_k}, \forall\, r \in \mathbb{N}, \tag{15.10}$$

$$(-zq;q)_\infty = 1 + \sum_{k=1}^{\infty} zq^k(-zq;q)_{k-1} \tag{15.11}$$

$$= 1 + \sum_{k=1}^{\infty} zq^k(-zq^{k+1};q)_\infty. \tag{15.12}$$

$$= 1 + \sum_{k=1}^{\infty} \frac{z^k q^{k(k+1)/2}}{(q;q)_k}. \tag{15.13}$$

Proof. The method of proof is to show that the coefficient of $z^m q^n$ in the expansion of each of the functions in (15.7)–(15.10) is $p(m,n)$, and likewise that the same coefficient in (15.11)–(15.13) is $Q(m,n)$. That this is true for the infinite products on the left side follows from Theorem 15.1.

By similar reasoning to that used in the proof of Theorem 15.1, the coefficient of $z^m q^n$ in the expansion of

$$\frac{zq^k}{(zq;q)_k}$$

equals the number of partitions of n into exactly m parts with largest part k, so that summing over all k gives that the coefficient of $z^m q^n$ in the expansion of the series on the right side of (15.7) equals the number of partitions of n into exactly m parts. This proves (15.7).

The proofs of (15.8)–(15.10) are similar, except that the k-th term in the series generates partitions, respectively, with smallest part k, with Durfee square of size k^2, and Durfee rectangle of size $k \times (k+r)$. In the case of (15.9), the $1/(zq;q)_k$ factor generates a sub-partition with parts from $1, 2, \ldots, k$ which is placed below the Durfee square (with the power of z generated tracking the number of additional parts), and the $1/(q;q)_k$ factor similarly generates a sub-partition with parts from $1, 2, \ldots, k$ which is placed to the right of the Durfee square. The interpretation of (15.10) is similar, with $1/(zq;q)_{k+r}$ factor generating a sub-partition with parts from $1, 2, \ldots, k+r$, which is placed below the side of the Durfee rectangle of length $k + r$.

Likewise, the k-th term in the series on the right side of (15.11) and (15.12) generates partitions into distinct parts with, respectively, largest part k, and smallest part k, and the proofs proceed similarly.

Finally, for (15.13), from the Ferrers graph of any partition into k distinct parts one may separate off a triangle of $1 + 2 + 3 + \cdots + k = k(k+1)/2$ dots, shown in Figure 15.3. The remainder of such a partition may be regarded as a sub-partition with parts from $1, 2, \ldots, k$ (generated by $1/(q;q)_k$) and placed to the right of the triangle. $\square$

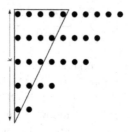

Figure 15.3: The Ferrers graph of a partition into k distinct parts, containing a triangle of $k(k+1)/2$ dots

15.2   Conjugation

If the Ferrers graph of a partition λ of a positive integer n is reflected across the main diagonal, a new partition of n, denoted by λ' and called the *conjugate* of λ, is obtained. For example, the conjugate of the partition $5 + 4 + 4 + 2 + 1$ is the partition $5 + 4 + 3 + 3 + 1$ — see Figure 15.4.

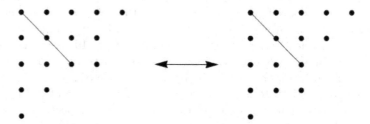

Figure 15.4: The conjugate of the partition $5 + 4 + 4 + 2 + 1$ is the partition $5 + 4 + 3 + 3 + 1$

Note that conjugation has the effect of switching the largest part and the number of parts. This means that if $p(l, m, n)$ denotes the number of partitions of n into exactly m parts and with largest part l, then

$$p(l, m, n) = p(m, l, n). \qquad (15.14)$$

The equality at (15.14) leads to the identity in the following theorem.

Theorem 15.3. *If s, z and q are complex numbers with $|q| < 1$, then*

$$\sum_{k=1}^{\infty} \frac{sz^k q^k}{(sq;q)_k} = \sum_{k=1}^{\infty} \frac{zs^k q^k}{(zq;q)_k}. \tag{15.15}$$

Proof. If each side is expanded as multi-series in the variables s, z and q, the coefficient of $s^m z^l q^n$ on the left side is $p(l, m, n)$, while the coefficient of $s^m z^l q^n$ on the right side is $p(m, l, n)$. $\qquad\square$

Remark: The identity at (15.15) was also stated in [208, Eq. 2.2.4].

Another example of conjugation providing a bijection between particular sets of partitions is the map between partitions into distinct parts and partitions in which every part from 1 to the largest part occurs at least once. A third example is the map between partitions into distinct odd parts and partitions in which the largest part occurs an odd number of times, and all parts smaller than the largest part occur an even number of times. Both of these bijections easily lead to basic hypergeometric identities. A number of the equalities in Theorem 15.2 may be extended, by keeping track of both the number of partitions and the largest part, in some cases also combined with conjugation — see Q15.2–Q15.5.

15.3   Other Bijections

15.3.1   Franklin's involution

If $\mathcal{P}_n$ denotes the set of all partitions of the positive integer n, conjugation is an example of an *involution* on $\mathcal{P}_n$, i.e. a map $\phi : \mathcal{P}_n \to \mathcal{P}_n$ such that $\phi(\phi(\lambda)) = \lambda$, $\forall\, \lambda \in \mathcal{P}_n$. One of the first and most important bijective proofs of a basic hypergeometric identity was Franklin's [119] proof of the pentagonal number theorem (see (6.14) below).

Franklin found an involution on the set of partitions of each positive integer n into distinct parts. Let $\mathcal{D}_n$ denote the set of partitions of n into distinct parts. For a given partition $\lambda \in \mathcal{D}_n$, let $s = s(\lambda)$ denote the smallest part of λ, and let $d = d(\lambda)$ denote the size of the largest decreasing sequence of consecutive parts, starting with the largest part (for ease of notation, the diagonal row of dots at the end of these rows in the corresponding Ferrers graph will be referred to as the "diagonal row"). Franklin's involution is a map $\phi : \mathcal{D}_n \to \mathcal{D}_n$ that acts on a partition $\lambda \in \mathcal{D}_n$ by moving some of the dots in the corresponding Ferrers graph around in a certain prescribed manner, as will be described next.

We first consider the case where bottom row of the Ferrers graph and the diagonal row do not intersect. If $s \leq d$, $\phi(\lambda)$ is the partition formed by removing the s dots at the bottom of the Ferrers graph and placing one dot at the end of each of the rows corresponding to the s largest parts. For example, the partition $\lambda = 9 + 8 + 7 + 6 + 3$ of the integer 33 has $s(\lambda) = 3$ and $d(\lambda) = 4$, so $\phi(\lambda) = 10 + 9 + 8 + 6$ is formed by moving the bottom row of 3 dots in the Ferrers graph of λ to the ends of the top 3 rows (see Figure 15.5).

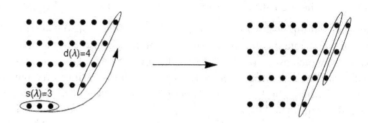

Figure 15.5: An example of Franklin's involution when $s(\lambda) \leq d(\lambda)$

If $s > d$ in the Ferrers graph of λ, $\phi(\lambda)$ is the partition formed by removing a dot from the right end each of the rows corresponding to the d largest parts of λ, and placing them at the bottom of the Ferrers graph to create a new smallest part. For example, the partition $\lambda = 10 + 9 + 8 + 6$ of the integer 33 has $s(\lambda) = 6$ and $d(\lambda) = 3$, so $\phi(\lambda) = 9 + 8 + 7 + 6 + 3$ is formed by moving the dot at the right end of the top three rows of the Ferrers graph of λ to the bottom of the Ferrers graph to create a new smallest part of size 3 (see Figure 15.6).

Figure 15.6: An example of Franklin's involution when $s(\lambda) > d(\lambda)$

When the bottom row and the diagonal row of the Ferrers graph of the partition λ do intersect, we consider a number of cases. If $s < d$ or $s > d + 1$, then the rule is the same as in the case that bottom row and the diagonal row of the Ferrers graph of the partition λ do not intersect. If $s = d$ or $s = d + 1$, then define $\phi(\lambda) := \lambda$. This latter situation occurs if $n = k^2 + k(k-1)/2 = k(3k-1)/2$ or $n = k^2 + k(k+1)/2 = k(3k+1)/2$, where $k(= d)$ is the number of rows in the Ferrers graph (see Figure 15.7).

Figure 15.7: Examples of partitions fixed by Franklin's involution

Thus $\phi(\phi(\lambda)) = \lambda$, $\forall \lambda \in \mathcal{D}_n$, and if $n \neq k(3k \pm 1)/2$, then n has as many partitions into an odd number of distinct parts as it has partitions into an even number of distinct parts, and if $n = k(3k\pm1)/2$, then n has one additional unmatched partition into k distinct parts. Thus, if $\mathcal{D}_n^o$ denotes the set of partitions of n into an *odd* number of distinct parts, and $\mathcal{D}_n^e$ denotes the set of partitions of n into an *even* number of distinct parts, then

$$|\mathcal{D}_n^e| - |\mathcal{D}_n^o| = \begin{cases} (-1)^k, & n = k(3k \pm 1)/2, \\ 0 & \text{otherwise.} \end{cases} \tag{15.16}$$

This conclusion leads to Franklin's combinatorial proof of the pentagonal number theorem.

Theorem 15.4.

$$(q;q)_\infty = \sum_{n=-\infty}^{\infty} (-1)^n q^{n(3n-1)/2} = 1 + \sum_{n=1}^{\infty} (-1)^n (q^{n(3n-1)/2} + q^{n(3n+1)/2}). \tag{15.17}$$

Proof. This follows immediately from (15.16) and the fact that

$$(q;q)_\infty = 1 + \sum_{n=1}^{\infty} (|\mathcal{D}_n^e| - |\mathcal{D}_n^o|)q^n. \tag{15.18}$$

□

15.3.2 Self-conjugate partitions and partitions into distinct odd parts

The identity in the next theorem follows easily as a special case of (15.13), but a different proof is given, to provide another example of the variety of combinatorial techniques used to prove basic hypergeometric identities.

A *self-conjugate* partition is a partition that is equal to its own conjugate. If the dots in first row and column are removed from the Ferrers graph of a self-conjugate partition, and combined into a single new part, then the new part will be odd. If this same step is applied to the Ferrers graph of the remaining partition, the new part formed will again be odd, but smaller than the first part. If this step is repeated until the Ferrers graph of the self-conjugate partition is used up, this partition will have been converted into a partition into distinct odd parts.

Figure 15.8: partitions into distinct odd parts $\Longleftrightarrow$ self-conjugate partitions

This process is also easily seen to be reversible. Further, noting that the size of the Durfee square of the self-conjugate partition equals the number of parts in the partition into distinct odd parts, what has been shown is that for any positive integers m and n,

\# self-conjugate partitions of n with Durfee square of side m
$$= \# \text{ of partitions of } n \text{ into } m \text{ distinct odd parts.}$$

Upon equating the respective generating functions, we obtain the following basic hypergeometric identity.

Theorem 15.5. *If $|q| < 1$, then*

$$\sum_{m=0}^{\infty} \frac{z^m q^{m^2}}{(q^2; q^2)_m} = (-zq; q^2)_\infty. \tag{15.19}$$

Proof. The coefficient of $z^m q^n$ on the right side is clearly the number of partitions of n into m distinct odd parts. One the left side, the factor $1/(q^2; q^2)_m$ may be regarded as generating pairs of identical sub-partitions with parts $\leq m$, and when these are placed above and to the right of the Durfee square, a self-conjugate partition with Durfee square of side m is produced. □

Remark: Theorem 15.5 could also have been proved by conjugation, since the conjugate of a partition into m distinct parts is a partition with m parts, in which the largest part occurs an odd number of times, and all smaller parts occur an even number of times and at least twice.

A more complicated identity deduced from examining partitions into distinct odd parts is stated in Q15.9.

15.3.3 An example of a more complicated bijection

Most bijective proofs involve more complicated operations than bijection or the relatively simple operation involved in Franklin's involution. Indeed, for reasons of space, most are beyond the scope of the present volume, so we restrict ourselves to giving one relatively simple example.

Theorem 15.6. *If $|q| < 1$, then*

$$1 + \sum_{k=1}^{\infty} \frac{sz^k q^k}{(sq^{k+1}; q)_k} = 1 + \sum_{k=1}^{\infty} \frac{s^k z q^{2k-1}}{(zq^k; q)_k}. \tag{15.20}$$

Proof. If the q-series on the left is expanded, the coefficient of $z^m s^r q^n$ is $p_1(m, r, n)$, the number of partitions of n with a unique smallest part m, r parts (counting multiplicity), and where the largest part is at most $2m$, and all parts in the interval $[m + 1, 2m]$ may occur (with repetition).

Likewise, when the series on the right is expanded, the coefficient of $z^m s^r q^n$ is $p_2(m, r, n)$, the number of partitions of n where the largest part is odd and equals $2r - 1$, there are m parts (counting multiplicity), and the smallest part is $\geq r$, and all parts in the interval $[r, 2r - 1]$ may occur (with repetition). We will exhibit a bijection between the set of partitions counted by $p_1(m, r, n)$ and that counted by $p_2(m, r, n)$, thus proving (15.20) above.

Consider a partition of the type counted by $p_2(m, r, n)$, with largest part $2r - 1$, exactly m parts, and smallest part $\geq r$, as indicated in the first Ferrers graph in Figure 15.9 (the partition shown has smallest part $= r$, and if the partition had smallest part $> r$, the corresponding partition counted

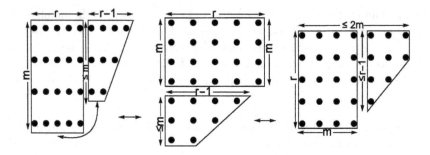

Figure 15.9: The two-step bijection that proves the identity in Theorem 15.6

by $p_1(m, r, n)$ would have largest part exactly $= 2m$). Next, separate the Ferrers graph into two pieces, an $m \times r$ rectangle and the sub-partition fitting inside a $(\leq m) \times (r - 1)$ rectangle to the right of the $m \times r$ rectangle. Move the part of the Ferrers graph inside the $\leq m \times r - 1$ rectangle to the bottom of the $m \times r$ rectangle to get an intermediate partition, as shown in the middle picture in Figure 15.9. Finally, take the conjugate of this intermediate partition to get a partition counted by $p_1(m, r, n)$, as shown in the final picture in Figure 15.9. This process is easily seen to be reversible, thus showing that $p_1(m, r, n) = p_2(m, r, n)$, and (15.20) follows.   □

Remark: The $s = z = 1$ special case of (15.20) was stated by Ramanujan in his last letter to Hardy [214, p. 354], and this special case was proved by Watson [260, p. 278]. A combinatorial proof of this same special case was given by Andrews [13, pp. 38–39].

15.4   Bijective proofs of identities involving Gaussian polynomials

As was remarked in the comments following (14.6), the coefficient of each power of q in the polynomial expansion of Gaussian polynomials have an interpretation in terms of certain restricted partition functions. This statement is made precise in the next theorem, and the proof of Andrews [30, Theorem 3.1, page 33] is given. This result is then used to provide combinatorial proofs of some basic hypergeometric identities. Recall that $g(n, r; q)$ is defined at (14.1).

Theorem 15.7. *For non-negative integers M N,*

$$h(N, M, q) := g(M + N, M; q) = \begin{bmatrix} M + N \\ M \end{bmatrix} = \sum_{n \geq 0} p(N, M, n)q^n, \quad (15.21)$$

where $p(N, M, n)$ denotes the number of partitions of the integer n into at most M parts, each $\leq N$.

Proof. From (14.1) it is clear that $h(N, 0; q) = h(0, M; q) = 1$, and from (14.9) that

$$h(N, M; q) = h(N, M - 1; q) + q^M h(N - 1, M; q). \quad (15.22)$$

Let $H(N, M, q)$ denote the right side of (15.21). Since $p(N, 0, n) = p(0, M, n) = 0$ for $n \geq 1$, it follows that $H(N, 0; q) = H(0, M; q) = 1$. It will be shown that $H(N, M, q)$ satisfies the same recurrence as $h(N, M, q)$, namely

$$H(N, M; q) = H(N, M - 1; q) + q^M H(N - 1, M; q). \quad (15.23)$$

Thus, since both sequences satisfy the same initial conditions and the same recurrence relation, it will follow that $h(N, M, q) = H(N, M; q)$ for all non-negative integers M and N, giving the result stated at (15.21).

Note that

$$p(N, 0; n) = p(0, M; n) = \begin{cases} 1 & N = M = n = 0, \\ 0 & \text{otherwise,} \end{cases}$$

$$p(N, M, n) = 0, \text{ if } n < 0 \text{ or } n > MN.$$

(In the case of $n = 0$, we consider there to be one partition into 0 parts of size 0, the empty partition.)

Partitions of an integer n into at most M parts, each part $\leq N$, fall into two groups, namely, those with exactly M parts, and those with at most $M - 1$ parts. For a partition into exactly M parts, delete each part of size 1 and reduce each part of size greater than 1 by 1, to get a partition of $n - M$ into at most M parts, each $\leq N - 1$. Hence

$$P(N, M; n) = P(N, M - 1, n) + P(N - 1, M; n - M) \quad (15.24)$$
$$\implies P(N, M; n)q^n = P(N, M - 1, n)q^n + q^M P(N - 1, M; n - M)q^{n-M}. \quad (15.25)$$

Upon summing over all n, (15.23), and hence (15.21), follows. $\square$

Remark: Andrews [30, Thm 3.1, page 33] used the notations $g(N, M, q)$ and $G(N, M, q)$ where we have used $h(N, M, q)$ and $H(N, M, q)$, which was done here to avoid confusion with our existing notation for the function $g(n, r; q)$ defined at (14.1).

The identity in (15.21) may be used to give combinatorial proofs of the two special cases of the q-binomial theorem (stated earlier in Corollary 3.2).

Corollary 15.1. *For each non-negative integer N,*

$$\sum_{m=0}^{N} \begin{bmatrix} N \\ m \end{bmatrix} z^m q^{m(m+1)/2} = (-zq; q)_N, \tag{15.26}$$

$$\sum_{m=0}^{\infty} q^m \begin{bmatrix} N + m - 1 \\ m \end{bmatrix} z^m = \frac{1}{(zq; q)_N}, \quad |z| < 1, \ |q| < 1. \tag{15.27}$$

Proof. We follow the proofs given by the authors in [45, Section 7.3], with some minor variations in explanation. The coefficient of $z^m q^n$ on the right side of (15.26) is the number of partitions of n into exactly m *distinct* parts, each $\leq N$, say $p'(m, \leq N, n)$. Thus the coefficient of z^m is

$$\sum_{n \geq m(m+1)/2} p'(m, \leq N, n) q^n.$$

By stripping off a triangle of $m(m+1)/2$ dots from the Ferrers graph of a partition of n counted by $p'(m, \leq N, n)$, one gets a partition of $n - m(m+1)/2$ containing at most m parts, each $\leq N - m$, or a partition of $n - m(m+1)/2$ counted by $p(N - m, m, n - m(m+1)/2)$, as defined at (15.21).

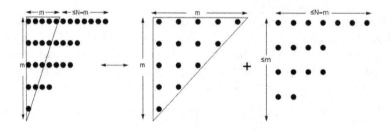

Figure 15.10: The bijection that proves the identity at (15.26)

This action is clearly reversible, giving a bijection between the two sets of partitions, giving that $p'(m, \leq N, n) = p(N - m, m, n - m(m+1)/2)$, and thus that

$$\sum_{n \geq m(m+1)/2} p'(m, \leq N, n)q^n$$

$$= q^{m(m+1)/2} \sum_{n \geq m(m+1)/2} p(N - m, m, n - m(m+1)/2)q^{n-m(m+1)/2}$$

$$= q^{m(m+1)/2} \sum_{k \geq 0} p(N - m, m, k)q^k = q^{m(m+1)/2} \begin{bmatrix} N \\ m \end{bmatrix}.$$

This completes the proof of (15.26).

Similarly, the coefficient of $z^m q^n$ on the right side of (15.27) is the number of partitions of n into exactly m parts, each $\leq N$, say $p^*(m, \leq N, n)$. Thus the coefficient of z^m is

$$\sum_{n \geq m} p^*(m, \leq N, n)q^n.$$

By stripping off the leftmost column of m dots from the Ferrers graph of a partition of n counted by $p^*(m, \leq N, n)$, one gets a partition of $n - m$ containing at most m parts, each $\leq N - 1$, or a partition of $n - m$ counted by $p(N - 1, m, n - m)$.

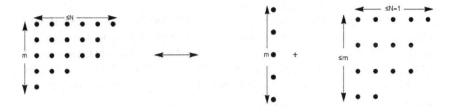

Figure 15.11: The bijection that proves the identity at (15.27)

This action is also reversible, so $p^*(m, \leq N, n) = p(N - 1, m, n - m)$ and

$$\sum_{n \geq m} p^*(m, \leq N, n)q^n = q^m \sum_{n \geq m} p(N - 1, m, n - m)q^{n-m}$$

$$= q^m \sum_{k \geq 0} p(N - 1, m, k)q^k = q^m \begin{bmatrix} N + m - 1 \\ m \end{bmatrix},$$

thus giving the result at (15.27). $\qquad \qquad \square$

15.5   From basic hypergeometric identities to partition identities

As remarked earlier, the main focus in this chapter is using combinatorial methods to prove basic hypergeometric identities. However, we conclude by remarking that it is also possible to go in the other direction, and give combinatorial interpretations to existing basic hypergeometric identities. The most famous such identities with combinatorial interpretations are the Rogers–Ramanujan identities (proved in Theorem 10.2).

$$\sum_{n=0}^{\infty} \frac{q^{n^2}}{(q;q)_n} = \frac{1}{(q,q^4;q^5)_\infty}, \tag{15.28}$$

$$\sum_{n=0}^{\infty} \frac{q^{n(n+1)}}{(q;q)_n} = \frac{1}{(q^2,q^3;q^5)_\infty}. \tag{15.29}$$

These two identities have the following combinatorial interpretations.
(1) By interpreting $1/(q;q)_n$ as generating parts from $\{1, 2, \ldots, n\}$ and interpreting n^2 as

$$n^2 = 1 + 1 + 2 + 2 + \cdots + (n-1) + (n-1) + n,$$

it may be seen that the coefficient of q^k in the expansion of the left side of (15.28) is equal to $A(k)$, the number of partitions of k in which all parts smaller than the largest part appear at least twice. The coefficient of q^k in the expansion of the right side of (15.28) is equal to $B(k)$, the number of partitions of k into parts $\equiv 1, 4 \pmod 5$, and thus $A(k) = B(k)$ for all positive integers k.
(2) In a similar manner, the coefficient of q^k in the expansion of the left- and right sides of (15.29) may be interpreted, respectively, as $C(k)$ and $D(k)$, where $C(k)$ is the number of partitions of k in which all parts from 1 to the largest part appear at least twice, and $D(k)$ is the number of partitions of k into parts $\equiv 2, 3 \pmod 5$, and thus that $C(k) = D(k)$ for all positive integers k.

Remark: The more usual interpretation of the series sides of (15.28)–(15.29) is in terms of the conjugates of the partitions described here (and thus of partitions in which the gap between consecutive parts is at least 2, and in the case of (15.29), no parts of size 1), but the present author finds the partition interpretations stated here easier to visualize.

Exercises

15.1 Prove the statements at (15.5)–(15.6).

15.2 Deduce, by combinatorial arguments, from (15.7) and (15.9) that

$$1 + \sum_{n=1}^{\infty} \frac{sz^n q^n}{(sq;q)_n} = 1 + \sum_{n=1}^{\infty} \frac{s^n z^n q^{n^2}}{(qz, sq; q)_n}. \tag{15.30}$$

15.3 (i) Prove combinatorially, for each positive integer r, that

$$\frac{1}{(sq;q)_r} = 1 + \sum_{n=1}^{r} \frac{sq^n}{(sq;q)_n}.$$

(ii) Deduce from (15.7) and (15.10) that

$$1 + \sum_{n=1}^{\infty} \frac{sz^n q^n}{(sq;q)_n} = 1 + \sum_{n=1}^{r} \frac{sz^n q^n}{(sq;q)_n} + \sum_{n=1}^{\infty} \frac{s^n q^{n^2+nr} z^{n+r}}{(qz;q)_n (sq;q)_{n+r}}. \tag{15.31}$$

15.4 By using the fact that the conjugate of a partition into odd parts is a partition in which the largest part occurs an odd number of times and all smaller parts occur an even number of times, prove that

$$1 + \sum_{k=1}^{\infty} \frac{z^{2k-1} q^{2k-1} s}{(sq;q^2)_k} = 1 + \sum_{k=1}^{\infty} \frac{zs^k q^k}{(z^2 q^2; q^2)_k}. \tag{15.32}$$

15.5 Deduce, by combinatorial arguments, from (15.11) and (15.13) that

$$1 + \sum_{k=1}^{\infty} (-zq;q)_{k-1} s^k z q^k = 1 + \sum_{k=1}^{\infty} \frac{s^k z^k q^{k(k+1)/2}}{(sq;q)_k}. \tag{15.33}$$

15.6 Let $S = \{n_k\}_{k \in I}$ be any collection, finite or infinite, of distinct positive integers such that $n_i > n_j$ if $i > j$. Give both analytic and combinatorial proofs of the fact that (with $|z|, |q| < 1$ when S is an infinite set)

$$1 + \sum_{k \in I} \frac{sz^{n_k}}{\prod_{j=1}^{k}(1 - sq^{n_j})} = 1 + \sum_{k \in I} \left(sz^{n_k} + q^{n_k} \sum_{j \geq k} \frac{s^2 z^{n_j}}{\prod_{i=k}^{j}(1 - sq^{n_i})} \right).$$

$$\tag{15.34}$$

15.7 Prove combinatorially the q-binomial theorem in the form

$$\sum_{k=0}^{\infty} \frac{(-aq;q)_k z^k q^k}{(q;q)_k} = \frac{(-azq^2;q)_\infty}{(zq;q)_\infty}, \qquad (15.35)$$

by considering the coefficient of $a^p z^k q^n$ in the expansion of the k-th term in the series on the left as counting the number of partitions of n with largest part k and with each of the p distinct parts coming from $(-aq;q)_k$ being placed below any parts of similar size generated by $(q;q)_k$ in the Ferrers graph, and with the right-most dots in these p rows being specially marked.

15.8 (Andrews [15]) Prove the Rogers–Fine identity in the form

$$\sum_{n=0}^{\infty} \frac{(-aq;q^2)_n bt^n q^{2n}}{(bq^2;q^2)_n}$$

$$= \sum_{N=0}^{\infty} \frac{(-aq;q^2)_N (-atq^3/b;q^2)_N (1 + atq^{4N+3}) b^{N+1} t^N q^{2N^2+2N}}{(bq^2;q^2)_N (tq^2;q^2)_{N+1}} \qquad (15.36)$$

combinatorially by interpreting the coefficient of $a^k b^l t^m q^n$ on the left hand side as the number of partitions of n with no repeated odd parts, where there are l even parts, k odd parts, and the largest part is $2m$. To get the right side, group these partitions according to the largest Durfee rectangle with $N + 1$ rows and $2N$ columns, consider separately the cases where there is and is not a dot in the Ferrers graph below and to the right of the Durfee rectangle, in each case interpreting the sub-partition below the Durfee rectangle in terms of $(-aq;q^2)_N/(bq^2;q^2)_N$ and the sub-partition to the right of the Durfee rectangle in terms of $(-atq^3/b;q^2)_N/(tq^2;q^2)_{N+1}$ (with an even part of size $2j$ being interpreted as two columns of height j, and an odd part $2j - 1$ being interpreted as a column of height j and a column of height $j - 1$).

15.9 By considering partitions into distinct odd parts and keeping track of the number of parts, the largest part and the smallest part, prove that

$$1 + \sum_{n=0}^{\infty} z s^{2n+1} q^{2n+1} \left(t^{2n+1} + \left(-zq;q^2\right)_n \sum_{j=1}^{n} \frac{z t^{2j-1} q^{2j-1}}{(-zq;q^2)_j} \right)$$

$$= 1 + \sum_{n=1}^{\infty} \frac{s^{2n-1} t z^n q^{n^2}}{(s^2 q^2;q^2)_{n-1} (1 - s^2 t^2 q^{2n})}. \qquad (15.37)$$

(It may also be interesting to give an analytic proof of this identity.)

15.10 Find a combinatorial proof of the identity

$$\sum_{m=0}^{\infty} \frac{z^m q^{m^2+m}}{(q^2;q^2)_m} = (-zq^2;q^2)_\infty. \tag{15.38}$$

(*Of course this identity also follows immediately from* (15.19), *upon replacing z with zq.*)

15.11 By interpreting the right side using Theorem 15.7 to interpret the right side, and them grouping the resulting partitions appropriately, give a combinatorial proof, for all integers m, $n \geq 1$ (the cases $m = 0$ or $n = 0$ being trivial), of the identity

$$\sum_{j=0}^{n} \begin{bmatrix} m+j \\ j \end{bmatrix} q^j = \begin{bmatrix} n+m+1 \\ m+1 \end{bmatrix}. \tag{15.39}$$

15.12 By using Theorem 15.7 to interpret the right side, and them grouping the resulting partitions by the size of the Durfee square, prove that for all integers $n \geq 0$,

$$\sum_{j=0}^{n} \begin{bmatrix} n \\ j \end{bmatrix}^2 q^{j^2} = \begin{bmatrix} 2n \\ n \end{bmatrix}. \tag{15.40}$$

15.13 Give a combinatorial proof of the identity (which holds for all integers $n \geq 0$)

$$\sum_{j=0}^{n} \begin{bmatrix} n \\ j \end{bmatrix} \frac{z^j q^{j^2}}{(zq;q)_j} = \frac{1}{(zq;q)_n}. \tag{15.41}$$

15.14 By mimicking the proof of Theorem 15.5 and using Theorem 15.7, show that if n is a non-negative integer, then

$$\sum_{j=0}^{n} \begin{bmatrix} n \\ j \end{bmatrix}_{q^2} z^j q^{j^2} = (-zq;q^2)_n. \tag{15.42}$$

Chapter 16

q-Continued Fractions

The best known representation of a q-hypergeometric function as a continued fraction is probably the expansion of the quotient of the Rogers–Ramanujan functions, or (via the Rogers–Ramanujan identities (10.17)–(10.18)) their corresponding infinite products, as the Rogers–Ramanujan continued fraction (see below for an explanation of the continued fraction notation on the right of (16.1)):

$$\frac{\displaystyle\sum_{n=0}^{\infty} \frac{q^{n^2}}{(q;q)_n}}{\displaystyle\sum_{n=0}^{\infty} \frac{q^{n^2+n}}{(q;q)_n}} = \frac{(q^2,q^3;q^5)_\infty}{(q,q^4;q^5)_\infty} = K(q) := 1 + \cfrac{q}{1 + \cfrac{q^2}{1 + \cfrac{q^3}{1 + \cfrac{q^4}{\ddots}}}}, \tag{16.1}$$

these identities holding for $|q| < 1$.

The subject of q-continued fractions has a long history (the first example possibly being the $a = 1$ case of (16.139) below, due to Gauss [126]), and there is a very extensive literature. There are numerous beautiful results, many of them due to Ramanujan (see [38], for example). In the present volume we can hope to give only a sample of these results.

Firstly, some general features of continued fractions, for example the odd- and even parts, more general contractions and extensions, the Bauer–Muir transform etc. are employed to derive some elementary identities.

Later in the chapter three general identities, in which ratios of basic hypergeometric series with several free parameters are represented as q-continued fractions, are proved. Several known continued fraction expansions then follow as special cases.

Heine's continued fraction, and other similar continued fractions that are derived via three-term recurrence relations are also considered. Finally, some connections between q-continued fractions and the regular continued fraction expansion of real numbers are briefly investigated.

16.1 Notation and Some Convergence Criteria

The modern understanding of continued fractions is in terms of composition of linear fractional transformations (see, for example, [183] or [184]). This approach is useful when investigating the general convergence of continued fractions, and various aspects of convergence theory such as tail sequences and value sets. Here we will adopt the simpler approach of regarding an infinite continued fraction as the limit of the sequence of its approximants (of course this interpretation is also a consequence of the linear fractional transformation viewpoint, but the full generality of the latter viewpoint is not needed here). In this section we introduce the standard notation for continued fractions, and some of their elementary properties, as may be found in any book on general continued fractions (see, for example [183] or [184]).

An infinite continued fraction has the form

$$b_0 + \cfrac{a_1}{b_1 + \cfrac{a_2}{b_2 + \cfrac{a_3}{b_3 + \dots}}}. \qquad (16.2)$$

Here $\{a_n\}_{n \geq 1}$ and $\{b_n\}_{n \geq 0}$ are sequences of real or complex numbers (we may also allow the a_n and b_n to be functions of one or more variables, with these variables taking values in $\mathbb{R}$ or $\mathbb{C}$). The term a_n is called the *n-th partial numerator*, while b_n is the *n-th partial denominator*. It will be assumed henceforth that $a_n \neq 0$, unless otherwise stated, otherwise the continued fraction terminates and becomes a finite continued fraction. As with other infinite objects, we ascribe a value to the object in (16.2) by defining this value to be the limit of a certain sequence. Define the sequence $\{f_n\}_{n \geq 0}$ by setting

$$f_0 : = b_0, \; f_1 := b_0 + \frac{a_1}{b_1} = \frac{b_0 b_1 + a_1}{b_1}, \qquad (16.3)$$

$$f_2 := b_0 + \cfrac{a_1}{b_1 + \cfrac{a_2}{b_2}} = \frac{a_2 b_0 + a_1 b_2 + b_1 b_2 b_0}{a_2 + b_1 b_2},$$

$$\vdots$$

$$f_n := b_0 + \cfrac{a_1}{b_1 + \cfrac{a_2}{b_2 + \cfrac{a_3}{\ddots \cfrac{a_{n-1}}{b_{n-1} + \cfrac{a_n}{b_n}}}}} =: \frac{A_n}{B_n},$$

$$\cdots$$

The term f_n is the *n-th approximant* of the continued fraction. When f_n is simplified to a rational function, with the numerator A_n and the denominator B_n being polynomials in the a_i's and b_i's, A_n is termed the *n-th canonical numerator* (or *n-th numerator convergent*) and B_n is termed the *n canonical denominator* (or *n-th denominator convergent*). If $\lim_{n \to \infty} f_n = f \in \hat{\mathbb{C}}$, then we say the continued fraction at (16.2) converges, and its value is f.

To save space, the continued fraction at (16.2) is sometimes written as

$$b_0 + \frac{a_1}{b_1} + \frac{a_2}{b_2} + \frac{a_3}{b_3} + \cdots$$

and as $b_0 + K_{n=1}^{\infty} a_n/b_n$ or even $b_0 + K(a_k/b_k)$, and the N-th approximant f_N may be expressed as

$$f_N = b_0 + \frac{a_1}{b_1} + \frac{a_2}{b_2} + \frac{a_3}{b_3} + \cdots + \frac{a_N}{b_N} = b_0 + K_{n=1}^{N} \frac{a_n}{b_n}.$$

It is easy to show by mathematical induction (see, for example, [183], p. 9, and Q16.1) that the A_n's and B_n's satisfy the following recurrence relations.

$$A_n = b_n A_{n-1} + a_n A_{n-2}, \tag{16.4}$$

$$B_n = b_n B_{n-1} + a_n B_{n-2},$$

with $A_0 = b_0$, $B_0 = 1$, and defining $A_{-1} := 1$ and $B_{-1} := 0$. It is also easy to show, again by mathematical induction, (see also [183], p. 9, and Q16.1) that for $n \geq 1$,

$$A_n B_{n-1} - A_{n-1} B_n = (-1)^{n-1} \prod_{i=1}^{n} a_i. \tag{16.5}$$

Another result that is easily derived (see Q16.1) is that, for all $n \geq 2$,

$$A_n B_{n-2} - A_{n-2} B_n = b_n (-1)^{n-2} \prod_{i=1}^{n-1} a_i. \tag{16.6}$$

A large body of the analytic theory of continued fractions is concerned with convergence questions, with determining criteria for the sequences $\{a_n\}$ and $\{b_n\}$ which guarantee the convergence of $K(a_k/b_k)$, and with properties of the sequence of approximants $\{f_n\}$ and, in the case of convergence of $K(a_k/b_k)$, with properties of the limit f. Since most q-continued fractions have the property that the $a_n \to 0$ and the b_n are bounded and bounded away from 0, at this stage we give just two simple convergence criteria, and do not overly concern ourselves with properties of the approximants $\{f_n\}$. The reader who is interested in understanding more of the convergence theory of continued fractions is directed to the books of Jones and Thron [162], Lorentzen and Waadeland [183,184] and Wall [252].

Since we are interested simply in convergence and divergence, our statements of the following two theorems are simpler than the usual versions.

Theorem 16.1. (*Śleszyński–Pringsheim Theorem [211, 239], see also [162, page 92]*) *The continued fraction $K(a_k/b_k)$ converges to a finite value if*

$$|b_k| \geq |a_k| + 1, \forall k \geq 1. \tag{16.7}$$

Under these conditions all the approximants satisfy $|f_n| < 1$ and the limit of the continued fractions satisfy $|f| \leq 1$.

Proof. Since

$$\left| \frac{a_1}{b_1} \right| = \frac{|a_1|}{|b_1|} \leq \frac{|a_1|}{|a_1| + 1} < 1,$$

and so $|f_1| < 1$. Suppose $|f_k| < 1$ for $k = 1, 2, \ldots n - 1$ for all continued fractions satisfying the conditions of the theorem. Consider $K_{k=1}^{n} a_k/b_k$. Then $|K_{k=2}^{n} a_k/b_k| < 1$ and

$$|f_n| = \left| K_{k=1}^{n} \frac{a_k}{b_k} \right| = \left| \frac{a_1}{b_1 + K_{k=2}^{n} \frac{a_k}{b_k}} \right| \leq \frac{|a_1|}{|b_1| - \left| K_{k=2}^{n} \frac{a_k}{b_k} \right|} < \frac{|a_1|}{|a_1| + 1 - 1} = 1.$$

Hence $|f_n| < 1$, $\forall\, n \geq 1$, and if the continued fraction converges, its value f satisfies $|f| \leq 1$.

Since $a_0 = 0$, then $A_0 = 0$ and $B_0 = 1$, so for each $n \geq 1$,

$$\frac{A_n}{B_n} = \frac{A_n}{B_n} - \frac{A_0}{B_0} = \sum_{k=1}^{n} \left(\frac{A_k}{B_k} - \frac{A_{k-1}}{B_{k-1}} \right)$$

$$= \sum_{k=1}^{n} \frac{A_k B_{k-1} - A_{k-1} B_k}{B_k B_{k-1}} = \sum_{k=1}^{n} \frac{(-1)^{k-1} \prod_{i=1}^{k} a_i}{B_k B_{k-1}},$$

where the last equality follows by (16.5). It will be shown that the last series converges absolutely as $n \to \infty$, so that it converges, and hence that $\lim_{n\to\infty} A_n/B_n$ exists and the continued fraction converges. By (16.4),

$$|B_k| = |b_k B_{k-1} + a_k B_{k-2}| \geq |b_k||B_{k-1}| - |a_k||B_{k-2}|$$

$$\geq (|a_k| + 1)|B_{k-1}| - |a_k||B_{k-2}|$$

$$\Longrightarrow |B_k| - |B_{k-1}| \geq |a_k|(|B_{k-1}| - |B_{k-2}|)$$

$$\geq |a_k||a_{k-1}|(|B_{k-2}| - |B_{k-3}|)$$

$$\cdots$$

$$\Longrightarrow |B_k| - |B_{k-1}| \geq \prod_{i=1}^{k} |a_i|(|B_0| - |B_{-1}|) = \prod_{i=1}^{k} |a_i| > 0.$$

Note that the last equality implies that $\{|B_k|\}$ is a strictly increasing sequence and since $|B_0| = 1$, $\lim_{n\to\infty} 1/|B_n|$ exists, and since

$$\sum_{k=1}^{n} \frac{\prod_{i=1}^{k} |a_i|}{|B_k B_{k-1}|} \leq \sum_{k=1}^{n} \left(\frac{1}{|B_{k-1}|} - \frac{1}{|B_k|} \right) = \frac{1}{|B_0|} - \frac{1}{|B_n|} = 1 - \frac{1}{|B_n|}.$$

Upon taking the limit as $n \to \infty$, the required absolute convergence of the series mentioned follows, and the theorem is proved. $\qquad\Box$

Theorem 16.2. (*Worpitzky [262], see [183, pp. 35–36]*) *For all $n \geq 1$, let*

$$|a_n| \leq \frac{1}{4}.$$

Then $K_{n=1}^{\infty} a_n/1$ converges. Under these conditions all the approximants satisfy $|f_n| < 1/2$ and the limit of the continued fraction satisfies $|f| \leq 1/2$.

Proof. After applying an equivalence transformation (see subsection 16.2.3), it can be seen that $K_{n=1}^{\infty} a_n/1$ has exactly the same sequence of approximants as

$$\frac{2a_1}{2} + K_{n=2}^{\infty} \frac{4a_n}{2}$$

Since the condition on the a_n implies $2 \geq |4a_n| + 1$ for $n \geq 2$, convergence follows by the Śleszyński–Pringsheim Theorem. If the continued fraction above is multiplied by 2, then $2 \geq |4a_n| + 1$ for $n \geq 1$, and by the Śleszyński–Pringsheim Theorem the approximants of this continued fraction, $\{f'_n\}$, satisfy $|f'_n| < 1$, and thus the approximants $\{f_n\}$ of $K_{n=1}^{\infty} a_n/1$ satisfy $|f_n| < 1/2$ and thus the limit f of the continued fraction satisfies $|f| \leq 1/2$. □

Note that a continued fraction $K_{n=1}^{\infty} a_n/b_n$ converges if its N-th tail $K_{n=N+1}^{\infty} a_n/b_n$ converges, for some positive integer N. This ensures, for example, the convergence (in $\hat{\mathbb{C}}$) for any continued fraction $K_{n=1}^{\infty} a_n/1$ in which $a_n \to 0$ as $n \to \infty$, which is the case for many q-continued fractions, including the Rogers–Ramanujan continued fraction $K(q)$ (see (16.1)). Observe also that the continued fraction $K_{n=1}^{\infty} a_n/b_n$ is equivalent to the continued fraction

$$\frac{a_1/b_1}{1} + K_{n=2}^{\infty} \frac{a_n/b_n b_{n-1}}{1},$$

so that Worpitzky's theorem is also applicable to any continued fraction in which $a_n/b_n b_{n-1} \to 0$ as $n \to \infty$.

16.2 Some Elementary Properties of Continued Fractions

In this section we briefly investigate a number of slightly more advanced aspects of continued fractions.

16.2.1 Constructing a continued fraction with numerators $\{A_n\}$ and denominators $\{B_n\}$

The question arises if, given two sequences $\{A_n\}_{n \geq -1}$ and $\{B_n\}_{n \geq -1}$, it is possible to construct a continued fraction in which, for $n \geq 0$, the n-th canonical numerator and n-th canonical denominator are given, respectively, by A_n and B_n. This question is not just of theoretical interest, as the formulae for determining the a_n and b_n in this continued fraction $b_0 + K_{n=1}^{\infty} a_n/b_n$ (when it is possible to construct it) will be useful in the proof of subsequent results.

Theorem 16.3. *([183, Theorem 7, page 69]) The sequences $\{A_n\}_{n \geq -1}$ and $\{B_n\}_{n \geq -1}$ of complex numbers are the canonical numerators and denominators of some continued fraction $b_0 + K_{n=1}^{\infty} a_n/b_n$ if and only if*

$$A_{-1} = B_0 = 1, \qquad B_{-1} = 0, \qquad \Delta_n := A_n B_{n-1} - A_{n-1} B_n \neq 0 \quad (16.8)$$

for all $n \in \mathbb{N} \cup \{0\}$. If (16.8) holds, then $b_0 + K_{n=1}^{\infty} a_n/b_n$ is uniquely determined by

$$b_0 = A_0, \qquad b_1 = B_1, \quad a_1 = A_1 - A_0 B_1,$$

$$a_n = -\frac{\Delta_n}{\Delta_{n-1}}, \qquad b_n = \frac{A_n B_{n-2} - A_{n-2} B_n}{\Delta_{n-1}} \; for \; n \geq 2. \quad (16.9)$$

Proof. If the continued fraction $b_0 + K_{n=1}^{\infty} a_n/b_n$ is given, then (16.8) follows from the statement after (16.4), and the "determinant" formula at (16.5) (recall that we are assuming that no $a_n = 0$, to avoid the continued fraction terminating).

If the sequences $\{A_n\}_{n \geq -1}$ and $\{B_n\}_{n \geq -1}$ with the properties stated in (16.8) are given, then the values for a_n and b_n follow from setting $b_0 = A_0/B_0$, $b_0 + a_1/b_1 = (b_0 b_1 + a_1)/b_1 = A_1/B_1$, and using the recurrence relations (16.4) to solve for a_n and b_n for $n \geq 2$. The solutions are unique for all $n \geq 2$, since the determinant of the system (16.4) is $\Delta_{n-1} \neq 0$. □

Remark: It will also be useful to write the continued fraction in the theorem above in terms of the sequence of approximants $\{f_n\}$, but this will be deferred until after a discussion of equivalent continued fractions.

16.2.2   The odd- and even part of a continued fraction

Let $\{n_k\}_{k \geq 0}$ be a subsequence of $\{n\}_{n \geq 0}$. We call $d_0 + K_{n=1}^{\infty} c_n/d_n$ a *canonical contraction* of $b_0 + K_{n=1}^{\infty} a_n/b_n$ if

$$C_k = A_{n_k}, \qquad D_k = B_{n_k} \qquad \text{for } k = 0, 1, 2, 3, \ldots,$$

where C_n, D_n, A_n and B_n are canonical numerators and denominators of $d_0 + K_{n=1}^{\infty} c_n/d_n$ and $b_0 + K_{n=1}^{\infty} a_n/b_n$ respectively. Of particular interest are the following contractions.

Theorem 16.4. *([183, Theorem 12, page 83]) The canonical contraction of $b_0 + K_{n=1}^{\infty} a_n/b_n$ with*

$$C_k = A_{2k} \qquad D_k = B_{2k} \qquad \text{for } k = 0, 1, 2, 3, \ldots,$$

exists if and only if $b_{2k} \neq 0$ for $k = 1, 2, 3, \ldots$, and in this case is given by

$$b_0 + \frac{b_2 a_1}{b_2 b_1 + a_2} - \frac{a_2 a_3 b_4/b_2}{a_4 + b_3 b_4 + a_3 b_4/b_2} - \frac{a_4 a_5 b_6/b_4}{a_6 + b_5 b_6 + a_5 b_6/b_4} + \cdots. \quad (16.10)$$

Proof. We will for the most part follow the proof in [183, pp. 83–84]. With the notation introduced above,

$$d_0 = C_0 = A_0 = b_0,$$
$$d_1 = D_1 = B_2 = b_2 b_1 + a_2,$$
$$c_1 = C_1 - C_0 D_1 = A_2 - A_0 B_2 = b_0 b_1 b_2 + b_0 a_2 + a_1 b_2 - b_0 (b_1 b_2 + a_2) = a_1 b_2.$$

For $n \geq 2$, c_n and d_n are given by (16.9):

$$c_n = -\frac{\Delta_n}{\Delta_{n-1}}, \text{ where}$$

$$\Delta_n = C_n D_{n-1} - C_{n-1} D_n = A_{2n} B_{2n-2} - A_{2n-2} B_{2n} = b_{2n} \prod_{i=1}^{2n-1} a_i,$$

$$d_n = \frac{C_n D_{n-2} - C_{n-2} D_n}{\Delta_{n-1}} = \frac{A_{2n} B_{2n-4} - A_{2n-4} B_{2n}}{\Delta_{n-1}}.$$

The claimed values for the c_n follow immediately upon simplifying the ration $-\Delta_n / \Delta_{n-1}$. For the d_n, use the recurrence relations (16.4) to write

$$A_{2n} = (b_{2n} b_{2n-1} + a_{2n}) A_{2n-2} + b_{2n} a_{2n-1} A_{2n-3},$$

$$B_{2n} = (b_{2n} b_{2n-1} + a_{2n}) B_{2n-2} + b_{2n} a_{2n-1} B_{2n-3},$$

substitute these values in the expression for d_n, and then use (16.8) (with n replaced with $2n - 3$) and (16.9) (with n replaced with $2n - 2$), and simplify. The details are left to the reader. □

The continued fraction (16.10) is called the *even* part of $b_0 + K_{n=1}^{\infty} a_n / b_n$.

Theorem 16.5. *([183, Theorem 13, page 85]) The canonical contraction of* $b_0 + K_{n=1}^{\infty} a_n / b_n$ *with* $C_0 = A_1 / B_1$

$$C_k = A_{2k+1} \qquad D_k = B_{2k+1} \qquad for \ k = 1, 2, 3, \dots,$$

exists if and only if $b_{2k+1} \neq 0$ *for* $k = 0, 1, 2, 3, \dots$, *and in this case is given by*

$$\frac{b_0 b_1 + a_1}{b_1} - \frac{a_1 a_2 b_3 / b_1}{b_1 (a_3 + b_2 b_3) + a_2 b_3} - \frac{a_3 a_4 b_5 b_1 / b_3}{a_5 + b_4 b_5 + a_4 b_5 / b_3}$$
$$- \frac{a_5 a_6 b_7 / b_5}{a_7 + b_6 b_7 + a_6 b_7 / b_5} - \frac{a_7 a_8 b_9 / b_7}{a_9 + b_8 b_9 + a_8 b_9 / b_7} + \cdots. \qquad (16.11)$$

Proof. The proof is similar to the proof of Theorem 16.4, and is left as an exercise (see Q16.2). □

The continued fraction (16.11) is called the *odd* part of $b_0 + K_{n=1}^\infty a_n/b_n$.

Remarks: (i) If a continued fraction converges, then clearly the odd- and even parts (and indeed any contraction) converge to the same limit.

(ii) A continued fraction $f_0 + K_{n=1}^\infty e_n/f_n$ is called an *extension* of the continued fraction $b_0 + K_{n=1}^\infty a_n/b_n$ if the latter is some contraction of the former.

The following corollary to Theorem 16.5 leads to an interesting extension of the Rogers–Ramanujan continued fraction $K(q)$ (see (16.1)).

Corollary 16.1. *(i) The odd part of the continued fraction*

$$\frac{c_1}{1} - \frac{c_2}{1} + \frac{c_2}{1} - \frac{c_3}{1} + \frac{c_3}{1} - \frac{c_4}{1} + \frac{c_4}{1} - \cdots \tag{16.12}$$

is

$$c_1 + \frac{c_1 c_2}{1} + \frac{c_2 c_3}{1} + \frac{c_3 c_4}{1} + \cdots. \tag{16.13}$$

(ii) Let $K(q)$ be as defined at (16.1), $|q| < 1$, $\alpha \in \mathbb{C}$ and let q^α be defined in the usual way by

$$q^\alpha = e^{\alpha \log q},$$

where $\log q$ is the principal logarithm of q. Then

$$K(q) = 1 - q^\alpha + \frac{q^\alpha}{1} - \frac{q^{1-\alpha}}{1} + \frac{q^{1-\alpha}}{1} - \frac{q^{1+\alpha}}{1} + \frac{q^{1+\alpha}}{1}$$
$$- \frac{q^{2-\alpha}}{1} + \frac{q^{2-\alpha}}{1} - \frac{q^{2+\alpha}}{1} + \frac{q^{2+\alpha}}{1} - \cdots. \tag{16.14}$$

Proof. In Theorem 16.5, set $b_0 = 0$, $a_1 = c_1$, and, for $k \geq 1$, $b_i = 1$, $a_{2i} = -c_{i+1}$ and $a_{2i+1} = c_{i+1}$ and the result in (i) follows.

For (ii), in (16.12) and (16.13) let $c_1 = q^\alpha$ and, for $k \geq 1$, let $c_{2k} = q^{k-\alpha}$ and $c_{2k+1} = q^{k+\alpha}$. This gives that the odd part of

$$\frac{q^\alpha}{1} - \frac{q^{1-\alpha}}{1} + \frac{q^{1-\alpha}}{1} - \frac{q^{1+\alpha}}{1} + \frac{q^{1+\alpha}}{1} - \frac{q^{2-\alpha}}{1} + \frac{q^{2-\alpha}}{1} - \frac{q^{2+\alpha}}{1} + \frac{q^{2+\alpha}}{1} - \cdots$$

is

$$q^\alpha + \frac{q}{1} + \frac{q^2}{1} + \frac{q^3}{1} + \cdots.$$

Since a tail of the left side of (16.14) converges (by Worpitzky's Theorem) and its odd part converges to $K(q)$, this proves the result. □

Remark: The $\alpha = 0$ case of the identity at (16.14) has a somewhat curious history. It was conjectured by Brillhart and Blecksmith in an email [80] to Bruce C. Berndt, was proved by Berndt and Yee in a paper [74] that appeared in 2003, yet first appeared (with proof) in the overlooked 1946 paper [118, Eq. (6.13)] of Evelyn Frank.

16.2.3 Equivalence transformations and equivalent continued fractions

Two continued fractions $b_0 + K_{n=1}^{\infty} a_n/b_n$ and $d_0 + K_{n=1}^{\infty} c_n/d_n$ are said to be *equivalent* if they have the same sequence of approximants $\{f_n\}$. If the N-th canonical numerator and denominator for $b_0 + K_{n=1}^{\infty} a_n/b_n$ are denoted by A_N and B_N, and those for $d_0 + K_{n=1}^{\infty} c_n/d_n$ by C_N and D_N, then $A_N/B_N = C_N/D_N$. It is not difficult to show (see [184, pp. 77–78]) that the equivalence of the two continued fractions implies the existence of a sequence of non-zero numbers r_n, $n \geq 0$, with $r_0 = 1$, and for $n \geq 1$,

$$c_n = r_n r_{n-1} a_n, \qquad d_n = r_n b_n,$$

$$C_n = A_n \prod_{k=0}^{n} r_k, \qquad D_n = B_n \prod_{k=0}^{n} r_k. \tag{16.15}$$

Note that the fact that the application of each step in an equivalence transformation leaves the approximants unchanged is essentially due to the fact that if α/β is any rational expression and r is any non-zero number, then $\alpha/\beta = (r\alpha)/(r\beta)$ (in what follows below, recall that $r_0 = 1$):

$$b_0 + \frac{a_1}{b_1} + \frac{a_2}{b_2} + \frac{a_3}{b_3} + \cdots$$
$$= b_0 + \frac{r_1 r_0 a_1}{r_1 b_1} + \frac{r_1 a_2}{b_2} + \frac{a_3}{b_3} + \frac{a_4}{b_4} + \cdots$$
$$= b_0 + \frac{r_1 r_0 a_1}{r_1 b_1} + \frac{r_2 r_1 a_2}{r_2 b_2} + \frac{r_2 a_3}{b_3} + \frac{a_4}{b_4} + \cdots$$
$$= b_0 + \frac{r_1 r_0 a_1}{r_1 b_1} + \frac{r_2 r_1 a_2}{r_2 b_2} + \frac{r_3 r_2 a_3}{r_3 b_3} + \frac{r_3 a_4}{b_4} + \cdots$$
$$= \cdots$$

16.2.4 From a sequence to a continued fraction

It is often more convenient to have a formula for generating a continued fraction whose n-th approximant is the n-th term in a single sequence $\{f_k\}$, rather than as the ratio of the n-th terms in two sequences, as was described in subsection 16.2.1. For example, it may be desired to represent an infinite product or an infinite series as a continued fraction in which the n-th approximant is equal to, respectively, the n-th partial product or the n-th partial sum. The following result, due originally to Daniel Bernoulli [76] follows from Theorem 16.3 via a sequence of equivalence transformations (see also [165, pp. 11–12], [184, page 51, Q24]).

Theorem 16.6. *Let $\{f_n\}_{n\geq 0}$ be a sequence of complex numbers such that $f_{i+1} \neq f_i$, $f_{i+2} \neq f_i$ for for $i \geq 0$. Then $\{f_n\}_{n\geq 0}$ is the sequence of approximants of the continued fraction*

$$f_0 + \frac{f_1 - f_0}{1} - \frac{f_2 - f_1}{f_2 - f_0} - \frac{(f_1 - f_0)(f_3 - f_2)}{f_3 - f_1} -$$
$$\cdots - \frac{(f_{n-2} - f_{n-3})(f_n - f_{n-1})}{f_n - f_{n-2}} - \ldots . \quad (16.16)$$

Proof. From (16.8) and (16.9) the continued fraction with n-th approximant $f_n = A_n/B_n$ is

$$A_0 + \frac{A_1 - A_0 B_1}{B_1} - \frac{\frac{A_2 B_1 - A_1 B_2}{A_1 B_0 - A_0 B_1}}{\frac{A_2 B_0 - A_0 B_2}{A_1 B_0 - A_0 B_1}} - \frac{\frac{A_3 B_2 - A_2 B_3}{A_2 B_1 - A_1 B_2}}{\frac{A_3 B_1 - A_1 B_3}{A_2 B_1 - A_1 B_2}} -$$
$$\cdots - \frac{\frac{A_n B_{n-1} - A_{n-1} B_n}{A_{n-1} B_{n-2} - A_{n-2} B_{n-1}}}{\frac{A_n B_{n-2} - A_{n-2} B_n}{A_{n-1} B_{n-2} - A_{n-2} B_{n-1}}} - \cdots$$

$$= f_0 + \frac{B_1(f_1 - f_0)}{B_1} - \frac{\frac{B_2(f_2 - f_1)}{(f_1 - f_0)}}{\frac{B_2(f_2 - f_0)}{B_1(f_1 - f_0)}} - \frac{\frac{B_3(f_3 - f_2)}{B_1(f_2 - f_1)}}{\frac{B_3(f_3 - f_1)}{B_2(f_2 - f_1)}} -$$
$$\cdots - \frac{\frac{B_n(f_n - f_{n-1})}{B_{n-2}(f_{n-1} - f_{n-2})}}{\frac{B_n(f_n - f_{n-2})}{B_{n-1}(f_{n-1} - f_{n-2})}} - \cdots$$

The result now follows after applying an equivalence transformation to clear denominators and eliminate the B_i.   □

The following special cases are useful

Corollary 16.2. *(i) Let $\sum_{i=0}^{\infty} a_i$ be a convergent infinite series such that $a_i \neq 0$ and $a_i + a_{i+1} \neq 0$ for any $i \geq 1$. Then*

$$\sum_{i=0}^{\infty} a_i = a_0 + \frac{a_1}{1} + \frac{-a_2}{a_2 + a_1} + \frac{-a_1 a_3}{a_3 + a_2} + \cdots + \frac{-a_{n-2} a_n}{a_n + a_{n-1}} + \cdots . \quad (16.17)$$

(ii) Let $\{b_i\}_{i\geq 0}$ be a sequence of numbers such that $b_i \neq 0$ for $i \geq 0$, $b_i \neq -1$ for $i \geq 2$, and $\sum_{n=0}^{\infty} \prod_{i=0}^{n} b_i$ converges. Then

$$\sum_{n=0}^{\infty} \prod_{i=0}^{n} b_i = b_0 + \frac{b_0 b_1}{1} + \frac{-b_2}{b_2 + 1} + \frac{-b_3}{b_3 + 1} + \cdots + \frac{-b_n}{b_n + 1} + \cdots. \quad (16.18)$$

(iii) If $\prod_{i=1}^{\infty} a_i$ is a convergent infinite product with no $a_i = 0$ or 1, and no $a_i a_{i+1} = 1$, then

$$\prod_{i=1}^{\infty} a_i = 1 + \frac{a_1 - 1}{1} - \frac{a_1(a_2 - 1)}{a_2 a_1 - 1} + K_{n=3}^{\infty} \frac{-a_{n-1}(a_{n-2} - 1)(a_n - 1)}{a_n a_{n-1} - 1}. \quad (16.19)$$

(iv) If $\prod_{i=1}^{\infty}(1 + b_i)$ is a convergent infinite product with no $b_i = 0$ or -1, and no $b_{i+1} = -b_i/(1 + b_i)$, then

$$\prod_{i=1}^{\infty}(1 + b_i) = 1 + \frac{b_1}{1} + K_{n=2}^{\infty} \frac{-(1 + b_{n-1})b_n/b_{n-1}}{b_n + b_n/b_{n-1} + 1}. \quad (16.20)$$

Proof. For (16.17), let $f_n = \sum_{i=0}^{n} a_i$ in (16.16). For (16.18), let $a_n = \prod_{i=0}^{n} b_i$ in (16.17) and apply an equivalence transformation to introduce ratios of the form a_n/a_{n-1} into the continued fraction. Next, for (16.19), apply an equivalence transformation to (16.16) to get it in the form

$$f_0 + \frac{f_1 - f_0}{1} - \frac{f_1/f_0(f_2/f_1 - 1)}{f_2/f_0 - 1}$$
$$- \frac{f_2/f_1(f_1/f_0 - 1)(f_3/f_2 - 1)}{f_3/f_1 - 1} -$$
$$\cdots - \frac{f_{n-1}/f_{n-2}(f_{n-2}/f_{n-3} - 1)(f_n/f_{n-1} - 1)}{f_n/f_{n-2} - 1} - \cdots,$$

and then set $f_n = \prod_{i=0}^{n} a_i$, with $a_0 = 1$ (this choice for a_0 turns out to be the most useful in practice). Finally, for (16.20) set $a_i = 1 + b_i$ in (16.19) and apply an equivalence transformation to clear denominators. □

These transformations allow elementary, yet still elegant, continued fraction expansions for various q-products and series.

Corollary 16.3. *If $|q| < 1$, then*

$$(q; q)_\infty = 1 - \frac{q}{1} + \frac{q^2 - q}{1 - q^2 + q} + \frac{q^3 - q}{1 - q^3 + q} + \frac{q^4 - q}{1 - q^4 + q} + \cdots, \quad (16.21)$$

$$\sum_{n=0}^{\infty} \frac{q^{n(n+1)/2}}{(q;q)_n} = 1 + \frac{q}{1-q} + \frac{q^3 - q^2}{1} + \frac{q^5 - q^3}{1} + \frac{q^7 - q^4}{1} + \cdots. \quad (16.22)$$

Proof. See Q16.3 and Q16.4 for generalizations of these two identities. □

16.3 The Bauer–Muir transformation of a continued fraction

The Bauer–Muir transform of a continued fraction $b_0 + K(a_n/b_n)$ with respect to a sequence $\{w_n\}$ from $\mathbb{C}$ is the continued fraction $d_0 + K(c_n/d_n)$ whose canonical numerators C_n and denominators D_n are given by

$$
\begin{aligned}
C_{-1} &= 1, & D_{-1} &= 0, & (16.23) \\
C_n &= A_n + w_n A_{n-1}, & D_n &= B_n + w_n B_{n-1}
\end{aligned}
$$

for $n = 0, 1, 2 \ldots$, where $\{A_n\}$ and $\{B_n\}$ are the canonical numerators and denominators of $b_0 + K(a_n/b_n)$. Bauer [66] (in 1872) and Muir [205] (in 1877) independently proved the following theorem:

Theorem 16.7. *The Bauer–Muir transform of $b_0 + K(a_n/b_n)$ with respect to $\{w_n\}$ from $\mathbb{C}$ exists if and only if*

$$\lambda_n := a_n - w_{n-1}(b_n + w_n) \neq 0 \text{ for } n = 1, 2, 3, \ldots. \quad (16.24)$$

If it exists, then it is given by

$$b_0 + w_0 + \frac{\lambda_1}{b_1 + w_1} + \frac{a_1 \lambda_2 / \lambda_1}{b_2 + w_2 - w_0 \lambda_2 / \lambda_1} + \frac{a_2 \lambda_3 / \lambda_2}{b_3 + w_3 - w_1 \lambda_3 / \lambda_2} + \cdots. \quad (16.25)$$

Proof. The claims in the theorem follow directly from Theorem 16.3, with C_i replacing A_i and D_i replacing B_i. The details are left as an exercise (see Q16.7). □

Note that it is not guaranteed that the sequences $\{A_n/B_n\}$ and $\{C_n/D_n\}$ converge to the same limit. In fact, solving

$$\frac{C_n}{D_n} = \frac{A_n + w_n A_{n-1}}{B_n + w_n B_{n-1}} \quad (16.26)$$

for w_n shows that the sequence $\{C_n/D_n\}$ may exhibit *any* behaviour by choosing the sequence $\{w_n\}$ appropriately. The sequence $\{C_n/D_n\}$ is the sequence of modified approximants of the continued fraction $K_{n=1}^{\infty} a_n/b_n$ (with respect to the sequence $\{w_n\}$) and thus the convergence of the Bauer–Muir transformation of $K_{n=1}^{\infty} a_n/b_n$ is thus related to the *general convergence* of $K_{n=1}^{\infty} a_n/b_n$ (see Jacobsen [155]). Jacobsen [158] investigated the question of when a continued fraction and its Bauer–Muir transformation converge to the same limit.

Theorem 16.8. *(Jacobsen, [158, page 497]) Let $b_0 + K(a_n/b_n)$; $a_n, b_n \in \mathbb{C}$, $a_n \neq 0$ converge to a value $f \in \hat{\mathbb{C}} = \mathbb{C} \cup \{\infty\}$. Let $\{g_n\}$ be a sequence of wrong tails; i.e., $g_n \in \hat{\mathbb{C}}$, $g_0 \neq f - b_0$, and*

$$g_{n-1} = \frac{a_n}{b_n + g_n} \quad for \ n = 1, 2, 3, \ldots.$$

Finally, let $\{w_n\}$; $w_n \in \mathbb{C}$ satisfy (16.24) for all $n \in \mathbb{N}$.

If $\liminf d(w_n, g_n) > 0$, where $d(.,.)$ denotes the chordal metric on the Riemann sphere $\hat{\mathbb{C}}$, then the Bauer–Muir transformation

$$b_0 + w_0 + \frac{\lambda_1}{b_1 + w_1} + \frac{a_1 \lambda_2/\lambda_1}{b_2 + w_2 - w_0 \lambda_2/\lambda_1} + \frac{a_2 \lambda_3/\lambda_2}{b_3 + w_3 - w_1 \lambda_3/\lambda_2} + \cdots$$

converges to the same value f.

This theorem may be difficult to use in practice for a particular continued fraction, unless some information is available about the limiting behaviour of wrong tail sequences.

The following theorem describes a continued fraction which is a common extension of the continued fraction $K_{n=1}^{\infty} a_n/b_n$ and its Bauer–Muir transformation with respect to a sequence $\{w_n\}$. If this extension converges, then it guarantees that $K_{n=1}^{\infty} a_n/b_n$ and its Bauer–Muir transformation with respect to the sequence $\{w_n\}$ converge to the same limit. It has a number of interesting applications, including another extension of the Rogers–Ramanujan continued fraction.

Theorem 16.9. *(Mc Laughlin and Wyshinski [198, Theorem 5])*
Let $\{w_n\}_{n=0}^{\infty}$ be a sequence from $\mathbb{C}$ such that $w_0 = 0$ and $w_n \neq 0$ for $n \geq 1$. We suppose further that $\{a_n\}_{n=1}^{\infty}$ and $\{b_n\}_{n=1}^{\infty}$ are sequences from $\mathbb{C}$ such that $a_n - w_{n-1}(b_n + w_n) \neq 0$ for $n = 1, 2, 3, \ldots$. Then the even part of the continued fraction

$$\frac{a_1}{b_1 + w_1} + \frac{-w_1}{1} + \frac{a_2/w_1}{b_2 + w_2 - a_2/w_1} + \frac{-w_2}{1} + \frac{a_3/w_2}{b_3 + w_3 - a_3/w_2} +$$

$$\cdots + \frac{-w_{n-1}}{1} + \frac{a_n/w_{n-1}}{b_n + w_n - a_n/w_{n-1}} + \frac{-w_n}{1} + \cdots \quad (16.27)$$

is $K_{n=1}^\infty a_n/b_n$ and its odd part is equal to the Bauer–Muir transform of $K_{n=1}^\infty a_n/b_n$ with respect to the sequence $\{w_n\}_{n=0}^\infty$.

Proof. The proofs are mostly just direct applications of Theorems 16.4 and 16.5. The theorem says that the odd part "is equal to" the Bauer–Muir transform of $K_{n=1}^\infty a_n/b_n$ with respect to the sequence $\{w_n\}_{n=0}^\infty$, rather than saying it "is" this continued fraction, so the odd part needs an application of the identity

$$\frac{E}{F} + \frac{G}{H + \alpha} = \frac{E}{F} - \frac{EG/F^2}{H + G/F + \alpha}. \quad (16.28)$$

to the beginning of the continued fraction to complete the proof. The details are once again left as an exercise (see Q16.8). □

Remark: Of course Theorem 16.9 may not be applied if any of the $w_n = 0$.

An application of this result is the following quite general extension of the Rogers–Ramanujan continued fraction $K(q)$.

Corollary 16.4. *(A) Let $\{\alpha_n\}_{n=1}^\infty$ be any sequence of complex numbers such that*

$$(i) \lim_{n\to\infty} Re(\alpha_n) = +\infty,$$

$$(ii) \lim_{n\to\infty} Re(n - \alpha_n) = +\infty.$$

Then

$$K(q) = 1 + q^{\alpha_1} - \frac{q^{\alpha_1}}{1} + \frac{q^{1-\alpha_1}}{1 + q^{\alpha_2} - q^{1-\alpha_1}} - \frac{q^{\alpha_2}}{1} + \frac{q^{2-\alpha_2}}{1 + q^{\alpha_3} - q^{2-\alpha_2}} -$$

$$\cdots - \frac{q^{\alpha_n}}{1} + \frac{q^{n-\alpha_n}}{1 + q^{\alpha_{n+1}} - q^{n-\alpha_n}} - \cdots. \quad (16.29)$$

$$(B)\ K(q) = 1 + q^{1/2} - \frac{q^{1/2}}{1} + \frac{q^{1/2}}{1} - \frac{q^{1/2}}{1} + \frac{q^{3/2}}{1} - \frac{q^{3/2}}{1} + \frac{q^{3/2}}{1} - \frac{q^{3/2}}{1}$$

$$+ \cdots + \frac{q^{(2n+1)/2}}{1} - \frac{q^{(2n+1)/2}}{1} + \frac{q^{(2n+1)/2}}{1} - \frac{q^{(2n+1)/2}}{1} + \cdots \quad (16.30)$$

Proof. The restrictions on the sequence $\{\alpha_n\}$ guarantee that the continued fraction converges by Worpitzky's theorem (after an equivalence transformation), and hence converges to the same limit as its odd part, $K(q)$. Part (B) follows the choices $\alpha_{2n-1} = \alpha_{2n} = (2n-1)/2$. Note that this latter continued fraction for $K(q)$ is similar, but not identical, to the $\alpha = 1/2$ case of the continued fraction given at (16.14) (there is a different pattern of $+/-$ signs). □

As another example of an application of Theorem 16.9, we illustrate how it may be used to show that a continued fraction and its Bauer–Muir transformation converge to the same limit, without having to rely on any information about the limiting behaviour of wrong tail sequences. In [41, pp. 13–14], the authors prove that the Bauer–Muir transformation of the continued fraction

$$1 - qx + \cfrac{q^3 x}{1 + q^2 - q^3 x} + \cfrac{q^7 x}{1 + q^4 - q^5 x} + \cfrac{q^{11} x}{1 + q^6 - q^7 x} + \cdots \qquad (16.31)$$

with respect to the sequence $w_k = -q^{2k}$ for $k \geq 1$ is the continued fraction

$$-qx + \cfrac{1}{1 - q^3 x} + \cfrac{q^5 x}{1 + q^2 - q^5 x} + \cfrac{q^9 x}{1 + q^4 - q^7 x} + \cdots. \qquad (16.32)$$

To prove that the first continued fraction and the sequence $\{w_k\}$ satisfy the requirements of Theorem 16.8, and thus that the two continued fractions converge to the same limit, the authors rely on results of one of the authors from another paper [156] on limit-periodic continued fractions to prove the required asymptotic behaviour of the wrong tail sequences. Here we will instead use Theorem 16.9 to derive a convergent continued fraction whose even part is the continued fraction at (16.31), and whose odd part equals the continued fraction at (16.32). Inserting the values $a_k = q^{4k-1} x$, $b_k = 1 + q^{2k} - q^{2k+1} x$ and $w_k = -q^{2k}$ into (16.27) (and pre-pending an initial $1 - qx$), it can be seen that this continued fraction is

$$1 - qx + \cfrac{q^3 x}{1 - q^3 x} + \cfrac{q^2}{1} - \cfrac{q^5 x}{1} + \cfrac{q^4}{1} - \cfrac{q^7 x}{1} + \cfrac{q^6}{1} - \cfrac{q^9 x}{1} + \cdots. \qquad (16.33)$$

This continued fraction is convergent by Worpitzky's theorem, so its odd and even parts converge to the same value. It is an easy check that, as Theorem 16.9 predicts, the even part of this last continued fraction is that at (16.31). The odd part, after a little elementary algebra is applied to some of the initial partial quotients, is found to be

$$- qx + \frac{1}{1 - q^3 x} - \frac{q^5 x/(1 - q^3 x)^2}{1 + q^2 - q^5 x + q^5 x/(1 - q^3 x)}$$
$$+ \frac{q^9 x}{1 + q^4 - q^7 x} + \frac{q^{13} x}{1 + q^6 - q^9 x} + \cdots,$$

and this is seen to be equal to the continued fraction at (16.32), after an application of (16.28) with $E = 1$, $F = 1-q^3 x$, $G = q^5 x$ and $H = 1+q^2-q^5 x$.

In some cases, an infinite sequence of Bauer–Muir transformations are applied to prove the equality of two continued fractions, as the following example shows.

Corollary 16.5. *Let x and q be real numbers with $|q| < 1$. Then*

$$1 - qx + \frac{q^3 x}{1 + q^2 - q^3 x} + \frac{q^7 x}{1 + q^4 - q^5 x} + \frac{q^{11} x}{1 + q^6 - q^7 x} + \cdots$$
$$= \lim_{n \to \infty} -qx + \frac{1}{-q^3 x} + \frac{1}{-q^5 x} + \frac{1}{-q^7 x} + \cdots + \frac{1}{-q^{2n+1} x + 1}. \quad (16.34)$$

Proof. Let $f(x; q)$ denote the continued fraction at (16.31). Then, as the authors in [41] point out, the equality of (16.31) and (16.32) imply that

$$f(x; q) = -qx + \frac{1}{f(q^2 x; q)}.$$

This expression may be iterated (or equivalently, a sequence of similar Bauer–Muir transformations may be applied to the successive tails of the continued fraction) to get that after $n + 1$ steps

$$f(x; q) = -qx + \frac{1}{-q^3 x} + \frac{1}{-q^5 x} + \frac{1}{-q^7 x} + \cdots + \frac{1}{-q^{2n+1} x} + \frac{1}{f(q^{2n} x; q)}.$$

After an equivalence transformation, Worpitzky's theorem gives that $f(q^{2n} x; q) \to 1$ as $n \to \infty$, giving the result. $\qquad \square$

16.3.1   A class of continued fractions that are easily shown to converge to their Bauer–Muir transforms

As remarked above, it is not necessarily the case that the Bauer–Muir transformation of a continued fraction with respect to a sequence $\{\omega_k\}$ converges, or if it does converge, that it converges to the same limit as the original continued fraction.

However, if the continued fractions is such that the numerator- and denominator convergents converge separately, and the sequence $\{\omega_k\}$ is sufficiently well-behaved (for example, if the sequence has a limit different from -1) then a continued fraction and its Bauer–Muir transform with respect to the sequence $\{\omega_k\}$ do converge to the same limit.

Let the n-th approximant of the continued fraction $b_0 + K_{n=1}^{\infty}(a_n/b_n)$ be denoted by A_n/B_n and let the n-th approximant of its transformation with respect to the sequence $\{\omega_n\}$ be denoted by C_n/D_n. From (16.26) it follows that for $n \geq 0$,

$$C_n = A_n + \omega_n A_{n-1}, \qquad D_n = B_n + \omega_n B_{n-1}, \qquad \implies \frac{C_n}{D_n} = \frac{A_n + \omega_n A_{n-1}}{B_n + \omega_n B_{n-1}}.$$

Now suppose that the sequences $\{A_n\}$ and $\{B_n\}$ converge separately, i.e.

$$\lim_{n \to \infty} A_n = A, \quad \lim_{n \to \infty} B_n = B, \quad \implies \lim_{n \to \infty} \frac{A_n}{B_n} = \frac{A}{B}, \quad \text{for some } A, B \in \mathbb{C}.$$

Suppose further that $\lim_{n \to \infty} \omega_n = \omega \neq -1$. Then

$$\lim_{n \to \infty} \frac{C_n}{D_n} = \lim_{n \to \infty} \frac{A_n + \omega_n A_{n-1}}{B_n + \omega_n B_{n-1}} = \frac{A + \omega A}{B + \omega B} = \frac{A}{B},$$

and thus the continued fraction $b_0 + K_{n=1}^{\infty}(a_n/b_n)$ and its Bauer–Muir transformation with respect to the sequence $\{\omega_n\}$ converge to the same limit.

This is of particular relevance for q-continued fractions, since it will be shown in the next section that many q-continued fractions do have the property that numerators and denominators do converge separately.

16.4 General Continued Fraction Identities

In this section we investigate three general continued fraction identities, forms of two of which were stated by Ramanujan (see [38, Chapter 6]), with a form equivalent to the third type having been stated by Hirschhorn (see [144]).

16.4.1 A first general continued fraction identity

To prove a group of less elementary continued fraction identities, stronger general results are necessary. Although the proofs of the following two theorems are a little long, even if they essentially involve only substitutions from the q-binomial theorem and rearrangement of series, they then provide as payoff straightforward proofs of many beautiful q-continued fraction identities.

For space-saving reasons, we occasionally use a notation employed by Ramanujan and the authors in [38, **Entry 6.2.1**], and define

$$G(a, \lambda; b; q) := \sum_{n=0}^{\infty} \frac{(-\lambda/a; q)_n a^n q^{n(n+1)/2}}{(-bq, q; q)_n}. \tag{16.35}$$

Noting that letting $a \to 0$ in (16.35) gives

$$G(0, \lambda; b; q) := \sum_{n=0}^{\infty} \frac{\lambda^n q^{n^2}}{(-bq, q; q)_n}. \tag{16.36}$$

For clarity, we prefer to use the *series* that defines $G(a, \lambda; b; q)$ instead of the "$G(a, \lambda; b; q)$" notation, unless the expression contains a quotient of such series, in which case we employ the latter notation.

Theorem 16.10. *Let* $|q| < 1$ *and define*

$$H(a, b, c, d, q) := \frac{1}{1} + \frac{-ab + cq}{a + b + dq} + \frac{-ab + cq^2}{a + b + dq^2} + \cdots + \frac{-ab + cq^n}{a + b + dq^n} + \cdots .$$

(i) Let $A_N := A_N(q)$ *and* $B_N := B_N(q)$ *denote the N-th numerator convergent and N-th denominator convergent, respectively, of $H(a, b, c, d, q)$. Then A_N and B_N are given explicitly by the formulae*

$$A_N =$$

$$\sum_{j,l,n \geq 0} a^j b^{N-1-n-j-l} c^l d^{n-l} q^{n(n+1)/2+l(l+1)/2} \begin{bmatrix} n+j \\ j \end{bmatrix} \begin{bmatrix} N-1-j-l \\ n \end{bmatrix} \begin{bmatrix} n \\ l \end{bmatrix}. \tag{16.37}$$

For $N \geq 2$,

$$B_N = A_N + (cq - ab) \times$$

$$\sum_{j,l,n \geq 0} a^j b^{N-2-n-j-l} c^l d^{n-l} q^{n(n+3)/2+l(l+1)/2} \begin{bmatrix} n+j \\ j \end{bmatrix} \begin{bmatrix} N-2-j-l \\ n \end{bmatrix} \begin{bmatrix} n \\ l \end{bmatrix}. \tag{16.38}$$

(ii) If $|a/b| < 1$ *and* $|q| < 1$ *then* $H(a, b, c, d, q)$ *converges and*

$$\frac{1}{H(a, b, c, d, q)} - 1 = (cq/b - a) \frac{G(dq/b, cq^2/b^2; -a/b; q)}{G(d/b, cq/b^2; -a/b; q)}. \tag{16.39}$$

(iii) If $|q| < 1$, $|a| < 1$ and $b = 1$, then the numerators and denominators converge separately and

$$\lim_{N \to \infty} A_N = \sum_{n=0}^{\infty} \frac{d^n q^{n(n+1)/2}(-cq/d)_n}{(a)_{n+1}(q)_n}, \tag{16.40}$$

$$\lim_{N \to \infty} B_N = \sum_{n=0}^{\infty} \frac{d^n q^{n(n+1)/2}(-cq/d)_n}{(a)_{n+1}(q)_n} + (cq - a) \sum_{n=0}^{\infty} \frac{d^n q^{n(n+3)/2}(-cq/d)_n}{(a)_{n+1}(q)_n}. \tag{16.41}$$

Remarks: (a) By symmetry the conditions on a and b in (i) and (ii) can be interchanged, in which case a and b are interchanged on the right sides. (b) The left side in (ii) is displayed as shown to simplify the representation on the right side.
(c) A similar continued fraction, namely,

$$1 + a + d + \frac{-a + cq}{a + 1 + dq} + \frac{-a + cq^2}{a + 1 + dq^2} + \cdots + \frac{-a + cq^n}{a + 1 + dq^n} + \cdots,$$

was investigated by Hirschhorn in [144] and [145], and he derived many of the well-known continued fraction identities as corollaries of his main result. A further continued fraction identity of a similar type was given by Bhargava and Adiga [77, Eq. II, page 14].

Proof. (i) The method of proof here is that of Hirschhorn in [144]. Set $F(t) = \sum_{N \geq 1} A_N t^n$ and $G(t) = \sum_{N \geq 1} B_N t^n$. From the recurrence relations at (16.4) it follows that

$$A_{N+1} = (a + b + dq^N)A_N + (-ab + cq^N)A_{N-1}$$

holds for $N \geq 1$. If this equation is multiplied by t^{N+1} and summed over $N \geq 1$, we get (noting that $A_1 = 1$ and $A_0 = 0$) that

$$F(t) - t = t(a + b)F(t) + tdF(tq) - abt^2 F(t) + ct^2 qF(tq).$$

The last equation can be rewritten to give the following recurrence relation for $F(t)$:

$$F(t) = \frac{t}{(1 - at)(1 - bt)} + \frac{t(d + ctq)}{(1 - at)(1 - bt)} F(tq).$$

Upon iterating this equation and using the fact that $F(0) = 0$, the following series for $F(t)$ results:

$$F(t) = \sum_{n \geq 1} \frac{t^n d^{n-1}(-ctq/d)_{n-1} q^{n(n-1)/2}}{(at)_n (bt)_n}. \tag{16.42}$$

The special cases of the q-binomial theorem at (3.5)–(3.6) are applied to the q-products in the expression above to give that

$$F(t) =$$

$$\sum_{\substack{n\geq 1,\\ j,k,l\geq 0}} t^{n+j+k+l} d^{n-1-l} a^j b^k c^l q^{\frac{n(n-1)}{2}+\frac{l(l+1)}{2}} \begin{bmatrix} n+j-1 \\ j \end{bmatrix} \begin{bmatrix} n+k-1 \\ k \end{bmatrix} \begin{bmatrix} n-1 \\ l \end{bmatrix}$$

$$= \sum_{j,k,l,n\geq 0} t^{n+1+j+k+l} d^{n-l} a^j b^k c^l q^{\frac{n(n+1)}{2}+\frac{l(l+1)}{2}} \begin{bmatrix} n+j \\ j \end{bmatrix} \begin{bmatrix} n+k \\ n \end{bmatrix} \begin{bmatrix} n \\ l \end{bmatrix}.$$

Finally, we organize $F(t)$ as a series in t by setting $N = n + j + k + l + 1$, substituting for k and then use the definition of $F(t)$ to get (16.37).

By a similar argument, it follows that

$$G(t) = \sum_{n\geq 1} \frac{t^n d^{n-1}(-ctq/d)_{n-1} q^{n(n-1)/2}}{(at)_n (bt)_n} (1 + (cq - ab)tq^{n-1})$$

$$= F(t) + (cq - ab) \sum_{n\geq 1} \frac{t^{n+1} d^{n-1}(-ctq/d)_{n-1} q^{(n-1)(n+2)/2}}{(at)_n (bt)_n}$$

$$= F(t) + (cq - ab) \sum_{n\geq 0} \frac{t^{n+2} d^n (-ctq/d)_n q^{n(n+3)/2}}{(at)_{n+1} (bt)_{n+1}}$$

$$= F(t) + (cq - ab) \sum_{j,k,l,n\geq 0} t^{n+2+j+k+l} d^{n-l} a^j b^k c^l q^{\frac{n(n+3)}{2}+\frac{l(l+1)}{2}}$$

$$\times \begin{bmatrix} n+j \\ j \end{bmatrix} \begin{bmatrix} n+k \\ n \end{bmatrix} \begin{bmatrix} n \\ l \end{bmatrix}.$$

We again reorganize the second series as a series in t by setting $N = n + j + k + l + 2$, and then substitute for k and use the definition of $G(t)$ to get (16.38).

(ii) The expression for A_N in (16.37) can be re-written as

$$A_N = b^{N-1} \sum_{n\geq 0} (d/b)^n q^{n(n+1)/2} \sum_{j\geq 0} \begin{bmatrix} n+j \\ j \end{bmatrix} (a/b)^j \tag{16.43}$$

$$\times \sum_{l\geq 0} \begin{bmatrix} N-1-j-l \\ n \end{bmatrix} q^{l(l-1)/2} \left(\frac{cq}{bd}\right)^l \begin{bmatrix} n \\ l \end{bmatrix}.$$

From the definition of the Gaussian polynomials at (14.1), j, l and n are restricted by $l \leq n$ and $l + j + n \leq N - 1$.

Likewise, the expression for B_N in (16.38) can be re-written as

$$B_N = A_N + b^{N-1}(cq/b - a)\sum_{n\geq 0}(d/b)^n q^{n(n+3)/2}\sum_{j\geq 0}\begin{bmatrix} n+j \\ j \end{bmatrix}(a/b)^j$$
$$\times \sum_{l\geq 0}\begin{bmatrix} N-2-j-l \\ n \end{bmatrix}q^{l(l-1)/2}\left(\frac{cq}{bd}\right)^l\begin{bmatrix} n \\ l \end{bmatrix}.$$

Thus

$$\lim_{N\to\infty}\frac{B_N - A_N}{b^{N-1}} = (cq/b - a)\sum_{n\geq 0}(d/b)^n q^{n(n+3)/2}\sum_{j\geq 0}\begin{bmatrix} n+j \\ j \end{bmatrix}(a/b)^j$$
$$\times \sum_{l=0}^{n}\frac{q^{l(l-1)/2}\left(\frac{cq}{bd}\right)^l}{(q)_n}\begin{bmatrix} n \\ l \end{bmatrix}$$
$$= (cq/b - a)\sum_{n=0}^{\infty}\frac{(d/b)^n q^{n(n+3)/2}(-cq/db)_n}{(a/b)_{n+1}(q)_n}.$$

Here (3.6) has been used on the inner sum over j, while (4.11) (with $c = 0$ and b replaced with $-cq/(bd)$) has been used on the sum over l.

Likewise the expression for A_N above gives that

$$\lim_{N\to\infty}\frac{A_N}{b^{N-1}} = \sum_{n=0}^{\infty}\frac{(d/b)^n q^{n(n+1)/2}(-cq/db)_n}{(a/b)_{n+1}(q)_n}.$$

Equation (16.39) now follows.

(iii) If the replacement $b = 1$ is now made in the limit above, it is an immediate consequence that

$$\lim_{N\to\infty}A_N = \sum_{n\geq 0}\frac{d^n q^{n(n+1)/2}(-cq/d)_n}{(a)_{n+1}(q)_n}.$$

This proves (16.40). The proof of (16.41) is similar. □

16.4.2 Continued fraction expansion of infinite products - I

One implication of (16.39) is that certain ratios of infinite q-products have elegant expansions as continued fractions. In particular, suppose each of the series

$$\sum_{n=0}^{\infty}\frac{(A;q)_n z^n q^{n(n+1)/2}}{(B,q;q)_n} \quad \text{and} \quad \sum_{n=0}^{\infty}\frac{(A;q)_n z^n q^{n(n+3)/2}}{(B,q;q)_n} \qquad (16.44)$$

becomes the series side of an identity of the Rogers–Ramanujan–Slater type for particular choices of A, B and z (and possibly also replacing q with q^2 or some higher power of q). If a, b, c and d are specialized in (16.39) so that the two series on the right side of this identity match the two series in the Slater-type identity (with q possibly also being replaced with q^2 or some higher power of q), then there results an identity in which the ratio of the corresponding infinite products are expressed as an infinite continued fraction. We give several examples, but before coming to these, we state some other special cases of (16.39) that result in ratios of series of slightly different type. We partly follow the exposition of Hirschhorn [144], although some particular identities were proved previously.

Corollary 16.6. *If* $|a|, |q| < 1$, *then*

$$1 + \frac{-a + cq}{a + 1} + \cdots + \frac{-a + cq^n}{a + 1} + \ldots = (1 - a)\frac{G(0, c; -a/q; q)}{G(0, cq; -a; q)}, \qquad (16.45)$$

$$1 + d + \frac{-a}{a + 1 + dq} + \cdots + \frac{-a}{a + 1 + dq^n} + \cdots = (1 - a)\frac{G(d/q, 0; -a/q; q)}{G(d, 0; -a; q)}, \tag{16.46}$$

$$1 + \frac{-a}{a + 1 + dq} + \cdots + \frac{-a}{a + 1 + dq^n} + \cdots = (1 - a)\frac{G(d, 0; -a/q; q)}{G(d, 0; -a; q)}. \tag{16.47}$$

Proof. For (16.45), let $d \to 0$ and set $b = 1$ in (16.39). Then add 1 to each side and simplify the resulting right side.

For (16.46), let $c \to 0$ and set $b = 1$ in (16.39). Then add $1 + d$ to each side and once again simplify the resulting right side.

The identity at (16.47) follows from (16.46) by subtracting d from both sides and simplifying the right side.

The details are once again left as exercise (see Q16.10).   □

The following identities are implications of the identities in Corollary 16.6.

Corollary 16.7. *If* $|q| < 1$, *then*

$$1 + \frac{q}{1} + \frac{q^2}{1} + \frac{q^3}{1} + \cdots = \frac{(q^2, q^3; q^5)_\infty}{(q, q^4; q^5)_\infty}; \tag{16.48}$$

$$1 + \frac{-q + q^2}{1 + q} + \frac{-q + q^4}{1 + q} + \frac{-q + q^6}{1 + q} + \cdots = \frac{(-q^3, -q^5; q^8)_\infty}{(-q, -q^7; q^8)_\infty}. \tag{16.49}$$

$$1 + q + \frac{-q}{1 + q + q^3} + \frac{-q}{1 + q + q^5} + \cdots = \frac{(q^8, q^{12}; q^{20})_\infty}{(q^4, q^{16}; q^{20})_\infty}. \tag{16.50}$$

Proof. For (16.48), let $a \to 0$ and $c \to 1$ in (16.45). The result follows after employing the Rogers–Ramanujan identities (10.17) and (10.18).

Similarly, to obtain (16.49), replace q with q^2, set $c = 1$ and $a = q$ in (16.45), and then use the identities at (10.48) and (10.49) to sum the resulting series.

Replace q with q^2 and set $a = d = q$ in (16.46). Then (16.50) follows after using (VII.1) and (VII.2) to sum the resulting series. $\square$

Remarks: (i) The identity at (16.48) is due originally to Rogers [217], and those at (16.49) and (16.50) are due to Gordon [129].

(ii) By using the same kind of argument as used in the proof of Corollary 16.6 (after setting $b = 1$), it can be shown that if $|a|, |q| < 1$, then

$$1 + a + d + \frac{-a + cq}{1 + a + dq} + \frac{-a + cq^2}{1 + a + dq^2} + \cdots + \frac{-a + cq^n}{1 + a + dq^n} + \cdots$$
$$= \frac{G(d/q, c; -a; q)}{G(d, cq; -a; q)}. \quad (16.51)$$

This is one of the main results of Hirschhorn in [144]. The proof is left as an exercise (see Q16.11).

16.4.3   A second general continued fraction identity

An equivalence transformation is necessary in the proof of the next theorem, which gives a second general formula for the limit of a q-continued fraction with several parameters.

Theorem 16.11. *Let a, b, c, d be complex numbers with $d \neq 0$ and $|q| < 1$. Define*

$$H_1(a, b, c, d, q) := \frac{1}{1} + \frac{-abq + c}{(a+b)q + d} + \cdots + \frac{-abq^{2n+1} + cq^n}{(a+b)q^{n+1} + d} + \cdots.$$

(i) Let $C_N := C_N(q)$ and $D_N := D_N(q)$ denote the N-th numerator convergent and N-th denominator convergent, respectively, of $H_1(a, b, c, d, q)$. Then C_N and D_N are given explicitly by the following formulae.

$$C_N = d^{N-1} \sum_{j, l, n \geq 0} a^j b^{n-j-l} c^l d^{-n-l} q^{n(n+1)/2 + l(l-1)/2}$$
$$\times \begin{bmatrix} N-1-n+j \\ j \end{bmatrix}_q \begin{bmatrix} N-1-j-l \\ n-j-l \end{bmatrix}_q \begin{bmatrix} N-1-n \\ l \end{bmatrix}_q. \quad (16.52)$$

For $N \geq 2$,

$$D_N = C_N + (c/bq - a) \sum_{j,l,n \geq 0} a^j b^{n+1-j-l} c^l d^{N-2-n-l} \times$$

$$q^{(n+1)(n+2)/2+l(l-1)/2} \begin{bmatrix} N-2-n+j \\ j \end{bmatrix}_q \begin{bmatrix} N-2-j-l \\ n-j-l \end{bmatrix}_q \begin{bmatrix} N-2-n \\ l \end{bmatrix}_q .$$

$$\tag{16.53}$$

(ii) If $|q| < 1$ then $H_1(a,b,c,d,q)$ converges and

$$\frac{1}{H_1(a,b,c,d,q)} - 1 = \frac{c - abq}{(d+aq)} \frac{G(bq/d, cq/d^2; aq/d; q)}{G(b/d, c/d^2; a/d; q)}. \tag{16.54}$$

(iii) If $|q| < 1$ and $d = 1$, then the numerators and denominators converge separately and

$$C_\infty := \lim_{N \to \infty} C_N = (-aq)_\infty \sum_{j=0}^\infty \frac{q^{j(j+1)/2} b^j (-c/b)_j}{(q)_j (-aq)_j}. \tag{16.55}$$

$$D_\infty := \lim_{N \to \infty} D_N = C_\infty + (c/q - ab)(-aq)_\infty \sum_{j=0}^\infty \frac{q^{(j+1)(j+2)/2} b^j (-c/b)_j}{(q)_j (-aq)_{j+1}}. \tag{16.56}$$

Proof. The continued fraction $H_1(a,b,c,d,q)$ is derived from $H(a,b,c,d,q)$ by making the substitution $q \to 1/q$ and then applying an equivalence transformation to clear the negative powers of q. With the notation above, the equivalence transformation is defined by setting $r_0 = 1$ and $r_n = q^{n-1}$ for $n \geq 1$, so that

$$\prod_{k=0}^N r_k = q^{N(N-1)/2}.$$

Thus, with $A_N(q)$ being as defined in Theorem 16.10, and employing the formula for C_N at (16.15),

$$C_N = q^{N(N-1)/2} A_N(1/q)$$

$$= \sum_{j,l,n \geq 0} a^j b^{N-1-n-j-l} c^l d^{n-l} q^{(N-n)(N-n-1)/2+l(l-1)/2}$$

$$\times \begin{bmatrix} n+j \\ j \end{bmatrix}_q \begin{bmatrix} N-1-j-l \\ n \end{bmatrix}_q \begin{bmatrix} n \\ l \end{bmatrix}_q$$

$$= \sum_{j,l,n\geq 0} a^j b^{n-j-l} c^l d^{N-1-n-l} q^{n(n+1)/2+l(l-1)/2}$$

$$\times \begin{bmatrix} N-1-n+j \\ j \end{bmatrix}_q \begin{bmatrix} N-1-j-l \\ n-j-l \end{bmatrix}_q \begin{bmatrix} N-1-n \\ l \end{bmatrix}_q$$

$$= d^{N-1} \sum_{n=0}^{N-1} \sum_{j=0}^{n} \sum_{l=0}^{\min(n-j,\,N-1-n)} a^j b^{n-j-l} c^l d^{-n-l} q^{n(n+1)/2+l(l-1)/2}$$

$$\times \begin{bmatrix} N-1-n+j \\ j \end{bmatrix}_q \begin{bmatrix} N-1-j-l \\ n-j-l \end{bmatrix}_q \begin{bmatrix} N-1-n \\ l \end{bmatrix}_q.$$

The transformation at (14.11) has been used to derive the second equality. For the next-to-last step, n has been replaced by $N-1-n$, and in the last step the upper limits on j, l and n come from the definition of the Gaussian polynomials at (14.1). This proves (16.52).

Similarly, with $B_N(q)$ also being as defined in Theorem 16.10,

$$D_N = q^{N(N-1)/2} B_N(1/q)$$

$$= C_N + (c/bq - a) \sum_{j,l,n\geq 0} a^j b^{N-1-n-j-l} c^l d^{n-l} q^{(N-n)(N-n-1)/2+l(l-1)/2}$$

$$\times \begin{bmatrix} n+j \\ j \end{bmatrix}_q \begin{bmatrix} N-2-j-l \\ n \end{bmatrix}_q \begin{bmatrix} n \\ l \end{bmatrix}_q$$

$$= C_N + (c/bq - a) \sum_{j,l,n\geq 0} a^j b^{n+1-j-l} c^l d^{N-2-n-l} q^{(n+1)(n+2)/2+l(l-1)/2}$$

$$\times \begin{bmatrix} N-2-n+j \\ j \end{bmatrix}_q \begin{bmatrix} N-2-j-l \\ n-j-l \end{bmatrix}_q \begin{bmatrix} N-2-n \\ l \end{bmatrix}_q$$

$$= C_N + (c/bq - a) \sum_{n=0}^{N-2} \sum_{j=0}^{n} \sum_{l=0}^{\min(n-j,\,N-2-n)} a^j b^{n+1-j-l} c^l d^{N-2-n-l} \times$$

$$q^{(n+1)(n+2)/2+l(l-1)/2} \begin{bmatrix} N-2-n+j \\ j \end{bmatrix}_q \begin{bmatrix} N-2-j-l \\ n-j-l \end{bmatrix}_q \begin{bmatrix} N-2-n \\ l \end{bmatrix}_q.$$

The second equality follows upon replacing n by $N-n-2$ and the bounds on j, l and n in the last equality follow, as above, from the definition of the Gaussian polynomials at (14.1). This proves (16.53).

From (16.53),

$$\lim_{N\to\infty} \frac{D_N - C_N}{d^{N-1}} = \frac{c/q - ab}{d} \sum_{j,l,n\geq 0} a^j b^{n-j-l} c^l d^{-n-l} \frac{q^{(n+1)(n+2)/2+l(l-1)/2}}{(q)_j (q)_{n-j-l} (q)_l}.$$

$$= \frac{c/q - ab}{d} \sum_{n \geq 0} (b/d)^n q^{(n+1)(n+2)/2} \sum_{j=0}^{n} \frac{(a/b)^j}{(q)_j (q)_{n-j}}$$

$$\times \sum_{l=0}^{n-j} q^{l(l-1)/2} (c/bd)^l \begin{bmatrix} n-j \\ l \end{bmatrix}$$

$$= \frac{c - abq}{dq} \sum_{n \geq 0} (b/d)^n q^{(n+1)(n+2)/2} \sum_{j=0}^{n} \frac{(a/b)^j (-c/bd)_{n-j}}{(q)_j (q)_{n-j}}$$

$$= \frac{c - abq}{dq} \sum_{n \geq 0} (b/d)^n q^{(n+1)(n+2)/2} \sum_{j=0}^{n} \frac{(a/b)^{n-j} (-c/bd)_j}{(q)_j (q)_{n-j}}$$

$$= \frac{c - abq}{dq} \sum_{j=0}^{\infty} \frac{(b/a)^j (-c/bd)_j}{(q)_j} \sum_{n \geq j} \frac{(a/d)^n q^{(n+1)(n+2)/2}}{(q)_{n-j}}$$

$$= \frac{c - abq}{dq} \sum_{j=0}^{\infty} \frac{(b/a)^j (-c/bd)_j}{(q)_j} \sum_{n \geq 0} \frac{(a/d)^{n+j} q^{n(n+3)/2+jn+(j+1)(j+2)/2}}{(q)_n}$$

$$= \frac{c - abq}{dq} \sum_{j=0}^{\infty} \frac{(b/d)^j (-c/bd)_j q^{(j+1)(j+2)/2}}{(q)_j} \sum_{n \geq 0} \frac{(aq^{j+2}/d)^n q^{n(n-1)/2}}{(q)_n}$$

$$= \frac{c - abq}{dq} \sum_{j=0}^{\infty} \frac{(b/d)^j (-c/bd)_j q^{(j+1)(j+2)/2}}{(q)_j} (-aq^{j+2}/d)_{\infty}$$

$$= \frac{c - abq}{dq} (-aq^2/d)_{\infty} \sum_{j=0}^{\infty} \frac{(b/d)^j (-c/bd)_j \, q^{(j+1)(j+2)/2}}{(q)_j (-aq^2/d)_j}.$$

The next-to-last equation follows from the special case (3.4) of the q-binomial theorem. It follows similarly from (16.52) that

$$\lim_{N \to \infty} \frac{C_N}{d^{N-1}} = \sum_{j,l,n \geq 0} a^j b^{n-j-l} c^l d^{-n-l} \frac{q^{n(n+1)/2+l(l-1)/2}}{(q)_j (q)_{n-j-l} (q)_l}$$

$$= (-aq/d)_{\infty} \sum_{j=0}^{\infty} \frac{(b/d)^j (-c/bd)_j \, q^{j(j+1)/2}}{(q)_j (-aq/d)_j}.$$

We omit the details. That (16.54) holds is now immediate.

That (16.55) and (16.56) hold follows immediately upon letting $d = 1$ in the limits above. $\quad\Box$

Remark: 1) A continued fraction identity equivalent to (16.54) was stated by Ramanujan (see [38, **Entry 6.4.1**, page 161]).

One implication of (16.54) is the following continued fraction identity.

Corollary 16.8. *If* $|a|, |q| < 1$, *then*

$$\frac{a}{1 - a + bq} + \frac{a}{1 - a + bq^2} + \cdots \frac{a}{1 - a + bq^n} \cdots$$

$$= a + \frac{-abq}{1 + (a + b)q} + \frac{-abq^3}{1 + (a + b)q^2} + \cdots + \frac{-abq^{2n-1}}{1 + (a + b)q^n} + \cdots. \quad (16.57)$$

Proof. Set $d = b$ in (16.46), replace a with $-a$, and take the reciprocal of both sides. Set $c = 0$ and $d = 1$ in (16.54), cancel $-abq$ on both sides, and then replace b with b/q and a with a/q. The quotient of series on the right sides of (16.46) and (16.54) are now identical, and thus the corresponding continued fractions must be equal. The result follows after some further elementary algebra. □

Suppose the same steps as above are initially followed in (16.54), namely, setting $c = 0$ and $d = 1$, cancelling $-abq$ on both sides, then replacing b with b/q and a with a/q. If next the reciprocal of both sides are taken and b is then subtracted from both sides, the following identity results.

Corollary 16.9. *If* $|q| < 1$, *then*

$$1 + a + \frac{-abq}{(a + b)q + 1} + \cdots + \frac{-abq^{2n-1}}{(a + b)q^n + 1} + \cdots$$

$$= (1 + a)\frac{G(b, 0; a/q; q)}{G(b, 0; a; q)}. \quad (16.58)$$

Proof. The details are left as an exercise (see Q16.14). □

By adding $1 + b$ to both sides of (16.54) and simplifying the resulting combination of series on the right side, the following identity results.

Corollary 16.10. *If* $|q| < 1$, *then*

$$1 + b + \frac{-abq + c}{(a + b)q + 1} + \cdots + \frac{-abq^{2n+1} + cq^n}{(a + b)q^{n+1} + 1} + \cdots$$

$$= \frac{G(b/q, c/q; a; q)}{G(b, c; a; q)}. \quad (16.59)$$

Proof. The details are once again left as an exercise (see Q16.15). □

Remark: This last identity is equivalent to a result of Ramanujan from the lost notebook (see [38, **Entry 6.4.1**, page 161], where this identity may be seen to follow from (16.59) upon replacing c with λq, interchanging a and b and then replacing a with aq):

$$\frac{G(a, \lambda; b; q)}{G(aq, \lambda q; b; q)}$$
$$= 1 + aq + \frac{\lambda q - abq^2}{1 + bq + aq^2} + \frac{\lambda q^2 - abq^4}{1 + bq^2 + aq^3} + \frac{\lambda q^3 - abq^6}{1 + bq^3 + aq^4} + \cdots \quad (16.60)$$

The symmetry in a and b in the continued fraction of Theorem 16.11 can be exploited to prove an identity of Ramanujan from the lost notebook [213] (see [38, **Entry 6.2.2**, p. 146], where an alternative proof to the one below is given).

Corollary 16.11. *Let a, b, c and q be complex numbers with $|q| < 1$ and $a, b \neq 0$. Then*

$$(-aq)_\infty \sum_{j=0}^{\infty} \frac{b^j(-c/b)_j\, q^{j(j+1)/2}}{(q)_j(-aq)_j} = (-bq)_\infty \sum_{j=0}^{\infty} \frac{a^j(-c/a)_j\, q^{j(j+1)/2}}{(q)_j(-bq)_j}. \quad (16.61)$$

Proof. Let $d = 1$ in Theorem 16.11. The symmetry in a and b in the continued fraction in Theorem 16.11 and the first equality in (16.55) give (16.61) immediately. □

Remark: By replacing c with λ we note for later use (see Q16.28) that this identity may be written as

$$(-aq)_\infty G(b, \lambda; a; q) = (-bq)_\infty G(a, \lambda; b; q). \quad (16.62)$$

16.4.4 Continued fraction expansion of infinite products - II

Before deriving some further continued fraction expansions for infinite q-products, we need some special cases of Watson's transformation for an $_8\phi_7$ series. In (5.26), replace all lower-case letters denoting parameters, apart from q, with the corresponding upper-case letters, and let $B, D \to \infty$ to get

$$\sum_{r \geq 0} \frac{(1 - Aq^{2r})(A, C, E; q)_r \left(-A^2/CE\right)^r q^{r(3r+1)/2}}{(1 - A)(Aq/C, Aq/E, q; q)_r}$$
$$= \frac{(Aq)_\infty}{(Aq/E)_\infty} \sum_{r \geq 0} \frac{(E)_r(-Aq/E)^r q^{r(r-1)/2}}{(Aq/C, q; q)_r}. \quad (16.63)$$

If we then set $A = c$, $C = -c/a$ and $E = -c/b$, we get from (16.55) that

$$C_\infty = (-aq)_\infty \sum_{j=0}^\infty \frac{b^j(-c/b)_j\, q^{j(j+1)/2}}{(q)_j(-aq)_j} \tag{16.64}$$

$$= \frac{(-aq)_\infty(-bq)_\infty}{(cq)_\infty} \sum_{r\geq 0} \frac{(1-cq^{2r})(-c/a, -c/b, c; q)_r\, (-ab)^r\, q^{r(3r+1)/2}}{(1-c)(-aq, -bq, q; q)_r}.$$

Similarly, upon setting $A = cq$, $C = -c/a$ and $E = -c/b$ in (16.63), it follows from (16.56) that

$$D_\infty = C_\infty + (c-abq)(-aq^2)_\infty \sum_{j=0}^\infty \frac{b^j(-c/b)_j\, q^{j(j+3)/2}}{(-aq^2, q; q)_j} \tag{16.65}$$

$$= C_\infty + (c-abq)\frac{(-aq^2)_\infty(-bq^2)_\infty}{(cq^2)_\infty}$$

$$\times \sum_{r\geq 0} \frac{(1-cq^{2r+1})(-c/a, -c/b, cq; q)_r\, (-abq^2)^r\, q^{r(3r+1)/2}}{(1-cq)(-aq^2, -bq^2, q; q)_r}.$$

The next continued fraction identity is also due to Ramanujan and can be found in his second notebook ([212, p. 290]). It has been proved (by different methods to that give below) in [41] and in [42], where the authors remark that this identity is "perhaps the deepest of Ramanujan's q-continued fractions".

Corollary 16.12. *If $|q| < 1$, then*

$$\frac{1}{1} - \frac{q}{1+q} - \frac{q^3}{1+q^2} - \cdots - \frac{q^{2n-1}}{1+q^n} - \cdots = \frac{(q^2; q^3)_\infty}{(q; q^3)_\infty}, \tag{16.66}$$

with the numerators converging to $1/(q; q^3)_\infty$ and the denominators converging to $1/(q^2; q^3)_\infty$.

Proof. Let $\omega = \exp(2\pi i/3)$ and set $a = -\omega$, $b = -\omega^2$, $c = 0$ and $d = 1$ in Theorem 16.11, so that the continued fraction in the theorem is the continued fraction in Corollary 16.12. By the second equality in (16.64),

$$C_\infty = (\omega q)_\infty(\omega^2 q)_\infty \sum_{r\geq 0} \frac{(-q^2)^r\, q^{3r(r-1)/2}}{(\omega q)_r(\omega^2 q)_r(q)_r}$$

$$= (\omega q)_\infty(\omega^2 q)_\infty \sum_{r\geq 0} \frac{(-q^2)^r\, (q^3)^{r(r-1)/2}}{(q^3; q^3)_r} = (\omega q)_\infty(\omega^2 q)_\infty(q^2; q^3)_\infty$$

$$= \frac{(\omega q)_\infty (\omega^2 q)_\infty (q)_\infty (q^2; q^3)_\infty}{(q)_\infty} = \frac{(q^3; q^3)_\infty (q^2; q^3)_\infty}{(q)_\infty} = \frac{1}{(q; q^3)_\infty}.$$

The third equality above follows from the q-binomial theorem (see (3.4)).
By the second equality in (16.65)

$$D_\infty - C_\infty = -q(\omega q^2)_\infty (\omega^2 q^2)_\infty \sum_{r \geq 0} \frac{(-1)^r q^{3r(r-1)/2+4r}}{(\omega q^2)_r (\omega^2 q^2)_r (q)_r}$$

$$= -q(\omega q)_\infty (\omega^2 q)_\infty \sum_{r \geq 0} \frac{(-1)^r q^{3r(r-1)/2+4r}(1 - q^{r+1})}{(\omega q)_{r+1}(\omega^2 q)_{r+1}(q)_{r+1}}$$

$$= q(\omega q)_\infty (\omega^2 q)_\infty \sum_{r \geq 1} \frac{(-1)^r q^{(3r^2-r)/2-1}(1 - q^r)}{(\omega q)_r (\omega^2 q)_r (q)_r}$$

$$= (\omega q)_\infty (\omega^2 q)_\infty \sum_{r \geq 1} \frac{(-1)^r q^{(3r^2-3r)/2}(q^r - q^{2r})}{(\omega q)_r (\omega^2 q)_r (q)_r}$$

$$= (\omega q)_\infty (\omega^2 q)_\infty \sum_{r \geq 0} \frac{(-1)^r (q^3)^{r(r-1)/2}(q^r - q^{2r})}{(q^3; q^3)_r}$$

$$= (\omega q)_\infty (\omega^2 q)_\infty ((q; q^3)_\infty - (q^2; q^3)_\infty).$$

Here again we have used the special case of the q-binomial theorem at (3.4).
From the third expression above for C_∞, it follows that

$$D_\infty = (\omega q)_\infty (\omega^2 q)_\infty (q; q^3)_\infty = \frac{(\omega q)_\infty (\omega^2 q)_\infty (q)_\infty (q; q^3)_\infty}{(q)_\infty}$$

$$= \frac{(q^3; q^3)_\infty (q; q^3)_\infty}{(q)_\infty} = \frac{1}{(q^2; q^3)_\infty}.$$

$\square$

The next continued fraction is also due to Ramanujan and may be found
in his second notebook ([212, page 373]). The first published proofs are due
to Watson [257] and Selberg [229]. Other proofs were given by Gordon [129],
Andrews [14] and Hirschhorn [146]. This continued fraction is referred to
in the literature as Ramanujan's cubic continued fraction, and some of its
properties were investigated by Chan in [94].

Corollary 16.13. *If $|q| < 1$, then*

$$S(q) := \frac{1}{1} + \frac{q + q^2}{1} + \frac{q^2 + q^4}{1} + \frac{q^3 + q^6}{1} + \cdots = \frac{(q, q^5; q^6)_\infty}{(q^3, q^3; q^6)_\infty}. \qquad (16.67)$$

Proof. Set $a = -1/q^{1/2}$, $b = 1/q^{1/2}$, $c = 1$ and $d = 1$ in Theorem 16.11, so that the continued fraction in the theorem is

$$\frac{1}{1 + 2S(q)}.$$

The remainder of the proof employs the identities at (16.64) and (16.65) as in the previous corollary, together with the Jacobi triple product identity (see (6.1)). The details are left as an exercise (see Q16.17). $\square$

16.4.5 A third general continued fraction identity

The next general continued fraction identity was also stated by Ramanujan, on page 41 of the lost notebook (see [38, **Entry 6.2.1**, page 144]). Proofs were given by Andrews [20] and Hirschhorn [146]. Here we will follow the first proof of the authors in [38], which in turn is based on the proof of Adiga and Bhargava [77]. Let $G(a, \lambda; b; q)$ be as defined at (16.35).

Theorem 16.12. *Let a, b, λ and q be complex numbers with $|q| < 1$. Then*

$$\frac{G(aq, \lambda q; b; q)}{G(a, \lambda; b; q)} = \frac{1}{1 +} \ \frac{aq + \lambda q}{1} \ + \ \frac{bq + \lambda q^2}{1} \ + \cdots$$
$$\cdots + \ \frac{aq^n + \lambda q^{2n-1}}{1} \ + \ \frac{bq^n + \lambda q^{2n}}{1} \ + \cdots. \quad (16.68)$$

The proof requires some auxiliary results. To that end define

$$P(a, b, \lambda) := (-bq; q)_\infty G(a, \lambda; b; q), \quad (16.69)$$

noting that this implies that the left side of (16.68) may be expressed as $P(aq, b, \lambda q)/P(a, b, \lambda)$.

Lemma 16.1. *Let $P(a, b, \lambda)$ be as defined at (16.69). Then*

$$P(a, b, \lambda) - P(aq, b, \lambda q) = aqP(aq, bq, \lambda q), \quad (16.70)$$
$$P(a, b, \lambda) - P(a, b, \lambda q) = \lambda qP(aq, bq, \lambda q^2), \quad (16.71)$$
$$P(a, b, \lambda) - P(a, bq, \lambda) = bqP(aq, bq, \lambda q), \quad (16.72)$$
$$P(a, b, \lambda) = P(aq, b, \lambda q) + (aq + \lambda q)P(aq, bq, \lambda q^2), \quad (16.73)$$
$$P(a, b, \lambda) = P(a, bq, \lambda q) + (bq + \lambda q)P(aq, bq, \lambda q^2). \quad (16.74)$$

Proof. Most of the details are left as an exercise (see Q16.16). For (16.70), just combine the two series on the left into a single series, using (16.69) and (16.35), together with the facts (easily proved) that

$$(-\lambda/a; q)_n - q^n(-\lambda/(aq); q)_n = \begin{cases} 0, & \text{if } n = 0, \\ (-\lambda/a; q)_{n-1}(1 - q^n), & \text{if } n > 0, \end{cases}$$

and, for $n > 0$,

$$\frac{(-bq; q)_\infty}{(-bq; q)_n} = \frac{(-bq^2; q)_\infty}{(-bq^2; q)_{n-1}}.$$

The identities at (16.71) and (16.72) follow similarly, after employing the identities

$$(-\lambda/a; q)_n - q^n(-\lambda q/a; q)_n = \begin{cases} 0, & \text{if } n = 0, \\ \dfrac{\lambda}{a}(-\lambda q/a; q)_{n-1}(1 - q^n), & \text{if } n > 0, \end{cases}$$

and

$$\frac{(-bq; q)_\infty}{(-bq; q)_n} - \frac{(-bq^2; q)_\infty}{(-bq^2; q)_n} = \frac{(-bq^2; q)_\infty}{(-bq^2; q)_n} bq^{n+1}.$$

To get (16.73), replace λ with λq in (16.70), and add the resulting identity to (16.71). Finally, (16.74) follows upon replacing λ with λq in (16.72), and adding the resulting identity to (16.71). □

The identities at (16.73) and (16.74) are now iterated to prove the theorem.

Proof of Theorem 16.12. Define

$$Q_n = \frac{P(aq^n, bq^n, \lambda q^{2n})}{P(aq^{n+1}, bq^n, \lambda q^{2n+1})},$$

$$Q'_n = \frac{P(aq^{n+1}, bq^n, \lambda q^{2n+1})}{P(aq^{n+1}, bq^{n+1}, \lambda q^{2n+2})}.$$

After replacing a with aq^n, b with bq^n and λ with λq^{2n} in (16.73) and replacing a with aq^{n+1}, b with bq^n and λ with λq^{2n+1} in (16.74) and rearranging the resulting identities, it follows that

$$Q_n = 1 + \frac{aq^{n+1} + \lambda q^{2n+1}}{Q'_n},$$

$$Q'_n = 1 + \frac{bq^{n+1} + \lambda q^{2n+2}}{Q_{n+1}},$$

$$\implies Q_n = 1 + \frac{aq^{n+1} + \lambda q^{2n+1}}{1} + \frac{bq^{n+1} + \lambda q^{2n+2}}{Q_{n+1}}. \tag{16.75}$$

Upon starting with $n = 0$ and iterating, it can be seen that for each integer N,

$$\frac{P(aq, b, \lambda q)}{P(a, b, \lambda)} = \frac{1}{Q_0} = \frac{1}{1} + \frac{aq + \lambda q}{1} + \frac{bq + \lambda q^2}{1} + \cdots$$
$$\cdots + \frac{aq^{N+1} + \lambda q^{2N+1}}{1} + \frac{bq^{N+1} + \lambda q^{2N+2}}{Q_{N+1}}.$$

Since $Q_N = 1 + o(1)$ as $N \to \infty$ (and likewise for Q'_N, so that odd and even parts converge to the same limit), the continued fraction on the right side of (16.68) converges (by Worpitzky's theorem — see Theorem 16.2) to the quotient of series on the left side of (16.68). $\square$

Corollary 16.14. *If $|a|, |q| < 1$, then*

$$1 + b + \frac{-abq + c}{(a + b)q + 1} + \cdots + \frac{-abq^{2n+1} + cq^n}{(a + b)q^{n+1} + 1} + \cdots \qquad (16.76)$$
$$= 1 - a + b + \frac{a + c}{1 - a + bq} + \frac{a + cq}{1 - a + bq^2} + \cdots + \frac{a + cq^{n-1}}{1 - a + bq^n} + \cdots$$
$$= 1 + \frac{b + c}{1} + \frac{aq + cq}{1} + \frac{bq + cq^2}{1} + \frac{aq^2 + cq^3}{1} + \cdots$$
$$\cdots + \frac{bq^n + cq^{2n}}{1} + \frac{aq^{n+1} + cq^{2n+1}}{1} + \cdots.$$

Proof. In (16.51), replace d with b, a with $-a$ and c with c/q. Then the right side of (16.51) becomes identical to the right side of (16.59), and so the corresponding continued fractions are equal, giving the equality of the first and second continued fractions.

In (16.68), invert both sides, replace (simultaneously) a with b/q, λ with c/q and b with a, and the equality of the third continued fraction with the other two follows similarly. $\square$

Remark: The condition $|a| < 1$ is not necessary for the equality of the first and third continued fractions above.

16.4.6 Continued fraction expansion of infinite products - III

Before considering implications of (16.68), we note that the equality of the continued fractions at (16.76) imply some new continued fraction expansions for some infinite series or products. For example, one implication is an expansion for $(q^2; q^3)_\infty/(q; q^3)_\infty$ that is different from that at (16.68), but is perhaps almost as elegant.

Corollary 16.15. *Let* $\omega := \exp(2\pi i/3)$ *and suppose* $|q| < 1$. *Then*

$$\frac{1}{-\omega} + \frac{-\omega^2}{1} + \frac{-\omega q}{1} + \frac{-\omega^2 q}{1} + \frac{-\omega q^2}{1} + \cdots$$
$$\cdots + \frac{-\omega^2 q^n}{1} + \frac{-\omega q^{n+1}}{1} + \cdots = \frac{(q^2; q^3)_\infty}{(q; q^3)_\infty}. \quad (16.77)$$

Proof. In the first and third continued fractions at (16.76), subtract b from each and then take the reciprocal of each. Set $a = -\omega$, $b = -\omega^2$, $c = 0$, and the equality of the continued fractions together with (16.66) gives the result, after setting $1 + \omega^2 = -\omega$ in the first partial numerator.   □

Remarks: 1) The continued fraction identity at (16.77) is also due to Ramanujan, is stated on page 27 of the lost notebook [213], and was proved by Lee and Sohn [177, Eq. (3.2)].

2) Note that the second continued fraction at (16.76) does not lead to another continued fraction for $(q^2; q^3)_\infty/(q; q^3)_\infty$, since $a = -\omega$ fails the $|a| < 1$ requirement.

Another identity of Ramanujan's from the lost notebook (see [38, **Corollary 6.2.2**, page 151]) follows in a similar manner from the identity of Carlitz [90] stated in Q16.12. Ramanujan also stated a special case of this identity, which results in a continued fraction expansion for $(-q^2; q^2)_\infty/(-q; q^2)_\infty$ (cancel a factor of $1 + cq$ on each side, and then set $c = 1/q$).

Corollary 16.16. *If* $|q| < 1$ *and* $c \neq q^{-2n}$ *for any positive integer* n, *then*

$$\frac{1 + cq}{1} + \frac{cq^2}{1} + \frac{q + cq^3}{1} + \frac{cq^4}{1} + \cdots$$
$$\cdots + \frac{q^{n-1} + cq^{2n-1}}{1} + \frac{cq^{2n}}{1} + \cdots = \frac{(-cq; q^2)_\infty}{(-cq^2; q^2)_\infty}. \quad (16.78)$$

Proof. Set $a = 0$ and $b = 1$ in (16.76). Replace c with cq, subtract 1 from each continued fraction and use (16.133) to demonstrate the equality of the infinite product and each of the two different continued fractions.   □

For the next result, we appeal directly to (16.68).

Corollary 16.17. *If* $|q| < 1$, *then*

$$\frac{(q, q^7; q^8)_\infty}{(q^3, q^5; q^8)_\infty} = \frac{1}{1} + \frac{q + q^2}{1} + \frac{q^4}{1} + \frac{q^3 + q^6}{1} + \cdots. \quad (16.79)$$

Proof. In (16.68), replace q with q^2, set $a = 1/q$, $b = 0$ and $\lambda = 1$. Use the Slater identities at (VII.5) and (VII.6) to sum the resulting series, and (16.79) follows directly. □

The continued fraction in (16.79) is known as the Ramanujan–Göllnitz–Gordon continued fraction. The identity at (16.79) was first proved by Selberg [229]. See (16.49) (replace q with $-q$) and (16.136) for other continued fraction expansions for this infinite product due to Andrews [14] and Gordon [129]. We also give another continued fraction for this infinite product in the next corollary.

Corollary 16.18. *If $|q| < 1$, then*

$$\frac{(q^3, q^5; q^8)_\infty}{(q, q^7; q^8)_\infty} = 1 + q + \frac{q^2}{1} + \frac{q^4 + q^3}{1} + \frac{q^6}{1} + \frac{q^8 + q^5}{1} + \dots ; \qquad (16.80)$$

$$\frac{(q^6, q^6; q^{12})_\infty}{(q^2, q^{10}; q^{12})_\infty} = 1 - q + \frac{q + q^2}{1} + \frac{-q^3 + q^4}{1} + \frac{q^3 + q^6}{1} + \frac{-q^5 + q^8}{1} + \dots \qquad (16.81)$$

$$= 1 + q + \frac{-q + q^2}{1 + q + q^3} + \frac{-q + q^4}{1 + q + q^5} + \frac{-q + q^6}{1 + q + q^7} + \dots ; \qquad (16.82)$$

$$\frac{(q^8, q^{12}; q^{20})_\infty}{(q^4, q^{16}; q^{20})_\infty} = 1 - q + \frac{q}{1} + \frac{-q^3}{1} + \frac{q^3}{1} + \frac{-q^5}{1} + \frac{q^5}{1} + \dots \qquad (16.83)$$

Proof. These identities follow from applying (16.76) to, respectively, (16.49) (with q replaced with $-q$), (16.67) (with q replaced with q^2), and (16.50). In each case, the parameters in (16.76) are specialized so that one of the continued fractions matches the continued fraction in the stated identities, and one of the other continued fractions in (16.76) then provides one of the identities in the present corollary. The details are left as an exercise (see Q16.21). □

Remark: Note that since the infinite products in (16.81) and (16.83) are unchanged when q is replaced with $-q$, so also are the continued fractions (in fact both sides of (16.83) remains unchanged when q is replaced with $-q$ or $\pm iq$). Note that (16.83) also follows from Corollary 16.1 and (16.48).

16.5   Heine's continued fraction and three-term recurrence relations

The function $G(a, \lambda; b; q)$ defined at (16.35) is easily seen to be a special case of the $_2\phi_1(a, b; c; q; z)$ function defined by

$$_2\phi_1(a, b; c; q; z) := \sum_{n=0}^{\infty} \frac{(a, b; q)_n}{(c, q; q)_n} z^n, \tag{16.84}$$

and the continued fraction expansions of ratios of the $G(a, \lambda; b; q)$ function described in the previous section may thus be regarded as expansions of particular cases of ratios of $_2\phi_1$ functions.

It is natural thus to consider continued fraction expansions of ratios of the more general $_2\phi_1$ function, rather than the special case $G(a, \lambda; b; q)$ (of course one reason for the interest in considering the $G(a, \lambda; b; q)$ function separately is that it leads to continued fraction expansions of infinite q-products via various identities of Rogers–Ramanujan–Slater type). In fact there are a number of such continued fractions, the first being due to Heine [140].

Before coming to Heine's continued fraction (and other similar ones), we prove a useful lemma.

Lemma 16.2. *Let $\{a_n\}_{n=1}^{\infty}$ be a sequence of non-zero complex numbers such that $\lim_{n\to\infty} a_n = 0$. Let A_n and B_n denote, respectively, the n-th numerator convergent and n-th denominator convergent of the continued fraction*

$$a_0 + \frac{a_1}{1} + \frac{a_2}{1} + \frac{a_3}{1} + \cdots. \tag{16.85}$$

(i) For each integer $n \geq 0$,

$$|A_n|, |B_n| \leq \prod_{i=0}^{n}(1 + |a_i|). \tag{16.86}$$

(ii) Suppose that in addition $\lim_{n\to\infty} \sum_{i=0}^{n} |a_i|$ exists. Then $\lim_{n\to\infty} A_n$ and $\lim_{n\to\infty} B_n$ exist in $\mathbb{C}$.

Proof. That (16.86) holds is a direct consequence of the recurrence relations at (16.4) and the fact that it holds for $n = 0$ and $n = 1$. The proof is given for A_n only, as the proof for B_n is similar.

Suppose (16.86) holds for $n = 0, 1, 2, \ldots, k$.

$$|A_{k+1}| = |A_k + a_{k+1}A_{k-1}| \le |A_k| + |a_{k+1}||A_{k-1}|$$

$$\le \prod_{i=0}^{k}(1 + |a_i|) + |a_{k+1}| \prod_{i=0}^{k-1}(1 + |a_i|)$$

$$\le \prod_{i=0}^{k}(1 + |a_i|) + |a_{k+1}| \prod_{i=0}^{k}(1 + |a_i|) = \prod_{i=0}^{k+1}(1 + |a_i|).$$

For (ii), observe that the convergence of $\sum_{i=0}^{\infty} |a_i|$ implies the convergence of $\prod_{i=0}^{\infty}(1 + |a_i|) =: P$, and thus that $|A_n|, |B_n| \le P$ for all integers $n \ge 0$. This, in turn, implies that $\{A_n\}$ (and $\{B_n\}$) is a Cauchy sequence, since (once again using the recurrence relations (16.4)),

$$|A_{m+n} - A_n| = |\sum_{k=n}^{m+n-1}(A_{k+1} - A_k)| \le \sum_{k=n}^{m+n-1}|A_{k+1} - A_k|$$

$$= \sum_{k=n}^{m+n-1}|a_{k+1}||A_{k-1}| \le P \sum_{k=n}^{m+n-1}|a_{k+1}|.$$

Hence, from the convergence of $\sum_{i=0}^{\infty}|a_i|$, $\{A_n\}$ ($\{B_n\}$) is Cauchy, and the claimed limits exist. □

The results in the above lemma are used in the proof of the next theorem (Heine's continued fraction).

Theorem 16.13. *Let* $|z|, |q| < 1$, *and assume* $c \ne q^{-j}$ *for any non-negative integer* j. *Then*

$$(1 - c)\frac{{}_2\phi_1(a, b; c; q; z)}{{}_2\phi_1(a, bq; cq; q; z)} = 1 - c + \frac{(1-a)(c-b)z}{1-cq} + \frac{(1-bq)(cq-a)z}{1-cq^2}$$

$$+ \frac{(1-aq)(cq-b)qz}{1-cq^3} + \frac{(1-bq^2)(cq^2-a)qz}{1-cq^4} + \cdots. \quad (16.87)$$

Proof. By comparing powers of z on both sides, it is easy to check that

$$_2\phi_1(a, b; c; q; z) = {}_2\phi_1(a, bq; cq; q; z) + \frac{(1-a)(c-b)z}{(1-c)(1-cq)} {}_2\phi_1(aq, bq; cq^2; q; z).$$

$$(16.88)$$

Upon noting that $_2\phi_1(e, f; g; q; z) = {}_2\phi_1(f, e; g; q; z)$ for all values of the parameters, rewriting (16.88) as

$$\frac{_2\phi_1(a, b; c; q; z)}{_2\phi_1(a, bq; cq; q; z)} = 1 + \frac{\dfrac{(1 - a)(c - b)z}{(1 - c)(1 - cq)}}{\dfrac{_2\phi_1(bq, a; cq; q; z)}{_2\phi_1(bq, aq; cq^2; q; z)}}, \qquad (16.89)$$

and iterating, it can be seen that, for each positive integer N,

$$\frac{_2\phi_1(a, b; c; q; z)}{_2\phi_1(a, bq; cq; q; z)} = 1 + \frac{a_1}{1} + \frac{a_2}{1} + \cdots + \frac{a_{2N-2}}{1} + \frac{a_{2N-1}}{b_{2N-1}} \qquad (16.90)$$

$$= 1 + \frac{a_1}{1} + \frac{a_2}{1} + \cdots + \frac{a_{2N-1}}{1} + \frac{a_{2N}}{b_{2N}},$$

where, for $n, N \geq 1$,

$$a_{2n-1} = \frac{(1 - aq^{n-1})(cq^{n-1} - b)q^{n-1}z}{(1 - cq^{2n-2})(1 - cq^{2n-1})}, \, a_{2n} = \frac{(1 - bq^n)(cq^n - a)q^{n-1}z}{(1 - cq^{2n-1})(1 - cq^{2n})},$$

$$b_{2N-1} = \frac{_2\phi_1(bq^n, aq^{n-1}; cq^{2n-1}; q; z)}{_2\phi_1(bq^n, aq^n; cq^{2n}; q; z)}, \, b_{2N} = \frac{_2\phi_1(aq^n, bq^n; cq^{2n}; q; z)}{_2\phi_1(aq^n, bq^{n+1}; cq^{2n+1}; q; z)}.$$

Since $a_n \to 0$ as $n \to \infty$, the continued fraction $1 + K_{n=1}^\infty(a_n/1)$ converges (by Worpitzky's theorem — see Theorem 16.2). Also, it can be seen that $b_n = 1 + o(1)$ as $n \to \infty$, so that each $b_n = 1 + \epsilon_n$, for some ϵ_n such that $\epsilon_n \to 0$ as $n \to \infty$. Note also that each a_n in (16.90) has the form

$$a_n = r_n q^{\lfloor n/2 \rfloor},$$

where the $|r_n| \leq M$ for some fixed number M. Hence the continued fraction $1 + K_{n=1}^\infty(a_n/1)$ satisfies the conditions of Lemma 16.2.

If A'_{2N}/B'_{2N} denotes the $2N$-th approximant of the second continued fraction at (16.90) (a similar argument works for the first), and A_{2N}/B_{2N} denotes the $2N$-th approximant of $1 + K_{n=1}^\infty(a_n/1)$, then since $b_{2N} = 1 + \epsilon_{2N}$,

$$\frac{A'_{2N}}{B'_{2N}} = \frac{A_{2N} + \epsilon_{2N}A_{2N-1}}{B_{2N} + \epsilon_{2N}B_{2N-1}}.$$

Since $\lim_{n \to \infty} A_n$ and $\lim_{n \to \infty} B_n$ exist in $\mathbb{C}$ by the lemma above,

$$\lim_{N \to \infty} \frac{A'_{2N}}{B'_{2N}} = \lim_{N \to \infty} \frac{A_{2N}}{B_{2N}}.$$

A similar argument shows the same statement is true when $2N$ is replaced with $2N + 1$. Hence

$$\frac{{}_2\phi_1(a, b; c; q; z)}{{}_2\phi_1(a, bq; cq; q; z)} = 1 + K_{n=1}^{\infty} \frac{a_n}{1}$$

and (16.87) follows upon multiplying both sides by $1 - c$ and applying an equivalence transformation (see subsection 16.2.3) to the continued fraction. $\square$

Remark: The above argument needs a little extra work if it were to happen that $\lim_{n\to\infty} A_n = \lim_{n\to\infty} B_n = 0$. Since $\lim_{n\to\infty} A_n/B_n := r$ exists, there is a tail of the sequence $\{B_n\}$ that has all terms non-zero. Set $A_n = (r + r_n)B_n$, where $r_n \to 0$ as $n \to \infty$, and it can again be shown that

$$\lim_{N\to\infty} \frac{A_N + \epsilon_N A_{N-1}}{B_N + \epsilon_N B_{N-1}} = \lim_{N\to\infty} \frac{A_N}{B_N}.$$

That the situation where $\lim_{n\to\infty} A_n = \lim_{n\to\infty} B_n = 0$ can arise naturally for a convergent continued fraction in which $\lim_{n\to\infty} A_n/B_n$ exists may be seen by considering equivalent continued fractions.

From (16.15) it can be seen that if the sequence $\{r_k\}$ is chosen so that $\prod_{k=0}^{n} r_k \to 0$ sufficiently fast as $n \to \infty$, then $\lim_{n\to\infty} C_n = \lim_{n\to\infty} D_n = 0$, and $\lim_{n\to\infty} C_n/D_n = \lim_{n\to\infty} A_n/B_n$.

As a non-trivial example of such a continued fraction, let C_n and D_n denote the n-th numerator convergent and n-th denominator convergent, respectively, of the continued fraction

$$(1 - c)q + (a - bq)z - \frac{(a - cq)(1 - bq)zq}{(1 - cq)q + (a - bq^2)z} - \frac{(a - cq^2)(1 - bq^2)zq}{(1 - cq^2)q + (a - bq^3)z}$$
$$- \frac{(a - cq^3)(1 - bq^3)zq}{(1 - cq^3)q + (a - bq^4)z} - \frac{(a - cq^4)(1 - bq^4)zq}{(1 - cq^4)q + (a - bq^5)z} - \cdots. \quad (16.91)$$

The continued fraction at (16.91) is derived from the continued fraction at (16.102) by applying an equivalence transformation in which each $r_i = q$. From (16.15), $C_n = q^{n+1} A_n$ and $D_n = q^{n+1} B_n$, where A_n and B_n denote the n-th numerator convergent and n-th denominator convergent, respectively, of the continued fraction at (16.102). Since it is shown below that A_n and B_n converge separately when $|az/q| < 1$, then $\lim_{n\to\infty} C_n = \lim_{n\to\infty} D_n = 0$ when $|az/q| < 1$ and the continued fraction does not terminate.

It would be interesting to see general criteria for determining from the sequences $\{a_n\}$ and $\{b_n\}$ when the continued fraction $K_{n=1}^{\infty}(a_n/b_n)$ has the property that $\lim_{n\to\infty} A_n = \lim_{n\to\infty} B_n = 0$.

The following beautiful continued fraction identity of Ramanujan [213, Second Notebook, Chapter 16, **Entry 12**] may be proved by employing Heine's continued fraction identity, as the authors in [2] did.

Corollary 16.19. *If* $|q|$, $|a| < 1$, *then*

$$\frac{(-a, b; q)_\infty - (a, -b; q)_\infty}{(-a, b; q)_\infty + (a, -b; q)_\infty} = \frac{a - b}{1 - q} - \frac{(a - bq)(b - aq)}{1 - q^3}$$
$$- \frac{(a - bq^2)(b - aq^2)q}{1 - q^5} - \frac{(a - bq^3)(b - aq^3)q}{1 - q^7} - \cdots. \quad (16.92)$$

Proof. In (16.87), replace q with q^2 and then replace a with bq/a, b with b/a, c with q and z with a^2. Then invert both sides and multiply both sides by $a - b$, so that the continued fraction on the right side of the resulting identity is the continued fraction on the right side of (16.92).

The left side of the resulting identity is B/A, where

$$A = {}_2\phi_1(bq/a, b/a; q; q^2; a^2) \qquad B = \frac{a - b}{1 - q} {}_2\phi_1(bq/a, bq^2/a; q^3; q^2; a^2).$$

It is an easy check that

$$A \pm B = \sum_{n=0}^\infty \frac{(b/a; q)_n}{(q; q)_n} (\pm a)^n = \frac{(\pm b; q)_\infty}{(\pm a; q)_\infty},$$

where the last equality follows from the q-binomial theorem (3.1). Hence the left side of (16.92) is

$$\frac{(b; q)_\infty/(a; q)_\infty - (-b; q)_\infty/(-a; q)_\infty}{(b; q)_\infty/(a; q)_\infty + (-b; q)_\infty/(-a; q)_\infty} = \frac{(A + B) - (A - B)}{(A + B) + (A - B)} = \frac{B}{A},$$

and (16.92) follows. □

Another continued fraction for a ratio of two ${}_2\phi_1$ functions was given by Lorentzen and Waadeland [183, Chapter VI, Theorem 11].

Theorem 16.14. *Let* $|z|$, $|q| < 1$, *and assume* $c \neq q^{-j}$ *for any non-negative integer* j. *Then*

$$(1 - c)\frac{{}_2\phi_1(a, b; c; q; z)}{{}_2\phi_1(aq, bq; cq; q; z)} = b_0 + K_{n=1}^\infty \frac{a_n}{b_n}, \quad (16.93)$$

where

$$a_n = (1 - aq^n)(1 - bq^n)(c - abzq^n)zq^{n-1}, \quad (16.94)$$
$$b_n = 1 - cq^n - (a + b - ab(1 + q)q^n)zq^n. \quad (16.95)$$

Proof. The method of proof is essentially the same as in the previous theorem, so most of the details are left as an exercise (see Q16.23). The starting point is the recurrence relation

$$\begin{aligned} {}_2\phi_1(a,b;c;q;z) = & \left[1 - \frac{a+b-ab(1+q)}{1-c}z\right] {}_2\phi_1(aq,bq;cq;q;z) \\ & + \frac{(1-aq)(1-bq)}{(1-c)(1-cq)}z(c-abqz)\,{}_2\phi_1(aq^2,bq^2;cq^2;q;z). \end{aligned} \quad (16.96)$$

When rewritten as

$$\begin{aligned} (1-c)\frac{{}_2\phi_1(a,b;c;q;z)}{{}_2\phi_1(aq,bq;cq;q;z)} = & [1-c-(a+b-ab(1+q))z] \\ & + \frac{(1-aq)(1-bq)z(c-abqz)}{(1-cq)\dfrac{{}_2\phi_1(aq,bq;cq;q;z)}{{}_2\phi_1(aq^2,bq^2;cq^2;q;z)}}, \end{aligned} \quad (16.97)$$

and this expression is iterated, (16.93) follows after the same kinds of arguments as used in the proof of Theorem 16.13. □

A third continued fraction was stated by the authors in [2] (Lemma 1, pages 17–18). The proof of this identity needs more machinery, and we examine the proof in some detail to illustrate the complexities that may arise in proving continued fraction identities, particularly with regard to convergence.

Before coming to the proof of the identity itself, we recall the following theorem of Thron [250, Theorem 2.1], which is needed to show that the numerator and denominator convergents converge separately in the *loxodromic* (see the proof of Theorem 16.17 for an explanation of this term) case.

Theorem 16.15. *Let $a_n(z)$, $b_n(z)$, $a(z)$ and $b(z)$ be holomorphic functions for $z \in \Delta$. Further assume that $a_n(z) \neq 0$ and*

$$\lim_{n\to\infty} a_n(z) = a(z), \quad \lim_{n\to\infty} b_n(z) = b(z) \text{ for } z \in \Delta.$$

Then

$$K_{n=1}^{\infty}\left(\frac{a_n(z)}{b_n(z)}\right) \quad (16.98)$$

is a limit periodic continued fraction for $z \in \Delta$. Set

$$a_n(z) = a(z) + \delta_n(z), \quad b_n(z) = b(z) + \eta_n(z).$$

Note that $\delta_n(z)$ and $\eta_n(z)$ are holomorphic and

$$\lim_{n\to\infty} \delta_n(z) = 0, \quad \lim_{n\to\infty} \eta_n(z) = 0$$

for $z \in \Delta$. Let $x_1(z)$ and $x_2(z)$ be the solutions of

$$\omega^2 + b(z)\omega - a(z) = 0$$

and assume that the solutions have been so numbered that

$$\left| \frac{x_1(z)}{x_2(z)} \right| < 1 \text{ for } z \in \Delta^* \subset \Delta.$$

Further assume that the series

$$\sum_{k=1}^{\infty} |\delta_k(z)| \text{ and } \sum_{k=1}^{\infty} |\eta_k(z)|$$

converge uniformly on compact subsets of $\Delta_0 \subset \Delta$ and that

$$|x_2(z)| > 0 \text{ for } z \in \Delta_0.$$

Finally, let $\Delta^{(\epsilon)}$ be such that

$$|x_1(z) - x_2(z)| > 2\epsilon \text{ for } z \in \Delta^{(\epsilon)}.$$

Then

$$\lim_{n\to\infty} \frac{A_n(z)}{(-x_2(z))^{n+1}} \text{ and } \lim_{n\to\infty} \frac{A_n(z)}{(-x_2(z))^{n+1}} \qquad (16.99)$$

both exist and are holomorphic in

$$\Delta^\dagger = \bigcup_{\epsilon > 0} \Delta^* \cap \Delta_0 \cap \Delta^{(\epsilon)}.$$

Here $A_n(z)$ and $B_n(z)$ are the numerator and denominator, respectively, of the n-th approximant of the continued fraction (16.98).

Also needed is a result of Lorentzen [182, Theorem 1.1], to deal with the *elliptic* (see the proof of Theorem 16.17 for an explanation of this term also) case. We modify the notation slightly to make it consistent with the notation of Theorem 16.15, and also omit the third statement of the theorem as stated at [182, Theorem 1.1] (it involves the tails of the continued fraction, and the information is not needed here).

Theorem 16.16. *Let $K(a_n/b_n)$ be a limit periodic continued fraction of elliptic type with*

$$a_n = a + \delta_n, \quad b_n = b + \eta_n, \quad \text{with} \quad \sum_{k=1}^{\infty} |\delta_k| < \infty, \quad \sum_{k=1}^{\infty} |\eta_k| < \infty.$$

Then
A. *$K(a_n/b_n)$ diverges.*
B. *Its (modified) approximants*

$$\frac{a_1}{b_1} + \frac{a_2}{b_2} + \cdots + \frac{a_n}{b_n + x} = \frac{A_n + xA_{n-1}}{B_n + xB_{n-1}}$$

converge to a limit $t_0 \in \hat{\mathbb{C}}$ as $n \to \infty$, if x is a fixed point of

$$T(\omega) = \frac{a}{b + \omega}.$$

Lorentzen actually proved more than what was stated in Theorem 16.16, namely that if x is a fixed point of $T(\omega)$ as above, then ([182, Eq. (1.6)])

$$\frac{A_n + xA_{n-1}}{(b+x)^n} \quad \text{and} \quad \frac{B_n + xB_{n-1}}{(b+x)^n} \tag{16.100}$$

both converge separately to finite values which are not both equal to zero. Further, there exists a constant M_1 such that for all integers $n \geq 1$

$$\left| \frac{B_n}{(-x)^n} \right| \leq M_1 \quad \text{and} \quad \left| \frac{A_n}{(-x)^n} \right| \leq M_1. \tag{16.101}$$

This follows from [182, Eq. (2.11)], together with the fact that $(K(a_n/b_n))^{-1}$ is also a continued fraction of elliptic type satisfying the conditions of Theorem 16.16.

As regards the proof of (16.102) below, Lorentzen [157] indicates in her discussion of the proof of an identity of Ramanujan ([213, Second Notebook, Chapter 16, **Entry 12**]) by the authors in [2], the proof of (16.102) in [2] is slightly insufficient. Indeed, the exact conditions on the parameters under which equality holds were not stated. The full correct proof of (16.102) is not given in [157], but instead a corrected version of the special case of **Entry 12** is given.

We will consider the difficulties involved in proving (16.102) in some detail, as they shed some interesting light on convergence questions.

Theorem 16.17. *Let* $|q|, |z| < 1$.
   (i) If $|az/q| < 1$, *then*

$$(1-c)\frac{{}_2\phi_1(a,b;c;q;z)}{{}_2\phi_1(a,bq;cq;q;z)} = (1-c) + (1-bq/a)az/q$$

$$-\frac{(1-cq/a)(1-bq)az/q}{(1-cq)+(1-bq^2/a)az/q} - \frac{(1-cq^2/a)(1-bq^2)az/q}{(1-cq^2)+(1-bq^3/a)az/q}$$

$$-\frac{(1-cq^3/a)(1-bq^3)az/q}{(1-cq^3)+(1-bq^4/a)az/q} - \frac{(1-cq^4/a)(1-bq^4)az/q}{(1-cq^4)+(1-bq^5/a)az/q} - \cdots.$$

$$(16.102)$$

(ii) If $|az/q| = 1$ *but* $az/q \neq 1$, *then*

$$(1-c)\frac{{}_2\phi_1(a,b;c;q;z)}{{}_2\phi_1(a,bq;cq;q;z)} = \lim_{n\to\infty}\frac{A_n - azA_{n-1}/q}{B_n - azB_{n-1}/q}, \qquad (16.103)$$

where A_n/B_n *denotes the n-th approximant of the continued fraction on the right side of* (16.102).
   (iii) If $|az/q| > 1$, *then*

$$\frac{az}{q}\left(1 - \frac{bq}{a}\right)\frac{{}_2\phi_1(q/a,c/a;bq/a;q;q/z)}{{}_2\phi_1(q/a,cq/a;bq^2/a;q;q/z)} = (1-c) + (1-bq/a)az/q$$

$$-\frac{(1-cq/a)(1-bq)az/q}{(1-cq)+(1-bq^2/a)az/q} - \frac{(1-cq^2/a)(1-bq^2)az/q}{(1-cq^2)+(1-bq^3/a)az/q}$$

$$-\frac{(1-cq^3/a)(1-bq^3)az/q}{(1-cq^3)+(1-bq^4/a)az/q} - \frac{(1-cq^4/a)(1-bq^4)az/q}{(1-cq^4)+(1-bq^5/a)az/q} - \cdots.$$

$$(16.104)$$

Remarks: (1) It seems that (16.102) holds when $az/q = 1$, but the proof would take us too far afield, so that case is not considered here.
(2) Note that the continued fractions on the right sides of (16.102) and (16.104) are the same.

The immediate difficulty stems from the fact that if the continued fraction at (16.102) is written in the form $c_0 + K_{n=1}^\infty c_n/1$, then $c_n \not\to 0$ as $n \to \infty$. This means that there is no immediate guarantee that the continued fraction converges, via Worpitzky's theorem (see Theorem 16.2 above), in contrast to the situation with the continued fractions in the previous two theorems. Instead, it is necessary to use results about *limit 1-periodic continued fractions*.

For space reasons, limit 1-periodic continued fractions will not be consider in detail, and the interested reader is referred to [183, Chapter III, Section 5].

Proof of Theorem 16.17.

$$a_n = -(a - cq^n)(1 - bq^n)z/q \qquad \Longrightarrow \lim_{n\to\infty} a_n = -az/q \qquad (16.105)$$

$$b_n = (1 - cq^n) + (a - bq^{n+1})z/q \qquad \Longrightarrow \lim_{n\to\infty} b_n = 1 + az/q.$$

The corresponding linear fractional transformation is

$$T(\omega) = \frac{-az/q}{1 + az/q + \omega}. \qquad (16.106)$$

The fixed points of T are $\omega_1 = -1$ and $\omega_2 = -az/q$. Thus the three cases that determine convergence are

$$\omega_1 = \omega_2 \qquad \Longrightarrow az/q = 1,$$
$$\omega_1 \neq \omega_2 \text{ and } |1 + az/q + \omega_1| = |1 + az/q + \omega_2| \qquad \Longrightarrow |az/q| = 1,$$
$$\omega_1 \neq \omega_2 \text{ and } |1 + az/q + \omega_1| \neq |1 + az/q + \omega_2| \qquad \Longrightarrow |az/q| \neq 1.$$

Limit 1-periodic continued fractions satisfying these conditions are said to be, respectively, of *parabolic, elliptic* and *loxodromic* type. From [183, Chapter III, Section 5] the continued fraction converges in the loxodromic and parabolic cases.

For the elliptic case, it will shown below that the continued fraction satisfies the conditions of Lorentzen's theorem (see Theorem 16.16 above), and hence diverges. Also, since $-az/q$ is one of the fixed points of the associated linear transformation, then (again by Theorem 16.16) the continued fraction converges in the modified sense, in that

$$\lim_{n\to\infty} \frac{A_n - azA_{n-1}/q}{B_n - azB_{n-1}/q}$$

exists.

The authors in [2] next show that the following recurrence relation holds (the tedious details of the proof are left to the reader):

$$(1 - c)(1 - cq)\left\{ {}_2\phi_1(a, b; c; q; z) - {}_2\phi_1(a, bq; cq; q; z) \right\}$$
$$= \frac{az}{q}\left\{ (1 - bq/a)(1 - cq)\, {}_2\phi_1(a, bq; cq; q; z) \right.$$
$$\left. - (1 - cq/a)(1 - bq)\, {}_2\phi_1(a, bq^2; cq^2; q; z) \right\}. \qquad (16.107)$$

If this is rewritten as

$$(1-c)\frac{2\phi_1(a,b;c;q;z)}{2\phi_1(a,bq;cq;q;z)}$$

$$= 1 - c + (a - bq)\,z/q - \frac{(a-cq)\,(1-bq)z/q}{(1-cq)\dfrac{2\phi_1(a,bq;cq;q;z)}{2\phi_1(a,bq^2;cq^2;q;z)}} \qquad (16.108)$$

and iterated n times, then the resulting continued fraction agrees with that at (16.102) to n terms, except instead of being $b_n = (1-cq^n)+(a-bq^{n+1})z/q$, the n-th partial denominator is

$$b_n' := (1-cq^n)\frac{2\phi_1(a,bq^n;cq^n;q;z)}{2\phi_1(a,bq^{n+1};cq^{n+1};q;z)} = (1-cq^n)(1+\epsilon_n),$$

where $\epsilon_n \to 0$ as $n \to \infty$, since the ratio of $2\phi_1$ functions in the formula for b_n' tends to 1 as $n \to \infty$.

If A_n/B_n denotes the n-th approximant of the continued fraction on the right side of (16.102), and A_n'/B_n' denotes the n-th approximant of the continued fraction resulting from the n-th iterate of (16.108), then

$$\frac{(1-c)\,2\phi_1(a,b;c;q;z)}{2\phi_1(a,bq;cq;q;z)} = \frac{A_n'}{B_n'} = \frac{A_n + \{(1-cq^n)\epsilon_n - (a-bq^{n+1})z/q\}\,A_{n-1}}{B_n + \{(1-cq^n)\epsilon_n - (a-bq^{n+1})z/q\}\,B_{n-1}}.$$

Recall that in the loxodromic case A_n and B_n converge separately by Theorem 16.15 ((16.99) with $x_2(z) = -1$). Recall also that in the elliptic case, A_n and B_n are bounded ((16.101) with $x = -1$) and $A_n - az/qA_{n-1}$ and $B_n - az/qB_{n-1}$ converge separately ((16.101) with $x = -az/q$ and $b + x = 1 + az/q + (-az/q) = 1$).

Thus

$$(1-c)\frac{2\phi_1(a,b;c;q;z)}{2\phi_1(a,bq;cq;q;z)} = \lim_{n\to\infty}\frac{A_n'}{B_n'} = \lim_{n\to\infty}\frac{A_n - az/qA_{n-1}}{B_n - az/qB_{n-1}}. \qquad (16.109)$$

Since A_n and B_n converge separately in the loxodromic case, $\lim_{n\to\infty} A_n'/B_n' = \lim_{n\to\infty} A_n/B_n$ and (16.102) holds.

We conclude by indicating how the continued fraction in (16.102) satisfy Theorem 16.15 and Theorem 16.16. With the notation of the former,

$$a(z) = az/q, \qquad\qquad b(z) = 1 + az/q,$$

$$\delta_n(z) = (ac + ab - bcq^n)zq^{n-1}, \qquad \eta_n(z) = (-c - bz)q^n,$$

$$x_1(z) = -az/q, \qquad\qquad x_2(z) = -1.$$

It is easy to check that the requirements of the theorem are met, and that the conclusions follow, in the region $|az/q| < 1$. Very conveniently, $(-x_2(z))^n = 1$, and $\lim_{n\to\infty} A_n$ and $\lim_{n\to\infty} B_n$ exist.

For the elliptic case $(az/q \neq 1$, but $|az/q| = 1)$ it can be seen that the conditions of Theorem 16.16 are also satisfied, and the claims made for the elliptic case are thus true.

The proof of (16.104) is left as an exercise (see Q16.24). $\qquad\square$

Remark: Since $\omega = -1$ is also a fixed point of the associated linear transformation, it follows from Theorem 16.16 that in the elliptic case,

$$\lim_{n \to \infty} \frac{A_n - A_{n-1}}{B_n - B_{n-1}}$$

also exists, and by a remark following Theorem 1.1 in [182], has a value different from the left side of (16.109).

The identities in Theorem 16.17 were key to the proof of the next corollary, which contains a continued fraction identity of Ramanujan [213, Second Notebook, Chapter 16, **Entry 12**]. The method of proof given here is closer to that given in [2] than that in [157], but is possibly shorter and more direct than either of these. Before coming to this, we prove a necessary lemma.

Lemma 16.3. *If* $|q| < 1$ *and* $|b| > 1$, *then*

$$\frac{{}_2\phi_1(a, b; a/(bq); q; -b^{-1})}{{}_2\phi_1(a, bq; a/b; q; -b^{-1})} - 1 = \frac{1}{1 - a/(bq)} \frac{(a, a/(b^2q); q^2)_\infty}{(aq, a/b^2; q^2)_\infty}. \qquad (16.110)$$

Proof. By the Bailey–Daum identity (4.10),

$${}_2\phi_1(a, bq; a/b; q; -b^{-1}) = \frac{(-q; q)_\infty (aq, a/b^2; q^2)_\infty}{(a/b, -1/b; q)_\infty},$$

so what remains to be shown is that

$${}_2\phi_1(a, b; a/(bq); q; -b^{-1}) - {}_2\phi_1(a, bq; a/b; q; -b^{-1})$$
$$= \frac{(-q; q)_\infty (a, a/(b^2q); q^2)_\infty}{(1 - a/(bq))(a/b, -1/b; q)_\infty}. \qquad (16.111)$$

However, after some simple algebra followed by a shift of summation index, the left side of (16.111) simplifies to

$$\frac{(1 - a)(1 - a/(b^2q))}{(1 - a/(bq))(1 - a/b)} \sum_{n=0}^{\infty} \frac{(aq, bq; q^2)_n}{(aq/b, q; q)_n} \left(\frac{-1}{b}\right)^n,$$

and the result follows after one further application of (4.10). $\qquad\square$

Corollary 16.20. *Let $|q| < 1$. Then*

$$1 - ab + \frac{(a-bq)(b-aq)}{(1-ab)(1+q^2)} + \frac{(a-bq^3)(b-aq^3)}{(1-ab)(1+q^4)} + \cdots$$

$$= \begin{cases} \dfrac{(a^2q, b^2q; q^4)_\infty}{(a^2q^3, b^2q^3; q^4)_\infty}, & |ab| < 1, \\[4mm] -ab\dfrac{(q/a^2, q/b^2; q^4)_\infty}{(q^3/a^2, q^3/b^2; q^4)_\infty}, & |ab| > 1. \end{cases} \qquad (16.112)$$

Proof. First suppose $|ab| < 1$. In (16.102), replace q with q^2 and then replace a with a^2q, b with $a/(bq)$, c with ab and z with $-bq/a$. The right side of the resulting identity is $1 - ab$ plus the left side of (16.112). Hence the proof of (16.112) will follow if it can be shown that

$$\frac{(1-ab)\,_2\phi_1(a^2q, a/(bq); ab; q^2; -qb/a)}{_2\phi_1(a^2q, aq/b; abq^2; q^2; -qb/a)} - (1-ab) = \frac{(a^2q, b^2q; q^4)}{(a^2q^3, b^2q^3; q^4)}. \qquad (16.113)$$

However, this follows from (16.110), after replacing q with q^2 and then a with a^2q and b with $a/(bq)$.

For the case $|ab| > 1$, factor out $-ab$ from the continued fraction on the left side of (16.112), apply an equivalence transformation with each $r_i = -1/ab$ (see subsection 16.2.3), and the result follows from the $|ab| < 1$ case, with $1/a$ instead of a and $1/b$ instead of b. $\square$

Remark: (1) The identity is also true when the continued fraction terminates, when $a = bq^{2k+1}$, some $k \in \mathbb{Z}$.

The next continued fraction identity was proved by Lee *et al.* in [178], and the proof from that paper is given to illustrate another method of deriving continued fraction identities.

Theorem 16.18. *If $|q|, |z|, |c/b| < 1$, then*

$$\frac{_2\phi_1(a, b; c; q; z)}{_2\phi_1(a, bq; cq; q; z)} = \frac{1-bz}{1-c} + \frac{(c-abz)(z-1)}{(1-c)(1-bzq)} + \frac{(1-c)(1-bq)(cq-a)z}{1-bzq^2}$$

$$+ \frac{(c-abzq)(zq-1)q}{1-bzq^3} + \frac{(1-bq^2)(cq^2-a)zq}{1-bzq^4} + \cdots. \qquad (16.114)$$

Proof. If the first iterate of Heine's transformation (4.3) is applied to each $_2\phi_1$ series on the left of (16.114), we obtain

$$\frac{_2\phi_1(a, b; c; q, z)}{_2\phi_1(a, bq; cq; q, z)} = \frac{1-bz}{1-c} \cdot \frac{_2\phi_1(abz/c, b; bz; q, c/b)}{_2\phi_1(abz/c, bq; bqz; q, c/b)}. \qquad (16.115)$$

Now apply (16.87) to (16.115) to get

$$(1-c)\frac{{}_2\phi_1(a,b;c;q,z)}{{}_2\phi_1(a,bq;cq;q,z)} = (1-bz)\frac{{}_2\phi_1(abz/c,b;bz;q,c/b)}{{}_2\phi_1(abz/c,bq;bqz;q,c/b)}$$

$$= 1 - bz + \frac{(c-abz)(z-1)}{1-bzq} + \frac{(1-bq)(cqz-az)}{1-bzq^2} +$$

$$\frac{(c-abzq)(zq-1)q}{1-bzq^3} + \frac{(1-bq^2)(cq^2z-az)q}{1-bzq^4} + \cdots,$$

which is equivalent to (16.114) □

Remark: It may be that (16.114) holds in at least some cases when $|c/b| \not< 1$.

One implication of (16.114) is another continued fraction representation for the quotient $G(a,\lambda;b;q)/G(aq,\lambda q;b;q)$ in addition to those at (16.60), (16.68), (16.143) and (16.144).

Corollary 16.21. *If $|q| < 1$, then*

$$\frac{G(a,b,\lambda)}{G(aq,b,\lambda q)} = 1 + aq + \frac{\lambda q - abq^2}{1+aq^2} + \frac{bq + \lambda q^2}{1+aq^3} + \frac{\lambda q^3 - abq^5}{1+aq^4} + \frac{bq^2 + \lambda q^4}{1+aq^5}$$

$$+ \frac{\lambda q^5 - abq^8}{1+aq^6} + \frac{bq^3 + \lambda q^6}{1+aq^7} + \cdots. \quad (16.116)$$

Proof. Set $c = 0$ and simultaneously replace a with $-\lambda/a$, b with $-\lambda/b$ and z with qab/λ in (16.114). The right side of (16.114) becomes the right side of (16.116), while Jackson's identity (4.12) gives that the left side of (16.114) is equal to $G(a,b,\lambda)/G(aq,b,\lambda q)$. □

Remark: This continued fraction was also proved by Bhatnagar [79], using similar methods.

The continued fraction at (16.114) also implies another continued fraction expansion for the first infinite product in Ramanujan's identity (16.112).

Corollary 16.22. *If $|q| < 1$ and $|bq| < |a| < 1/|b|$, then*

$$\frac{(a^2q, b^2q; q^4)_\infty}{(a^2q^3, b^2q^3; q^4)_\infty} = 1 + ab - \frac{(a+bq)(b+aq)}{(1+q^2)} + \frac{(a-bq)(b-aq)q^2}{(1+q^4)}$$

$$- \frac{(a+bq^3)(b+aq^3)q^2}{(1+q^6)} + \frac{(a-bq^3)(b-aq^3)q^4}{(1+q^8)} - \cdots \quad (16.117)$$

Proof. Replace q with q^2 in (16.114), multiply both sides by $1 - c$ and then replace a with $a^2 q$, b with $a/(bq)$, c with ab and z with $-bq/a$ so that the left side of (16.114) becomes the left side of (16.113), and thus equals the right side of (16.113). The same changes on the right side of (16.114) leads to the right side of (16.117). □

Remark: The restrictions on a and b in the corollary are to ensure that the requirements of (16.113) and (16.114) are met, but it may be the case that these restrictions may be relaxed and (16.117) will still hold.

Bhargava *et al.* [78] have described many other continued fraction expansions for ratios of $_2\phi_1$ functions that are derived from three-term recurrence relations.

16.6 Regular continued fractions from q-continued fractions

A *regular continued fraction* has the form

$$b_0 + \cfrac{1}{b_1} + \cfrac{1}{b_2} + \cfrac{1}{b_3} + \ldots := [b_0; b_1, b_2, b_3, \ldots], \qquad (16.118)$$

where b_0 is an integer, and b_i is a positive integer if $i \geq 1$.

If the above continued fraction terminates, then it converges to a rational number, and if it does not terminate, then it converges to an irrational number. Each irrational number has a unique expansion as a non-terminating regular continued fraction, and a rational number has two representations (the final partial quotient b_n may also be written as $b_n - 1 + 1/1$ if $b_n > 1$, and if $b_n = 1$ then $b_{n-1} + 1/1$ may be replaced with $b_{n-1} + 1$).

Regular continued fractions have numerous connections with number theory (Pell's equation, rational approximation, cryptography, transcendence questions, patterns in the regular continued fraction expansion of famous constants, classes of numbers with predictable patterns in their regular continued fraction expansions, ergodic theory to name a few). Here we will simply be interested in families of regular continued fractions which may be evaluated in terms of infinite q-series or infinite q-products via q-continued fraction identities.

Before coming to consider this general class of what have been termed "Tasoevian" continued fractions, we briefly consider another class of continued fractions, which is possibly the motivation for the former class. This latter class, *Hurwitzian* continued fractions, have the form

$$[b_0; b_1, \cdots, b_k, f_1(1), \cdots, f_n(1), f_1(2), \cdots, f_n(2), \cdots]$$
$$=: [b_0; b_1, \cdots, b_k, \overline{f_1(m), \cdots, f_n(m)}]_{m=1}^\infty,$$

where b_0 is an integer, $b_1, \ldots, b_k$ are positive integers, and the $f_i(x)$ are polynomials with rational coefficients taking only positive integral values for integral $x \geq 1$, with at least one polynomial non-constant. A sub-class consists of those Hurwitzian continued fractions in which all of the non-constant polynomials are linear, so that the quasi-periodic part consists of interlaced arithmetic progressions, interspersed with constant terms. One family of such expansions is given by

$$e^{2/(2n+1)} = [1; n, 12n + 6, 5n + 2,$$
$$\overline{1, 1, (6m+1)n + 3m, (24m+12)n + 12m + 6, (6m+5)n + 3m + 2}]_{m=1}^\infty.$$

D. H. Lehmer [179] found closed forms for the numbers represented by regular continued fractions whose partial quotients were either terms in an arithmetic progression,

$$[0; a, a + c, a + 2c, a + 3c, \cdots],$$

or terms in two interlaced arithmetic progressions,

$$[0; a, b, a + c, b + d, a + 2c, b + 2d, \cdots].$$

Lehmer gave the following example of a continued fraction of the first type:

$$[1; 2, 3, 4, 5, \cdots] = \frac{\sum_{m=0}^\infty \frac{1}{(m!)^2}}{\sum_{m=0}^\infty \frac{1}{m!(m+1)!}}$$

Tasoev [248], [249] introduced a new type of continued fraction of the form

$$[a_0; \underbrace{a, \cdots, a}_{m}, \underbrace{a^2, \cdots, a^2}_{m}, \underbrace{a^3, \cdots, a^3}_{m}, \cdots], \qquad (16.119)$$

where $a_0 \geq 0$, $a \geq 2$ and $m \geq 1$ are integers. This type was further investigated by Komatsu in a series of papers which included [166], [167], [168], [169], [170], [171], [172] and [173], by Mc Laughlin and Wyshinski [199], and by Mc Laughlin [192].

It is this class of regular continued fractions, namely those containing geometric progressions, or interlaced geometric progressions, possibly interspersed with constant terms, that are examined in this section. As will be seen, many q-continued fractions transform easily, via equivalence operations and other algebraic manipulations, to continued fractions of this type. We allow here for modified geometric progressions, i.e. those in which the n-th terms has the form $ba^n + c$, instead of the simple a^n considered by Tasoev, as exhibited in (16.119).

As a first simple example, consider the Rogers–Ramanujan continued fraction $K(q)$ at (16.1). The application of an equivalence transformation (see (16.15)), in which $r_{2i-1} = r_{2i} = 1/q^i$ for $i \geq 1$, gives

$$K(q) = 1 + \frac{1}{1/q} + \frac{1}{1/q} + \frac{1}{1/q^2} + \frac{1}{1/q^2} + \frac{1}{1/q^3} + \frac{1}{1/q^3} + \cdots$$

Next, setting $q = 1/a$, where $a \geq 2$ is a positive integer, gives, using the notation at (16.120), that

$$1 + [0; a, a, a^2, a^2, a^3, a^3, \cdots] = \frac{G(0,1;0;1/a)}{G(0,1/a;0;1/a)}$$

$$= \frac{(1/a^2; 1/a^5)_\infty (1/a^3; 1/a^5)_\infty}{(1/a; 1/a^5)_\infty (1/a^4; 1/a^5)_\infty}.$$

The left side is exactly the $m = 2$ case of Tasoev's continued fraction at (16.119), with $a_0 = 1$.

Before coming to the next example, we note how to remove 0's and negative integers from "almost" regular continued fraction expansions (i.e. regular apart from these zero- and negative integers where there should be positive integers). Indeed, one can easily check (see also [251]) that $[m, n, 0, p, \alpha] = [m, n + p, \alpha]$ and $[m, -n, \alpha] = [m - 1, 1, n - 1, -\alpha]$.

We start with the $a = 0$ case of (16.45) (with c replaced with b):

$$1 + \frac{bq}{1} + \frac{bq^2}{1} + \cdots + \frac{bq^n}{1} + \cdots = \frac{G(0,b;0;q)}{G(0,bq;0;q)} =: g(b,q), \qquad (16.120)$$

where

$$G(0, \lambda; 0; q) = \sum_{n=0}^{\infty} \frac{\lambda^n q^{n^2}}{(q; q)_n}.$$

Theorem 16.19. *Let $a \geq 2$ be a positive integer and suppose c and d are rationals such that $ca, da \in \mathbb{Z}^+$. Then*

$$[0; \overline{da^k, ca^k}]_{k=1}^{\infty} = \frac{1}{ad} \frac{G(0, 1/(cda^2); 0; 1/a)}{G(0, 1/(cda); 0; 1/a)}. \qquad (16.121)$$

$$[0; \overline{ca^k}]_{k=1}^{\infty} = \frac{1}{ac} \frac{G(0, 1/(c^2a^3); 0; 1/a^2)}{G(0, 1/(c^2a); 0; 1/a^2)}. \tag{16.122}$$

If $ca > 1$, then

$$[0; ca - 1, \overline{1, ca^{k+1} - 2}]_{k=1}^{\infty} = \frac{1}{ac} \frac{G(0, -1/(c^2a^3); 0; 1/a^2)}{G(0, -1/(c^2a); 0; 1/a^2)}. \tag{16.123}$$

If $da > 1$ and $ca > 2$, then

$$[0; da - 1, 1, ca - 2, \overline{1, da^{k+1} - 2, 1, ca^{k+1} - 2}]_{k=1}^{\infty}$$
$$= \frac{1}{ad} \frac{G(0, -1/(cda^2); 0; 1/a)}{G(0, -1/(cda); 0; 1/a)}. \tag{16.124}$$

If $c > 1$ is an integer and $da > 1$, then

$$[c - 1; \overline{1, da^k - 1, ca^k - 1}]_{k=1}^{\infty} = c \frac{G(0, 1/(cd); 0; -1/a)}{G(0, -1/(cda); 0; -1/a)}. \tag{16.125}$$

Proof. On the left side of (16.120), let $b = 1/(cd)$, so that $g(b,q)$ can be written as

$$g\left(\frac{1}{cd}, q\right) := 1 + \frac{1}{c}\left(\frac{1}{d/q} + \frac{1}{c/q} + \frac{1}{d/q^2} + \frac{1}{c/q^2} + \cdots\right).$$

Now set $q = 1/a$ and let c and d be rationals such that $ca, da \in \mathbb{Z}^+$ then

$$1 + \frac{1}{c}[0; \overline{da^k, ca^k}]_{k=1}^{\infty} = \frac{\sum_{n=0}^{\infty} \dfrac{1}{(cd)^n a^{n^2} (1/a; 1/a)_n}}{\sum_{n=0}^{\infty} \dfrac{1}{(cd)^n a^{n^2+n} (1/a; 1/a)_n}}. \tag{16.126}$$

After some manipulation, (16.121) follows.

Next, replace a by a^2 and d by c/a in Equation 16.121, and we get the identity at (16.122).

As noted above, it is possible to allow the parameters a, c and d in Equations 16.126, 16.121 and 16.122 to take negative values, and then transform the resulting continued fraction so that it is again regular. Thus, if we replace c by $-c$ and a by $-a$ in (16.122), for example, and repeatedly apply the second of the above conditions, we have that

$$[0; ca, -ca^2, ca^3, -ca^4, ca^5, -ca^6, \cdots]$$
$$= [0; ca - 1, 1, ca^2 - 1, -ca^3, ca^4, -ca^5, ca^6, \cdots]$$
$$= [0; ca - 1, 1, ca^2 - 2, 1, ca^3 - 1, -ca^4, ca^5, -ca^6, \cdots]$$
$$= [0; ca - 1, 1, ca^2 - 2, 1, ca^3 - 2, 1, ca^4 - 1, -ca^5, ca^6, \cdots]$$
$$= [0; ca - 1, 1, ca^2 - 2, 1, ca^3 - 2, 1, ca^4 - 2, 1, ca^5 - 1, -ca^6, \cdots] \text{ etc.}$$

Finally, we have, for $ca > 1$, the identity at (16.123)

Likewise, if c is replaced by $-c$ in 16.121 and the resulting continued fraction similarly manipulated, one gets (16.124). The identity at (16.125) is derived from Equation 16.126 by multiplying across by c, replacing a by $-a$ and similarly manipulating the continued fraction to remove the negatives.

□

Other q-continued fractions may be treated in ways similar to those above (for example, the $\lambda = 0$ case of (16.68)) to produce other regular continued fraction expansions that are summable. In addition, the general identity in part (i) of Corollary 16.1 may be applied to all of these q-continued fraction to produce still more such regular continued fractions. The even parts of some regular continued fractions correspond to certain summable q-continued fractions (see, for example, Theorem 4 in [192]). The interested reader is encouraged to experiment, or to see the papers [199] and [192].

Exercises

16.1 Prove the formulae at (16.4), (16.5) and (16.6).

16.2 Prove the formula for the odd part of the continued fraction $b_0 + K_{n=1}^{\infty} a_n/b_n$ stated in Theorem 16.5.

16.3 Generalize (16.21) to show that

$$(-zq; q)_\infty = 1 + \frac{zq}{1} - \frac{zq^2 + q}{1 + zq^2 + q} - \frac{zq^3 + q}{1 + zq^3 + q} - \frac{zq^4 + q}{1 + zq^4 + q} - \cdots$$
$$(16.127)$$

16.4 Generalize (16.22) to show that

$$\sum_{n=0}^{\infty} \frac{z^n q^{n(n+1)/2}}{(q; q)_n} = 1 + \frac{zq}{1 - q} + \frac{z(q^3 - q^2)}{(z - 1)q^2 + 1} + \frac{z(q^5 - q^3)}{(z - 1)q^3 + 1}$$
$$+ \frac{z(q^7 - q^4)}{(z - 1)q^4 + 1} + \cdots \quad (16.128)$$

Remark: Since the left sides of (16.127) and (16.128) are equal by a special case of the q-binomial theorem (see (3.4)), then the continued fractions on the right sides are equal also.

16.5 Use one of the transformations in Corollary 16.2 to show that

$$\sum_{n=0}^{\infty} q^{n^2} z^n = 1 + \frac{zq}{1} - \frac{zq^3}{zq^3 + 1} - \frac{zq^5}{zq^5 + 1} - \frac{zq^7}{zq^7 + 1} - \cdots \quad (16.129)$$

16.6 Prove that the even part of

$$d_0 + \cfrac{c_1}{d_1 - c_2} + \cfrac{c_2}{1} + \cfrac{-1}{d_2 - c_3 + 1} + \cfrac{c_3}{1} + \cfrac{-1}{d_3 - c_4 + 1} + \cfrac{c_4}{1} + \cdots \qquad (16.130)$$

and the even part of

$$d_0 + \cfrac{c_1}{d_1 + 1} + \cfrac{-1}{1} + \cfrac{c_2}{d_2 - c_2 + 1} + \cfrac{-1}{1} + \cfrac{c_3}{d_3 - c_3 + 1} + \cfrac{-1}{1} + \cdots \qquad (16.131)$$

are each equal to

$$d_0 + \cfrac{c_1}{d_1} + \cfrac{c_2}{d_2} + \cfrac{c_3}{d_3} + \cfrac{c_4}{d_4} + \cdots \qquad (16.132)$$

16.7 Fill in the details to complete the proof of Theorem 16.7.

16.8 Fill in the details to complete the proof of Theorem 16.9.

16.9 By applying the transformation at (16.28) iteratively to the continued fraction $b_0 + K_{n=1}^{\infty} a_n/b_n$, derive a continued fraction equivalent to its odd part (16.11).

16.10 Complete the proof of the identities at (16.45), (16.46) and (16.47).

16.11 Prove the identity at (16.51).

16.12 By specializing the parameters in (16.51) and (4.10), prove the following identity of Carlitz ([90]):

$$1 + \cfrac{cq}{1 + q} + \cfrac{cq^2}{1 + q^2} + \cfrac{cq^3}{1 + q^3} + \cdots = \frac{(-cq; q^2)_\infty}{(-cq^2; q^2)_\infty}. \qquad (16.133)$$

16.13 By specializing the parameters in (16.47) and using the identities at (VII.3) and (VII.4), prove the identity

$$1 + \cfrac{q}{1 - q + q^2} + \cfrac{q}{1 - q + q^4} + \cfrac{q}{1 - q + q^6} + \cdots = \frac{(q^2, q^3; q^5)_\infty}{(q, q^4; q^5)_\infty}. \qquad (16.134)$$

16.14 Complete the proof of the identity at (16.58).

16.15 Complete the proof of Ramanujan's continued fraction identity at (16.59).

16.16 Complete the proofs of the identities in Lemma 16.1.

16.17 Complete the proof of the continued fraction identity at (16.67).

16.18 By employing (16.47) in conjunction with the identities of Rogers (see [217]) at (VII.3) and (VII.4), prove the identity

$$1 + \cfrac{-q}{1 + q + q^2} + \cfrac{-q}{1 + q + q^4} + \cdots = \frac{(q, q^9; q^{10})_\infty (q^8, q^{12}; q^{20})_\infty}{(q^3, q^7; q^{10})_\infty (q^4, q^{16}; q^{20})_\infty}. \qquad (16.135)$$

16.19 By replacing q with q^2 in (16.55) and (16.56), specializing a, b and c in the resulting continued fraction and series, and employing the analytic versions of the Göllnitz–Gordon identities at (VII.5) and (VII.6), prove Gordon's continued fraction identity [129, Theorem 1] (a proof of this identity was also given by Andrews [14]):

$$1 + q + \frac{q^2}{1 + q^3} + \frac{q^4}{1 + q^5} + \cdots = \frac{(q^3, q^5; q^8)_\infty}{(q, q^7; q^8)_\infty}. \tag{16.136}$$

16.20 By employing (16.135) in conjunction with the identities at (16.76), prove that

$$\frac{(q, q^9; q^{10})_\infty (q^8, q^{12}; q^{20})_\infty}{(q^3, q^7; q^{10})_\infty (q^4, q^{16}; q^{20})_\infty} = -q + \frac{1}{1} + \frac{-q^3}{1} + \frac{q^2}{1} + \frac{-q^5}{1} + \frac{q^4}{1} + \cdots$$

$$= 1 - q + \frac{q^3}{1 + (1-q)q^2} + \frac{q^7}{1 + (1-q)q^4} + \frac{q^{11}}{1 + (1-q)q^6} + \cdots . \tag{16.137}$$

Remark: Ismail and Stanton [149, Theorem 4.1] give another continued fraction expansion for the infinite product at (16.137).

16.21 Complete the proof of the identities in Corollary 16.18, by supplying the missing details.

16.22 i) By specializing the parameters in (4.12), prove the identity

$$(aq; q)_\infty \sum_{n=0}^\infty \frac{b^n q^{n^2}}{(aq, q; q)_n} = \sum_{n=0}^\infty \frac{(b/a; q)_n (-a)^n q^{n(n+1)/2}}{(q; q)_n}. \tag{16.138}$$

ii) By making the appropriate choices for a, b and λ in (16.68) and transforming the resulting series $G(aq, \lambda q; b; q)$ and $G(a, \lambda; b; q)$ by specializing a and b appropriately in (16.138), derive the identity

$$\sum_{n=0}^\infty (-a)^n q^{n(n+1)/2} = \frac{1}{1} + \frac{aq}{1} + \frac{a(q^2 - q)}{1} + \frac{aq^3}{1} + \frac{a(q^4 - q^2)}{1} + \cdots . \tag{16.139}$$

Remark. This continued fraction was first written down by Eisenstein [113], although the $a = 1$ case was stated previously by Gauss [126].

16.23 Complete the proof of Theorem 16.14 by filling in the missing details.

16.24 Prove (16.104) by applying an equivalence transformation to the continued fraction on the right side of (16.104), transforming it to a continued fraction satisfying the conditions of (16.102), so that it can thus be summed as a constant times a ratio of two $_2\phi_1$ functions.

16.25 By starting with (16.83), prove that if $m > 1$ is an integer, then

$$[1; \overline{m-1, 1, m^{2k}-1}]_{k=1}^{\infty} = \frac{1}{m} + \frac{\left(m^{-8}, m^{-12}; m^{-20}\right)_{\infty}}{\left(m^{-4}, m^{-16}; m^{-20}\right)_{\infty}}. \tag{16.140}$$

16.26 By starting with (16.66), prove that if $a > 1$ is an integer, then

$$[1; \overline{a^{2k-1}-1, a^{2k}+1}]_{k=1}^{\infty} = \frac{(a^{-5}; a^{-6})_{\infty}(a^{-4}; a^{-12})_{\infty}}{(a^{-1}; a^{-6})_{\infty}(a^{-8}; a^{-12})_{\infty}}. \tag{16.141}$$

$$[1; \overline{a^{k}-1, 1}]_{k=1}^{\infty} = \frac{(a^{-2}; a^{-3})_{\infty}}{(a^{-1}; a^{-3})_{\infty}}. \tag{16.142}$$

16.26 (Lee, Mc Laughlin and Sohn [178]) Show that Ramanujan's continued fraction identity at (16.68) may be derived from Heine's continued fraction identity (16.87), by relabelling the parameters in the latter and then using Jackson's transformation (4.12).

16.27 Prove that Hirschhorn's identity (which holds for $|b| < 1$)

$$\frac{G(a, \lambda; b; q)}{G(aq, \lambda q; b; q)} = 1 - b + aq + \frac{\lambda q + b}{1 - b + aq^2} + \frac{\lambda q^2 + b}{1 - b + aq^3} + \cdots \tag{16.143}$$

may be derived from (16.102) by following similar steps to those described in Q16.26.

16.28 By simultaneously replacing b with aq and a with b/q in (16.143) and then employing (16.62), followed by (16.69) combined with (16.70) and (16.72) (or otherwise), prove Ramanujan's identity (which holds for $|aq| < 1$)

$$\frac{G(a, \lambda; b; q)}{G(aq, \lambda q; b; q)} = 1 + \frac{aq + \lambda q}{1 - aq + bq} + \frac{aq + \lambda q^2}{1 - aq + bq^2} + \frac{aq + \lambda q^3}{1 - aq + bq^3} + \cdots. \tag{16.144}$$

16.29 By treating (16.87) in a way similar to the treatment of (16.114) in Corollary 16.22, prove the following alternative continued fraction expansion for the infinite product in Ramanujan's identity (16.112) (with the identity below being valid at least for $|qb/a| < 1$):

$$\frac{(a^2 q, b^2 q; q^4)_{\infty}}{(a^2 q^3, b^2 q^3; q^4)_{\infty}} = \frac{(1 - a^2 q)(1 - b^2 q)}{(1 - abq^2)} + \frac{(a - bq)(b - aq)q^2}{(1 - abq^4)}$$
$$+ \frac{(1 - a^2 q^3)(1 - b^2 q^3)q^2}{(1 - abq^6)} + \frac{(a - bq^3)(b - aq^3)q^4}{(1 - abq^8)} + \cdots. \tag{16.145}$$

16.30 By applying the method of proof of (16.92) to (16.102), prove that if $|a^2|$, $|ab/q| < 1$, then

$$\frac{(-a, b; q)_\infty - (a, -b; q)_\infty}{(-a, b; q)_\infty + (a, -b; q)_\infty} = \frac{(a - b)q}{(ab + q)(1 - q)} - \frac{(a - bq^2)(b - aq^2)q}{(ab + q)(1 - q^3)}$$

$$- \frac{(a - bq^4)(b - aq^4)q}{(ab + q)(1 - q^5)} - \frac{(a - bq^6)(b - aq^6)q}{(ab + q)(1 - q^7)} - \cdots . \quad (16.146)$$

16.31 By similarly applying the method of Q16.30 to (16.114), prove that if $|a^2|$, $|aq/b| < 1$, then

$$\frac{(-a, b; q)_\infty - (a, -b; q)_\infty}{(-a, b; q)_\infty + (a, -b; q)_\infty} = \frac{(a - b)}{1 - ab}$$

$$- \frac{(1 - a^2)(1 - b^2)q}{1 - abq^2} - \frac{(a - bq^2)(b - aq^2)q}{1 - abq^4}$$

$$- \frac{(1 - a^2q^2)(1 - b^2q^2)q^3}{1 - abq^6} - \frac{(a - bq^4)(b - aq^4)q^3}{1 - abq^8} - \cdots . \quad (16.147)$$

Chapter 17

Lambert Series

The topic of this chapter, Lambert series, has connections with several areas, including number theory, Eisenstein series, theta functions and mock theta functions.

A *Lambert series* is a series of the form

$$\sum_{n=1}^{\infty} \frac{a_n q^n}{1 - q^n}. \tag{17.1}$$

If $|q| < 1$, then

$$\sum_{n=1}^{\infty} \frac{a_n q^n}{1 - q^n} = \sum_{n=1}^{\infty} a_n q^n \sum_{m=0}^{\infty} q^{mn} = \sum_{n=1}^{\infty} a_n \sum_{m=1}^{\infty} q^{mn} = \sum_{k=1}^{\infty} \left(\sum_{n|k} a_n \right) q^k.$$

This in turn leads to a number of identities for various arithmetic functions, including

$$\sum_{n=1}^{\infty} \frac{\phi(n) q^n}{1 - q^n} = \sum_{k=1}^{\infty} k q^k = \frac{q}{(1 - q)^2}. \tag{17.2}$$

Here ϕ is the Euler ϕ function, and the first equality follows from the relation $\sum_{n|k} \phi(n) = k$, and second from differentiating the identity $\sum_{k=1}^{\infty} q^k = (1 - q)^{-1}$. To consider the full range of identities that exist, it is necessary to consider more general series of the form

$$\sum_{n=1}^{\infty} \frac{a_n q^{jn}}{1 - b_n q^{kn}}, \tag{17.3}$$

for positive integers j and k, and also to allow the a_n to be functions of q, in the most general setting. In particular, bilateral sums are sometimes considered, and the a_n may contain factors of the form q^{rn^2}, for rational r.

There are a number of ways to derive Lambert series identities, the most obvious being to specialize the parameters in existing basic hypergeometric identities, so that a basic hypergeometric series becomes a (sum of) Lambert series. For example, suppose a series contains a factor

$$\frac{(b,c;q)_n q^n}{(aq/b, aq/c; q)_n}.$$

Upon making the specialization $b = a/c$, this product becomes

$$\frac{(a/c, c; q)_n q^n}{(cq, aq/c; q)_n} = \frac{(1-a/c)(1-c)}{(1-cq^n)(1-aq^n/c)}$$

$$= \frac{(1-a/c)(1-c)}{a-c^2}\left[\frac{aq^n}{1-a/cq^n} - \frac{c^2 q^n}{1-cq^n}\right]. \quad (17.4)$$

This was essentially the method used by the authors in [48], in one of their two proofs of the next theorem.

17.1   Lambert series identities deriving from Bailey's $_6\psi_6$ summation formula

We will stick more closely to the idea mentioned above than the authors in [48] did, just to make the method a little more explicit.

Theorem 17.1. *([48, Andrews, Lewis, Liu]) Suppose $a, b, c \neq q^n$ (for any $n \in \mathbb{Z}$) are non-zero complex numbers with $abc \neq q^n$. Then*

$$\frac{(ab, ac, bc, q/(ab), q/(ac), q/(bc), q, q; q)_\infty}{(a, b, c, q/a, q/b, q/c, abc, q/(abc); q)_\infty}$$

$$= 1 + \sum_{n=0}^{\infty} \left(\frac{aq^n}{1-aq^n} + \frac{bq^n}{1-bq^n} + \frac{cq^n}{1-cq^n} - \frac{abcq^n}{1-abcq^n}\right)$$

$$+ \sum_{n=1}^{\infty} \left(\frac{-q^n/a}{1-q^n/a} + \frac{-q^n/b}{1-q^n/b} + \frac{-q^n/c}{1-q^n/c} + \frac{q^n/(abc)}{1-q^n/(abc)}\right). \quad (17.5)$$

Proof. In Bailey's $_6\psi_6$ identity (8.1) set $c = a/b$ and $e = a/d$ to get

$$\frac{(aq, q, aq/bd, dq/b, bq/d, bdq/a, q, q, q/a; q)_\infty}{(aq/b, bq, aq/d, dq, q/b, bq/a, q/d, dq/a, q; q)_\infty}$$

$$= \sum_{n=-\infty}^{\infty} \frac{(q\sqrt{a}, -q\sqrt{a}, b, a/b, d, a/d; q)_n}{(\sqrt{a}, -\sqrt{a}, aq/b, bq, aq/d, dq; q)_n} q^n$$

$$= \frac{(1-b)(1-a/b)(1-d)(1-a/d)}{1-a}$$

$$\times \sum_{n=-\infty}^{\infty} \frac{(1-aq^{2n})q^n}{(1-aq^n/b)(1-bq^n)(1-aq^n/d)(1-dq^n)} \quad (17.6)$$

Now use the partial fraction expansions

$$\frac{(1-aq^{2n})q^n}{(1-aq^n/b)(1-bq^n)(1-aq^n/d)(1-dq^n)}$$

$$= \frac{b}{a(1-b/d)(1-bd/a)} \left(\frac{aq^n/b}{1-aq^n/b} - \frac{aq^n/d}{1-aq^n/d} + \frac{bq^n}{1-bq^n} - \frac{dq^n}{1-dq^n} \right)$$

$$= \frac{b}{a(1-b/d)(1-bd/a)}$$

$$\times \left(\frac{-bq^{-n}/a}{1-bq^{-n}/a} + \frac{dq^{-n}/a}{1-dq^{-n}/a} - \frac{q^{-n}/b}{1-q^{-n}/b} + \frac{q^{-n}/d}{1-q^{-n}/d} \right)$$

to replace terms in the final series of (17.6), with the first expansion being used to replace terms with $n \geq 0$ and the second expansion for terms with $n < 0$. Multiply both sides of the resulting identity by

$$\frac{a(1-a)(1-b/d)(1-bd/a)}{b(1-b)(1-a/b)(1-d)(1-a/d)}$$

to get

$$\frac{a(q, q, q/a, a, b/d, qd/b, bd/a, aq/bd; q)_\infty}{b(b, q/b, a/b, qb/a, d, q/d, a/d, qd/a; q)_\infty}$$

$$= \sum_{n=0}^{\infty} \left(\frac{aq^n/b}{1-aq^n/b} - \frac{aq^n/d}{1-aq^n/d} + \frac{bq^n}{1-bq^n} - \frac{dq^n}{1-dq^n} \right)$$

$$+ \sum_{n=1}^{\infty} \left(\frac{-bq^n/a}{1-bq^n/a} + \frac{dq^n/a}{1-dq^n/a} - \frac{q^n/b}{1-q^n/b} + \frac{q^n/d}{1-q^n/d} \right). \quad (17.7)$$

Finally, (17.5) follows after simultaneously replacing a with ab, c with $-1/(abc)$ and d with $1/c$, combined with a little elementary algebra. $\square$

The following implications of (17.5) are used later. Slightly different proofs may be found in [48].

Corollary 17.1. *If* $|q| < 1$ *and* $a, b \neq q^n$ *(for any* $n \in \mathbb{Z}$*) are non-zero complex numbers. Then*

$$\frac{b(ab, q/(ab), a/b, qb/a; q)_\infty (q; q)_\infty^4}{(a, q/a, b, q/b; q)_\infty^2}$$

$$= \sum_{n=-\infty}^{\infty} \left(\frac{bq^n}{(1 - bq^n)^2} - \frac{aq^n}{(1 - aq^n)^2} \right); \quad (17.8)$$

$$\frac{a(a^2, q/a^2; q)_\infty (q; q)_\infty^6}{(a, q/a; q)_\infty^4} = \frac{a(1 + a)}{(1 - a)^3} + \sum_{n=1}^{\infty} \frac{(a^n - a^{-n})n^2 q^n}{1 - q^n}. \quad (17.9)$$

Proof. It is an easy check that

$$\frac{-q^n/a}{1 - q^n/a} + \frac{-q^n/b}{1 - q^n/b} + \frac{cq^n}{1 - cq^n} + \frac{q^n/(abc)}{1 - q^n/(abc)}$$

$$= -\frac{abc}{q^n - abc} + \frac{a}{q^n - a} + \frac{b}{q^n - b} - \frac{1/c}{q^n - 1/c},$$

and thus, after a little algebra, that (17.5) may be rewritten as

$$\frac{(ab, ac, bc, q/(ab), q/(ac), q/(bc), q, q; q)_\infty}{(a, b, c, q/a, q/b, q/c, abc, q/(abc); q)_\infty}$$

$$= \sum_{n=-\infty}^{\infty} \left(\frac{aq^n}{1 - aq^n} + \frac{bq^n}{1 - bq^n} - \frac{q^n/c}{1 - q^n/c} - \frac{abcq^n}{1 - abcq^n} \right)$$

$$= \sum_{n=-\infty}^{\infty} \left(\frac{b(1 - ac)q^n}{(1 - bq^n)(1 - abcq^n)} - \frac{(1 - ac)q^n}{c(1 - aq^n)\left(1 - \frac{q^n}{c}\right)} \right). \quad (17.10)$$

Now divide both sides by $1 - ac$, let $c \to 1/a$, and (17.8) follows.

For (17.9), rewrite the right side of (17.8) as

$$\sum_{n=-\infty}^{\infty} \frac{(b - a)q^n \left(1 - abq^{2n}\right)}{(1 - aq^n)^2 (1 - bq^n)^2}.$$

Now divide both sides of (17.8) by $b - a$, let $b \to a$ and multiply both sides by a. The left side of (17.8) becomes the left side of (17.9), and the right side of (17.8) becomes, after a little simplification,

$$\sum_{n=-\infty}^{\infty} \frac{aq^n \left(1 + aq^n\right)}{(1 - aq^n)^3}$$

$$= \frac{a(1 + a)}{(1 - a)^3} + \sum_{n=1}^{\infty} \frac{aq^n \left(1 + aq^n\right)}{(1 - aq^n)^3} - \sum_{n=1}^{\infty} \frac{(1/a)q^n \left(1 + (1/a)q^n\right)}{(1 - (1/a)q^n)^3},$$

and (17.9) follows after two applications of (17.49). □

The following corollary uses the following special case (sometimes with q replaced with q^k, where $k > 1$ is a positive integer) of the Jacobi triple product identity (6.1) (see (6.8)):

$$\phi(q) = \sum_{n=-\infty}^{\infty} q^{n^2} = (-q, -q, q^2; q^2)_\infty. \tag{17.11}$$

Corollary 17.2. *If* $|q| < 1$, *then*

$$\sum_{m,n=-\infty}^{\infty} q^{m^2+n^2} = 1 + 4 \sum_{n=0}^{\infty} \frac{q^{4n+1}}{1-q^{4n+1}} - 4 \sum_{n=0}^{\infty} \frac{q^{4n+3}}{1-q^{4n+3}}; \tag{17.12}$$

$$\sum_{m,n=-\infty}^{\infty} q^{m^2+2n^2} = 1 + 2 \left(\sum_{n=0}^{\infty} \frac{q^{8n+1}}{1-q^{8n+1}} + \sum_{n=0}^{\infty} \frac{q^{8n+3}}{1-q^{8n+3}} \right)$$

$$- 2 \left(\sum_{n=0}^{\infty} \frac{q^{8n+5}}{1-q^{8n+5}} + \sum_{n=0}^{\infty} \frac{q^{8n+7}}{1-q^{8n+7}} \right); \tag{17.13}$$

$$\sum_{m,n=-\infty}^{\infty} q^{m^2+3n^2} = 1 + 2 \left(\sum_{n=0}^{\infty} \frac{q^{3n+1}}{1-q^{3n+1}} - \sum_{n=0}^{\infty} \frac{q^{3n+2}}{1-q^{3n+2}} \right)$$

$$+ 4 \left(\sum_{n=0}^{\infty} \frac{q^{12n+4}}{1-q^{12n+4}} - \sum_{n=0}^{\infty} \frac{q^{12n+8}}{1-q^{12n+8}} \right). \tag{17.14}$$

Proof. In each case, write the sum

$$\sum_{m,n=-\infty}^{\infty} q^{m^2+kn^2} = \sum_{m=-\infty}^{\infty} q^{m^2} \sum_{n=-\infty}^{\infty} q^{kn^2},$$

and use (17.11) to write the double sum as an infinite product.

By manipulating the resulting infinite product, show that it equals the infinite product that results on the left side from making the changes indicated below in (17.5) (in the case of (17.14), after also multiplying both sides of the identity that results from (17.5) by 2).

Similarly show that the sum of Lambert series that is produced by making the same changes on the right of (17.5) is the sum of Lambert series on the corresponding right sides of each of the identities.

For (17.12), replace q with q^4, and set $a = b = c = q$. For (17.13), replace q with q^8, and set $a = b = q$ and $c = q^3$. Lastly, for (17.14), replace q with q^6, and set $a = q$, $b = -q^2$ and $c = q^3$. The details in each case are left as exercises. □

The identity at (17.12) was also stated by Ramanujan (see [69, **Entry 8.** (i), page 114] and is equivalent to Jacobi's two squares theorem (that the number of representations of an integer n as a sum of two squares is four times the difference between the number of divisors of n congruent to 1 modulo 4 and the number of divisors of n congruent to 3 modulo 4). The identities (17.13) and (17.14) also have similar number theoretical interpretations relating representations of an integer n by certain binary quadratic forms and the number of divisors of n in various congruence classes (see, for example, [70, pp. 74–75]).

The identity at (17.12) expresses

$$\phi^2(q) = \left(\sum_{n=-\infty}^{\infty} q^{n^2} \right)^2$$

as a sum of Lambert series. The authors in [48] also prove expressions for each of $\phi^{2k}(q)$, $k = 2, 3, 4$ as a sum of Lambert series (these identities were first stated by Jacobi [154]).

Corollary 17.3. *If* $|q| < 1$, *then*

$$\phi^4(q) = 1 + 8 \sum_{n=1}^{\infty} \frac{nq^n}{1 - q^{2n}} - 8 \sum_{n=1}^{\infty} \frac{(-1)^n nq^{2n}}{1 - q^{2n}}; \tag{17.15}$$

$$\phi^6(q) = 1 + 16 \sum_{n=1}^{\infty} \frac{n^2 q^n}{1 + q^{2n}} + 4 \sum_{n=1}^{\infty} \frac{(-1)^n (2n-1)^2 q^{2n-1}}{1 - q^{2n-1}}; \tag{17.16}$$

$$\phi^8(q) = 1 + 16 \sum_{n=1}^{\infty} \frac{(-1)^n n^3 q^{2n}}{1 - q^{2n}} + 16 \sum_{n=1}^{\infty} \frac{n^3 q^n}{1 - q^{2n}}. \tag{17.17}$$

Proof. For (17.15), apply (17.48) (with $a = 1$) to (8.6) to get

$$\phi(-q)^4 = 1 + 8 \sum_{n=1}^{\infty} \frac{(-1)^n nq^n}{1 + q^n}$$

$$= 1 + 8 \sum_{n=1}^{\infty} \frac{(-1)^n n(q^n - q^{2n})}{1 - q^{2n}}.$$

Replace q with $-q$ and (17.15) follows.

For (17.16), first set $a = i$ in (17.9) to get, after a little simplification and manipulation of infinite products, that

$$1 + 4 \sum_{n=1}^{\infty} \frac{(-1)^n (2n-1)^2 q^{2n-1}}{1 - q^{2n-1}} = (q^2; q^2)_\infty^6 (-q; q^2)_\infty^4 (q; q^2)_\infty^8. \tag{17.18}$$

Next, continuing to follow the proof in [48] and noting that

$$\frac{a(1+a)}{(1-a)^3} = \sum_{n=1}^{\infty} n^2 a^n,$$

the identity at (17.9) may be written as

$$\frac{a(a^2, q/a^2; q)_{\infty}(q; q)_{\infty}^6}{(a, q/a; q)_{\infty}^4} = \sum_{n=1}^{\infty} \frac{n^2(a^n - a^{-n}q^n)}{1-q^n}. \tag{17.19}$$

Now replace q with q^4, set $a = q$ and multiply both sides by 16 to get, again after some infinite product manipulation, that

$$16 \sum_{n=1}^{\infty} \frac{n^2 q^n}{1+q^{2n}} = 16q(q^2; q^2)_{\infty}^6 (-q; q^2)_{\infty}^4 (-q^2; q^2)_{\infty}^8. \tag{17.20}$$

Now add (17.18) and (17.20) to get

$$1 + 4\sum_{n=1}^{\infty} \frac{(-1)^n (2n-1)^2 q^{2n-1}}{1-q^{2n-1}} + 16 \sum_{n=1}^{\infty} \frac{n^2 q^n}{1+q^{2n}}$$
$$= (q^2; q^2)_{\infty}^6 (-q; q^2)_{\infty}^4 \left[(q; q^2)_{\infty}^8 + 16q(-q^2; q^2)_{\infty}^8 \right]$$
$$= (q^2; q^2)_{\infty}^6 (-q; q^2)_{\infty}^4 (-q; q^2)_{\infty}^8 = (-q, -q, q^2; q^2)_{\infty}^6 = \phi(q)^6,$$

where the second equality follows directly from (6.24), thus completing the proof.

The proof of (17.17) is similar to that of (17.15). Apply (17.53) (with $a = -1$) to (8.7) to get

$$\phi(-q)^8 = 1 + 16 \sum_{n=1}^{\infty} \frac{(-1)^n n^3 q^n}{1-q^n}$$
$$= 1 + 16 \sum_{n=1}^{\infty} \frac{(-1)^n n^3 (q^n + q^{2n})}{1-q^{2n}},$$

and finally replace q once again with $-q$. $\qquad\qquad\square$

17.2 Lambert series identities deriving from Ramanujan's $_1\psi_1$ summation formula

Bilateral Lambert series have already made an appearance in the chapter on Ramanujan's $_1\psi_1$ summation formula, at (7.8), and replacing x with q^i,

y with q^j and q with q^m (where $1 \leq i, j < m$ are positive integers) leads to

$$\sum_{n=-\infty}^{\infty} \frac{q^{in}}{1-q^{mn+j}} = \frac{(q^m, q^m, q^{i+j}, q^{m-i-j}; q^m)_\infty}{(q^j, q^{m-j}, q^i, q^{m-i}; q^m)_\infty}. \qquad (17.21)$$

Special cases of Fine's identity (9.16) and the identity stated by Agarwal (9.17) are also useful. In particular, it follows from (9.17) and (17.21) that

$$\sum_{n=-\infty}^{\infty} \frac{q^{in}}{1-q^{mn+j}}$$
$$= \sum_{n=-\infty}^{\infty} \frac{(1-q^{2mn+i+j})q^{mn^2+n(i+j)}}{(1-q^{mn+i})(1-q^{mn+j})} = \frac{(q^m, q^m, q^{i+j}, q^{m-i-j}; q^m)_\infty}{(q^j, q^{m-j}, q^i, q^{m-i}; q^m)_\infty},$$
$$(17.22)$$

and if $i = j$, then

$$\sum_{n=-\infty}^{\infty} \frac{q^{in}}{1-q^{mn+i}}$$
$$= \sum_{n=-\infty}^{\infty} \frac{(1+q^{mn+i})q^{mn^2+2ni}}{1-q^{mn+i}} = \frac{(q^m, q^m, q^{2i}, q^{m-2i}; q^m)_\infty}{(q^i, q^{m-i}, q^i, q^{m-i}; q^m)_\infty}. \quad (17.23)$$

As Agarwal points out in [3], many of Ramanujan's identities involving Lambert series are special cases of one of the summation formulae above, and gives a list of eleven examples. Denis [110, 111] also stated (17.22) and gave many examples of its use. This formula was also used by Andrews and Berndt in [38, Section 4.4] to prove **Entry 4.4.1 (p. 47)** (which is stated and proved below), the ten entries in **Entry 4.4.2 (p. 47)** and the three entries in **Entry 4.4.2 (p. 47)** from Ramanujan's lost notebook [213].

The identity at (17.23) may be used to prove the following identity of Ramanujan from the lost notebook, and which was first proved by Andrews [20].

Proposition 17.1. *If* $|q| < 1$, *then*

$$\left\{ \frac{(q, q^4; q^5)_\infty}{(q^2, q^3; q^5)_\infty} \right\}^3 = \frac{\displaystyle\sum_{n=0}^{\infty} \frac{q^{2n}}{1-q^{5n+2}} - \sum_{n=0}^{\infty} \frac{q^{3n+1}}{1-q^{5n+3}}}{\displaystyle\sum_{n=0}^{\infty} \frac{q^n}{1-q^{5n+1}} - \sum_{n=0}^{\infty} \frac{q^{4n+3}}{1-q^{5n+4}}}.$$

$$\sum_{n=0}^{\infty} \frac{1+q^{5n+2}}{1-q^{5n+2}} q^{5n^2+4n} - \sum_{n=0}^{\infty} \frac{1+q^{5n+3}}{1-q^{5n+3}} q^{5n^2+6n+1}$$
$$= \frac{}{\sum_{n=0}^{\infty} \frac{1+q^{5n+1}}{1-q^{5n+1}} q^{5n^2+2n} - \sum_{n=0}^{\infty} \frac{1+q^{5n+4}}{1-q^{5n+4}} q^{5n^2+8n+3}}. \quad (17.24)$$

Proof. First set $m = 5$ and $i = 1$ in (17.23) to get

$$\sum_{n=-\infty}^{\infty} \frac{q^n}{1-q^{5n+1}}$$

$$= \sum_{n=-\infty}^{\infty} \frac{(1+q^{5n+1})q^{5n^2+2n}}{1-q^{5n+1}} = \frac{(q^5, q^5, q^2, q^3; q^5)_{\infty}}{(q, q^4, q, q^4; q^5)_{\infty}}, \quad (17.25)$$

and then set $m = 5$ and $i = 2$ to get

$$\sum_{n=-\infty}^{\infty} \frac{q^{2n}}{1-q^{5n+2}}$$

$$= \sum_{n=-\infty}^{\infty} \frac{(1+q^{5n+2})q^{5n^2+4n}}{1-q^{5n+2}} = \frac{(q^5, q^5, q^4, q; q^5)_{\infty}}{(q^2, q^3, q^2, q^3; q^5)_{\infty}}. \quad (17.26)$$

Now divide each expression in (17.26) by its corresponding expression in (17.25), and (17.24) follows after a simple manipulation of each bilateral series. $\qquad \square$

Remark: The expression inside the braces at (17.24) is equal to the reciprocal of the Rogers–Ramanujan continued fraction (see (16.1)).

Another useful special case of (7.8), used to prove some of the results collected in the next corollary, follows upon replacing q with q^2, x with q and y with zq, and renumbering the terms of negative index to get

$$\sum_{n=0}^{\infty} \left[\frac{q^n}{1-zq^{2n+1}} - \frac{q^n/z}{1-q^{2n+1}/z} \right] = \frac{(q^2, q^2, 1/z, q^2 z; q^2)_{\infty}}{(q, q, zq, q/z; q^2)_{\infty}}. \quad (17.27)$$

Recall also that $\left(\frac{n}{3}\right)$ denotes the Legendre symbol and that the Ramanujan theta function $\psi(q)$ (see (6.12)) has the representation

$$\psi(q) = \frac{(q^2; q^2)_{\infty}}{(q; q^2)_{\infty}} = (-q, -q, q; q)_{\infty}. \quad (17.28)$$

Corollary 17.4. *If $|q| < 1$, then*

$$\sum_{n=0}^{\infty} \frac{(2n+1)q^n}{1-q^{2n+1}} = \psi^4(q); \tag{17.29}$$

$$\sum_{n=0}^{\infty} \frac{q^n}{1+q^{2n+1}} = \psi^2(q^2); \tag{17.30}$$

$$\sum_{n=1}^{\infty} \left(\frac{n}{3}\right) \frac{q^n}{1-q^{2n}} = \frac{q\psi^3(q^3)}{\psi(q)}; \tag{17.31}$$

$$\sum_{n=0}^{\infty} \left(\frac{-n+1}{3}\right) \frac{q^n}{1-q^{2n+1}} = \psi(q)\psi(q^3); \tag{17.32}$$

$$\frac{1}{16}\sum_{n=0}^{\infty} \frac{(2n+1)^2 q^n}{1+q^{2n+1}} - \frac{1}{16}\sum_{n=0}^{\infty} \frac{(2n+1)^2(-1)^n q^n}{1-q^{2n+1}} = q\psi(q^2)^6; \tag{17.33}$$

$$\sum_{n=1}^{\infty} \frac{n^3 q^n}{1-q^{2n}} = q\psi(q)^8. \tag{17.34}$$

Proof. In (17.27) combine the two terms in the sum on the left together, divide both sides by $1 - 1/z$ and then let $z \to 1$. The right side of (17.27) becomes $\psi^4(q)$ and the left side becomes

$$\sum_{n=0}^{\infty} \frac{(1+q^{2n+1})q^n}{(1-q^{2n+1})^2} = \sum_{n=0}^{\infty} \frac{(2n+1)q^n}{1-q^{2n+1}},$$

where the last equality follows from differentiating

$$\sum_{n=0}^{\infty} \frac{aq^n}{1-a^2 q^{2n+1}}$$

with respect to a (in two different ways) and then setting $a = 1$.

For (17.30), replace q with q^2 and z with q in (17.27), multiply both sides by q and simplify, and finally replace q with $-q$.

Similarly, for (17.31), replace q with q^3 and z with q in (17.27), multiply both sides by q^2 and simplify. For (17.32), replace q with q^3 and z with q^2, multiply both sides by q^2 and similarly simplify.

To get (17.33), replace q with $-q$ in (17.19) and let $a = \sqrt{q}$ to derive, once again after some simplification and manipulation of infinite products, that

$$\sum_{n=0}^{\infty} \frac{(2n+1)^2 q^n}{1+q^{2n+1}} = (q^2; q^2)_\infty^6 (-q^2; q^2)_\infty^4 (-q; q^2)_\infty^8. \tag{17.35}$$

Replace q with $-q$ in (17.35) and subtract this new equation from (17.35) so that

$$\sum_{n=0}^{\infty} \frac{(2n+1)^2 q^n}{1+q^{2n+1}} - \sum_{n=0}^{\infty} \frac{(2n+1)^2 (-1)^n q^n}{1-q^{2n+1}}$$
$$= (q^2;q^2)_\infty^6 (-q^2;q^2)_\infty^4 \left[(-q;q^2)_\infty^8 - (q;q^2)_\infty^8 \right]$$
$$= (q^2;q^2)_\infty^6 (-q^2;q^2)_\infty^4 \left[16q(-q^2;q^2)_\infty^8 \right]$$
$$= 16q(-q^2,-q^2,q^2;q^2)_\infty^6 = 16q\psi^6(q^2),$$

where the last equality follows, as in the proof of (17.17), from (6.24).

Lastly, to prove (17.34), replace q with q^2 in (17.19), divide both sides by $1-q^2/a^2$, and then let $a \to q$. $\qquad\square$

Remark: All of the identities above were stated by Ramanujan (most can be found in [38, Section 18.2]).

The identity at (7.18) may also be specialized to produce a number of other Lambert series stated by Ramanujan (once again, see [38, Section 18.2]). Specifically, setting $\alpha = q^m/\beta$ and shifting the summation index in the series to start at $n = 0$ leads to

$$1 + 2\sum_{n=0}^{\infty} (-1)^n \left(\frac{\beta q^{mn}}{1-\beta q^{mn}} + \frac{q^{m(n+1)}/\beta}{1-q^{m(n+1)}/\beta} \right) = \frac{(-q^m/\beta, -\beta, q^m, q^m; q^m)_\infty}{(q^m/\beta, \beta, -q^m, -q^m; q^m)_\infty}.$$
$$(17.36)$$

Corollary 17.5. *If $|q| < 1$, then*

$$1 - 2\sum_{n=1}^{\infty} \left(\frac{n}{3} \right) \frac{q^n}{1-(-q)^n} = \frac{\phi^3(q^3)}{\phi(q)}; \qquad (17.37)$$

$$1 + 2\sum_{n=1}^{\infty} \left(\frac{n}{3} \right) \frac{q^n}{1+(-q)^n} = \phi(q^3)\phi(q); \qquad (17.38)$$

$$1 + 2\sum_{n=0}^{\infty} (-1)^n \left(\frac{q^{4n+1}}{1-q^{4n+1}} + \frac{q^{4n+3}}{1-q^{4n+3}} \right) = \frac{\phi(-q^4)^2 \phi(-q^2)}{\phi(-q)}. \qquad (17.39)$$

Proof. For (17.37), set $m = 3$, $\beta = q$ in (17.36) so that right side becomes $\phi^3(-q^3)/\phi(-q)$, and then replace q with $-q$. The identity at (17.38) follows similarly, upon setting $m = 3$ and $\beta = -q$, and that at (17.39) upon setting $m = 4$ and $\beta = q$. The details of the proofs are left to the reader. $\qquad\square$

17.3 Lambert series identities arising via differentiation

Differentiation of infinite q-products lead naturally to Lamberts series via logarithmic differentiation, using the calculus rule $f'(z) = f(z)(\ln(f(z)))'$, so that, for example, if $f(z) = (tz; q)_\infty$, then

$$\frac{d\,(tz;q)_\infty}{d\,z} = (tz;q)_\infty \sum_{n=0}^{\infty} \frac{tq^n}{1 - ztq^n}.$$

Applying this calculus rule to the Jacobi triple product identity leads to the following transformation.

Proposition 17.2. *Let m be a positive integer and let p and $t > 0$ be integers or half-integers such that none of the denominators in (17.40) vanish. If $|q| < 1$ then*

$$1 + m \sum_{n=0}^{\infty} \frac{(\pm z^m)q^{(2n+1)t+p}}{1 + (\pm z^m)q^{(2n+1)t+p}} - m \sum_{n=0}^{\infty} \frac{q^{(2n+1)t-p}/(\pm z^m)}{1 + q^{(2n+1)t-p}/(\pm z^m)}$$

$$= \frac{\sum_{n=-\infty}^{\infty}(mn+1)(\pm z)^{mn}q^{tn^2+pn}}{(\mp z^m q^{t+p}, \mp q^{t-p}/z^m, q^{2t}; q^{2t})_\infty}. \tag{17.40}$$

Proof. This follows directly from differentiating the identity

$$\sum_{n=-\infty}^{\infty} (\pm z^m q^p)^n q^{tn^2} = (\mp z^m q^{t+p}, \mp q^{t-p}/z^m, q^{2t}; q^{2t})_\infty, \tag{17.41}$$

(which is true by the Jacobi triple product identity (6.1)) with respect to z, multiplying both sides by $\pm z$, and then adding (17.41) to the derivative. Finally, divide both sides by $(\mp z^m q^{t+p}, \mp q^{t-p}/z^m, q^{2t}; q^{2t})_\infty$. $\square$

Special cases of (17.40) provide proofs of a number of Lambert series identities of Ramanujan, including (17.12), and two other identities not encountered above. The identities in the next corollary, all due to Ramanujan, were all proved by Hirschhorn [147], and his method of proof is essentially the one used here.

Corollary 17.6. *If $|q| < 1$, then*

$$1 + 4\sum_{n=0}^{\infty} \frac{q^{4n+1}}{1 - q^{4n+1}} - 4\sum_{n=0}^{\infty} \frac{q^{4n+3}}{1 - q^{4n+3}} = \phi^2(q); \tag{17.42}$$

$$1 - 6\sum_{n=1}^{\infty} \left(\frac{n}{3}\right)\frac{q^n}{1+q^n} = \frac{\phi^3(-q)}{\phi(-q^3)}; \qquad (17.43)$$

$$1 + 3\sum_{n=0}^{\infty}\left(\frac{q^{6n+1}}{1-q^{6n+1}} - \frac{q^{6n+5}}{1-q^{6n+5}}\right) = \frac{\psi^3(q)}{\psi(q^3)}. \qquad (17.44)$$

Proof. For (17.42), set $m = 4$, $t = 2$, $p = 1$ and $z = 1$ in the "−" case of (17.40) to get

$$1 + 4\sum_{n=0}^{\infty}\left(\frac{-q^{4n+3}}{1-q^{4n+3}} + \frac{q^{4n+1}}{1-q^{4n+1}}\right) = \frac{\sum_{n=-\infty}^{\infty}(4n+1)(-1)^{mn}q^{2n^2+n}}{(q^3, q, q^4; q^4)_\infty}.$$

The result follows after using (6.4)

$$\sum_{n=0}^{\infty}(-1)^n(2n+1)q^{n(n+1)/2} = \sum_{n=-\infty}^{\infty}(4n+1)q^{2n^2+n} = (q;q)_\infty^3,$$

with q replaced with $-q$, to sum the series, and then simplifying the infinite products.

The second identity, (17.43), follows similarly upon setting $m = 6$, $t = 3/2$, $p = 1/2$ and $z = 1$ in the "+" case of (17.40) to get

$$1 + 6\sum_{n=0}^{\infty}\left(\frac{q^{3n+2}}{1+q^{3n+2}} - \frac{q^{3n+1}}{1+q^{3n+1}}\right) = \frac{\sum_{n=-\infty}^{\infty}(6n+1)q^{(3n^2+n)/2}}{(-q^2, -q, q^3; q^3)_\infty},$$

and then using (6.25) to sum the series, and once again manipulating the infinite products.

Finally, for (17.43) set $m = t = 3$, $p = 2$ and $z = 1$ in the "−" case of (17.40) to get

$$1 + 3\sum_{n=0}^{\infty}\left(\frac{-q^{6n+5}}{1-q^{6n+5}} - \frac{-q^{6n+1}}{1-q^{6n+1}}\right) = \frac{\sum_{n=-\infty}^{\infty}(3n+1)(-1)^n q^{3n^2+2n}}{(q^5, q, q^6; q^6)_\infty},$$

and then using (6.26) (with q replaced with $-q$) to sum the series, and again manipulating the infinite products. □

Remark: All three identities in the above corollary were also stated by Ramanujan (see, for example, [38, Section 18.2]).

Collections of Lambert series identities were given by Lam [175] and Shen [231, 232]. Lambert series will appear again in chapter 18, in identities involving various mock theta functions.

Exercises

17.1 Prove that if $|q|, |x| < 1$, then

$$\sum_{n=1}^{\infty} \frac{q^n x^{n^2}}{1 - qx^n} + \sum_{n=1}^{\infty} \frac{q^n x^{n^2+n}}{1 - x^n} = \sum_{n=1}^{\infty} \frac{q^n x^n}{1 - x^n}. \tag{17.45}$$

17.2 Fill in the details in the proof of Corollary 17.2.

17.3 Prove that if $|q| < 1$, then

$$\sum_{n=1}^{\infty} \frac{q^n}{1 + q^{2n}} = \sum_{n=1}^{\infty} \frac{(-1)^{n-1} q^{2n-1}}{1 - q^{2n-1}}. \tag{17.46}$$

17.4 If $|q|, |aq| < 1$, by differentiating appropriate series (see for example, (9.18)) in two different ways, prove that

$$\sum_{n=1}^{\infty} \frac{q^n}{(1 - aq^n)^2} = \sum_{n=1}^{\infty} \frac{n a^{n-1} q^n}{1 - q^n}; \tag{17.47}$$

$$\sum_{n=1}^{\infty} \frac{(-q)^n}{(1 + aq^n)^2} = \sum_{n=1}^{\infty} \frac{n(-1)^n a^{n-1} q^n}{1 + q^n}; \tag{17.48}$$

$$\sum_{n=1}^{\infty} \frac{q^n(1 + aq^n)}{(1 - aq^n)^3} = \sum_{n=1}^{\infty} \frac{n^2 a^{n-1} q^n}{1 - q^n}; \tag{17.49}$$

$$\sum_{n=1}^{\infty} \frac{(-1)^n q^n (1 - aq^n)}{(1 + aq^n)^3} = \sum_{n=1}^{\infty} \frac{(-1)^n n^2 a^{n-1} q^n}{1 + q^n}; \tag{17.50}$$

$$\sum_{n=1}^{\infty} \frac{2q^{2n}}{(1 - aq^n)^3} = \sum_{n=1}^{\infty} \frac{n(n-1) a^{n-2} q^n}{1 - q^n}; \tag{17.51}$$

$$\sum_{n=1}^{\infty} \frac{2(-1)^{n-1} q^{2n}}{(1 + aq^n)^3} = \sum_{n=1}^{\infty} \frac{(-1)^n n(n-1) a^{n-2} q^n}{1 + q^n}; \tag{17.52}$$

$$\sum_{n=1}^{\infty} \frac{q^n(1 + 4aq^n + a^2 q^{2n})}{(1 - aq^n)^4} = \sum_{n=1}^{\infty} \frac{n^3 a^{n-1} q^n}{1 - q^n}; \tag{17.53}$$

$$\sum_{n=1}^{\infty} \frac{(-1)^n q^n(1 - 4aq^n + a^2 q^{2n})}{(1 + aq^n)^4} = \sum_{n=1}^{\infty} \frac{(-1)^n n^3 a^{n-1} q^n}{1 + q^n}; \tag{17.54}$$

17.5 (Agarwal [3]) Define

$$F(z) = \sum_{n=1}^{\infty} \frac{(q; q)_{n-1} z^n}{(1 - q^n)(z; q)_n}.$$

By employing a particular (limiting) case of Heine's q-Gauss summation formula (4.8), or otherwise, show that

$$F(z) - F(qz) = \frac{z}{(1-z)^2}.$$

Hence deduce that

$$F(z) = \sum_{n=1}^{\infty} \frac{nz^n}{1 - q^n}.$$

17.6 By letting a and b be appropriate fifth roots of unity in (17.8), prove Ramanujan's identity

$$\frac{(q;q)_\infty^5}{(q^5;q^5)_\infty} = 1 - 5\sum_{n=1}^{\infty}\left(\frac{(5n+1)q^{5n+1}}{1-q^{5n+1}} - \frac{(5n+2)q^{5n+2}}{1-q^{5n+2}}\right.$$
$$\left. - \frac{(5n+3)q^{5n+3}}{1-q^{5n+3}} + \frac{(5n+4)q^{5n+4}}{1-q^{5n+4}}\right). \quad (17.55)$$

17.7 (Ramanujan [213], see also [38, Eq. 5 (4.4.15), (4.4.16)]) Prove that if $|q| < 1$ then

$$\frac{(q^5, q^5; q^5)_\infty}{(q, q^4; q^5)_\infty} = \sum_{n=-\infty}^{\infty} \frac{q^{4n}}{1 - q^{10n+1}},$$

$$\frac{(q^5, q^5; q^5)_\infty}{(q^2, q^3; q^5)_\infty} = \sum_{n=-\infty}^{\infty} \frac{q^{2n}}{1 - q^{10n+3}}.$$

17.8 Prove that if $|q| < 1$ then

$$\frac{(q^3; q^3)_\infty^3}{(q; q)_\infty} = \sum_{n=-\infty}^{\infty} \frac{q^n}{1 - q^{3n+1}}.$$

17.9 (Chan [96]) By replacing q with q^3, letting $b = \omega$, a primitive third root of unity, and $a = q$ in (17.8), or otherwise, prove the identity of Borwein and Garvan [81]

$$1 + 3\sum_{n=1}^{\infty} \frac{nq^n}{1 - q^n} - 27\sum_{n=1}^{\infty} \frac{nq^{9n}}{1 - q^{9n}} = \frac{(q^3; q^3)_\infty^{10}}{(q; q)_\infty^3 (q^9; q^9)_\infty^3}. \quad (17.56)$$

17.10 Prove that

$$\sum_{n=0}^{\infty} \frac{n(a; q)_n z^n}{(q; q)_n} = \frac{(az; q)_\infty}{(z; q)_\infty} \sum_{n=0}^{\infty}\left(\frac{zq^n}{1 - zq^n} - \frac{azq^n}{1 - azq^n}\right). \quad (17.57)$$

Hence show that

$$\sum_{n=0}^{\infty} \frac{(4n+1)(q^2;q^4)_n q^n}{(q^4;q^4)_n} = \frac{(q^3;q^4)_\infty}{(q;q^4)_\infty}\phi^2(q);$$ (17.58)

$$\sum_{n=0}^{\infty} \frac{(6n+1)(q;q^3)_n(-q)^n}{(q^3;q^3)_n} = \frac{(-q^2;q^3)_\infty}{(-q;q^3)_\infty}\frac{\phi^3(-q)}{\phi(-q^3)};$$ (17.59)

$$\sum_{n=0}^{\infty} \frac{(3n+1)(q^4;q^6)_n q^n}{(q^6;q^6)_n} = \frac{(q^5;q^6)_\infty}{(q;q^6)_\infty}\frac{\psi^3(q)}{\psi(q^3)}.$$ (17.60)

17.11 Prove that

$$\sum_{n=-\infty}^{\infty} \frac{n(a;q)_n z^n}{(b;q)_n} = \frac{(q,b/a,az,q/(az);q)_\infty}{(b,q/a,z,b/(az);q)_\infty}$$
$$\times \sum_{n=0}^{\infty} \left(\frac{zq^n}{1-zq^n} - \frac{azq^n}{1-azq^n} + \frac{q^{n+1}/(az)}{1-q^{n+1}/(az)} - \frac{q^n b/(az)}{1-q^n b/(az)} \right).$$ (17.61)

Hence show that

$$\sum_{n=-\infty}^{\infty} \frac{(2n+1)q^n}{1-q^{8n+4}} = \frac{(q^3,q^5,q^8,q^8;q^8)_\infty}{(q,q^4,q^4,q^7;q^8)_\infty}\phi(q)\phi(q^2);$$ (17.62)

$$\sum_{n=-\infty}^{\infty} \frac{(4n+1)q^n}{1-q^{8n+2}} = \frac{(q^3,q^5,q^8,q^8;q^8)_\infty}{(q,q^2,q^6,q^7;q^8)_\infty}\phi(q)^2;$$ (17.63)

$$\sum_{n=-\infty}^{\infty} \frac{(3n+1)q^n}{1-q^{12n+4}} = \frac{(q^5,q^7,q^{12},q^{12};q^{12})_\infty}{(q,q^4,q^8,q^{11};q^{12})_\infty}\frac{\psi(q)^3}{\psi(q^3)};$$ (17.64)

$$\sum_{n=-\infty}^{\infty} \frac{(6n+1)q^n}{1+q^{6n+1}} = \frac{(-q^2,-q^4,q^6,q^6;q^6)_\infty}{(q^2,q^{10};q^{12})_\infty}\frac{\phi(q)^3}{\phi(q^3)}.$$ (17.65)

17.12 By employing Lambert series identities in the present chapter in conjunction with identities from Section 13.9, prove that

$$\phi^2(q) = 1 + 4\sum_{n=1}^{\infty} \left(\frac{q^{4n^2+n}}{1+q^{2n}} + \frac{q^{4n^2-3n}}{1-q^{4n-3}} - \frac{q^{4n^2-n}}{1-q^{4n-1}} \right).$$

Chapter 18

Mock Theta Functions

Mock theta functions were introduced by Ramanujan in his last letter to
G.H. Hardy [214, pp. 354–355], [73, pp. 220–223], in which he also gave
examples of mock theta functions of orders three, five and seven. It is still
not clear what exactly Ramanujan meant by a "mock theta function" or its
"order". We will not be concerned here with the definitions of these terms,
focusing instead on basic hypergeometric identities satisfied by Ramanujan's
mock theta functions, and those found subsequently by later investigators.

However, for completeness sake, we briefly mention current understanding of these terms. Andrews and Hickerson [47] interpreted Ramanujan's
statements to mean that a mock theta function is a function $f(q)$ defined
by a q-series which converges for $|q| < 1$ and which satisfies the following
two conditions:
(0) For every root of unity ζ, there is a θ-function $\theta_\zeta(q)$ such that the difference $f(q) - \theta_\zeta(q)$ is bounded as $q \to \zeta$ radially.
(1) There is no single θ-function which works for all ζ; i.e., for every θ-
function $\theta(q)$ there is some root of unity ζ for which $f(q) - \theta(q)$ is unbounded
as $q \to \zeta$ radially.

Note that it was not until the appearance of the paper by Griffin, Ono,
and Rolen [134] was it shown that Ramanujan's mock theta functions indeed
satisfied condition (1) above.

A similar definition was given by Gordon and McIntosh [130, 132], where
they also distinguish between a mock theta function and a "strong" mock
theta function. Zwegers [263, 264] has shown that the mock theta functions
are holomorphic parts of certain harmonic weak Maass forms. Note that
Rhoades [215] has shown that the original definition of Ramanujan and
the modern definition coming from the work of Zwegers are not equivalent.

Gordon and McIntosh interpret the "order" of a mock theta function by its behavior under the action of the modular group (see [131, Section 4] for details).

Many important papers have appeared in recent years dealing with subjects related to mock theta functions. Apart from papers dealing with basic hypergeometric aspects of these to be mentioned below, we note the following. Dragonette [112] initiated the study of the coefficients in the Taylor series expansions of the mock theta functions, using the circle method, with subsequent improvements being made by Andrews [12] and Bringmann and Ono [86, 88]. The Mock Theta Conjectures were proven by Hickerson [141], building on prior work by Andrews [23, 26], and Andrews and Garvan [46]. The transformation of mock theta functions was first investigated by Watson [259], later work was carried out by Dragonette [112] and Andrews [12], and more recently by Gordon and McIntosh [131, 132]. Important papers that exhibit the connections between mock theta functions and Maass forms include those of Bringmann and Ono [86–88], Folsom, Ono and Rhoades [117], Ono [206, 207], Zwegers [263, 264], and others (see the bibliography in the survey paper of Gordon and McIntosh [132] for further references).

As may possibly be gathered from the previous paragraphs, the mock theta functions comprise a very extensive research area. In the present chapter we may hope to just sample something of their nature and properties as basic hypergeometric series. The starting point for the approach taken in this chapter is the observation that many of the mock theta functions are special cases of one "side" ($n \geq 0$ or $n < 0$) of certain general bilateral series which arise as special cases of the $2\psi_2$ series

$$\sum_{n=-\infty}^{\infty} \frac{(a, c; q)_n}{(b, d; q)_n} z^n = \sum_{n=0}^{\infty} \frac{(a, c; q)_n}{(b, d; q)_n} z^n + \sum_{n=1}^{\infty} \frac{(q/b, q/d; q)_n}{(q/a, q/c; q)_n} \left(\frac{bd}{acz} \right)^n.$$

The mock theta functions which arise in this way include:

Third order - all 9 (Ramanujan, Watson, Gordon and McIntosh);

Fifth order - 8 of 10 (Ramanujan);

Sixth order - 8 (Ramanujan);

Eighth order - 4 of 8 (Gordon and McIntosh).

Many of the known properties of the mock theta functions may then be shown to derive from transformation- and summation formulae for these more general functions.

18.1 The Third Order Mock Theta Functions

The third order mock theta functions are the following basic hypergeometric series:

$$f(q) = \sum_{n=0}^{\infty} \frac{q^{n^2}}{(-q, -q; q)_n}, \qquad \phi(q) = \sum_{n=0}^{\infty} \frac{q^{n^2}}{(-q^2; q^2)_n},$$

$$\chi(q) = \sum_{n=0}^{\infty} \frac{q^{n^2}(-q; q)_n}{(-q^3; q^3)_n}, \qquad \psi(q) = \sum_{n=1}^{\infty} \frac{q^{n^2}}{(q; q^2)_n},$$

$$\omega(q) = \sum_{n=0}^{\infty} \frac{q^{2n^2+2n}}{(q, q; q^2)_{n+1}}, \qquad \nu(q) = \sum_{n=0}^{\infty} \frac{q^{n^2+n}}{(-q; q^2)_{n+1}},$$

$$\rho(q) = \sum_{n=0}^{\infty} \frac{q^{2n^2+2n}(q; q^2)_{n+1}}{(q^3; q^6)_{n+1}}, \qquad \xi(q) = 1 + 2\sum_{n=1}^{\infty} \frac{q^{6n^2-6n}}{(q, q^5; q^6)_n},$$

$$\sigma(q) = \sum_{n=1}^{\infty} \frac{q^{3n^2-3n}}{(-q, -q^2; q^3)_n}.$$

The first four third order mock theta functions listed above were stated by Ramanujan ([214, pp. 354–355], [73, pp. 220–223]), the next three by Watson [259], and the final two by Gordon and McIntosh [131].

Hickerson and Mortenson [143, Eqs. (5.4)–(5.10)] showed that all of the third order mock theta functions of Ramanujan and Watson may be expressed in terms of their function $g(x, q)$, where

$$g(x, q) := \sum_{n=0}^{\infty} \frac{q^{n^2+n}}{(x, q/x; q)_{n+1}} = x^{-1}\left(-1 + \sum_{n=0}^{\infty} \frac{q^{n^2}}{(x; q)_{n+1}(q/x; q)_n}\right). \tag{18.1}$$

This function was also defined by Gordon and McIntosh [132], where it was labelled "$g_3(x, q)$", and it is not difficult to see that their functions $\xi(q)$ and $\sigma(q)$ above may be similarly represented. Two of the sixth order mock theta functions, $\beta(q)$ and $\gamma(q)$, also have representations in terms of $g(x, q)$, which, as Gordon and McIntosh indicate in [132], possibly means that these should be reclassified as being of the third order. See Question 2.10 for the equality of the two series at (18.1). The function $g(x, q) = g_3(x, q)$ may also be considered as a bridge linking the mock theta functions of orders three and five, as not only may all of the order three mock theta functions be represented as special cases of $g(x, q)$ (in some cases with q replaced with q^2), special cases of $g(x, q)$ appear in the "mock theta conjectures" involving the mock theta functions of order five.

We initially instead consider the series

$$G_3(s,t,q) := 1 + \sum_{n=1}^{\infty} \frac{s^n t^n q^{n^2}}{(sq, tq; q)_n}, \tag{18.2}$$

which was defined in [101, Eq. (7)], where it appeared in a result (Theorem 18.5 below) which in turn played an important part in the main result in [117]. Note that the connection with the third order mock theta functions is that

$$G_3(x, q/x, q) = (1-x)(1-q/x)g(x,q). \tag{18.3}$$

We first collect together a number of other representations of $G_3(s,t,q)$ which in turn may be used to give different representations for the various third order mock theta functions.

Proposition 18.1. *Let $G_3(s,t,q)$ be as defined at (18.2) above. Then*

$$G_3(s,t,q) = \frac{1}{(tq;q)_\infty} \sum_{r=0}^{\infty} \frac{(s;q)_r(-t)^r q^{r(r+1)/2}}{(sq,q;q)_r}; \tag{18.4}$$

$$= \frac{(q;q)_\infty}{(sq,tq;q)_\infty} \sum_{r=0}^{\infty} \frac{(s,t;q)_r q^r}{(q;q)_r}; \tag{18.5}$$

$$= \frac{1}{(stq;q)_\infty} \sum_{r=0}^{\infty} \frac{(1-stq^{2r})(s,t,st;q)_r(-st)^r q^{(3r^2+r)/2}}{(1-st)(sq,tq,q;q)_r}; \tag{18.6}$$

$$= -\sum_{r=1}^{\infty}(s^{-1},t^{-1};q)_r q^r + \frac{(q/s,q/t;q)_\infty}{(sq,tq;q)_\infty} \sum_{r=-\infty}^{\infty}(s,t;q)_r q^r; \tag{18.7}$$

$$= -\sum_{r=1}^{\infty}(s^{-1},t^{-1};q)_r q^r + \frac{(q/t;q)_\infty}{(sq,q;q)_\infty} \sum_{r=-\infty}^{\infty} \frac{(t;q)_r(-s)^r q^{r(r+1)/2}}{(tq;q)_r}; \tag{18.8}$$

$$= -\sum_{r=1}^{\infty}(s^{-1},t^{-1};q)_r q^r$$

$$+ \frac{(q/s,q/t;q)_\infty}{(stq,q/(st),q;q)_\infty} \sum_{r=-\infty}^{\infty} \frac{(1-stq^{2r})(s,t;q)_r(st)^{2r}q^{2r^2}}{(1-st)(sq,tq;q)_r}. \tag{18.9}$$

Proof. For (18.4), let $a,b \to \infty$ and set $c = q$, $d = sq$ and $e = tq$ in (4.17). The identity at (18.5) follows after making the same substitutions in (4.18).

The identity at (18.6) follows from (5.27), after letting $d, e \to \infty$ and setting $a = q$, $f = sq$ and $g = tq$.

For (18.7) and (18.8), we first prove two more general identities. Replace z with zq/ac, let $a, c \to \infty$ and set $b = sq$ and $d = tq$ in, respectively, (7.20) and (7.21), to get that

$$\sum_{n=-\infty}^{\infty} \frac{z^n q^{n^2}}{(sq, tq; q)_n} = \frac{(sq/z, tq/z; q)_\infty}{(sq, tq; q)_\infty} \sum_{r=-\infty}^{\infty} (z/s, z/t; q)_r \left(\frac{stq}{z}\right)^r, \quad (18.10)$$

$$= \frac{(qs/z; q)_\infty}{(sq, stq/z; q)_\infty} \sum_{r=-\infty}^{\infty} \frac{(z/s; q)_r (-s)^r q^{r(r+1)/2}}{(tq; q)_r}. \quad (18.11)$$

Lastly, replace z with st, and use (2.5) on the terms of negative index in the new series on the left sides.

We also prove a generalization of (18.9) first, by letting $e, f \to \infty$ in (8.9), and then replacing a with z, c with z/s and d with z/t, to get

$$\sum_{n=-\infty}^{\infty} \frac{z^n q^{n^2}}{(sq, tq; q)_n}$$

$$= \frac{(sq/z, tq/z; q)_\infty}{(zq, q/z, stq/z; q)_\infty} \sum_{r=-\infty}^{\infty} \frac{(1 - zq^{2r})(z/s, z/t; q)_r (zst)^r q^{2r^2}}{(1 - z)(sq, tq; q)_r}. \quad (18.12)$$

The identity at (18.9) follows after replacing z with st. $\qquad\square$

The identity at (18.6) was stated by Ramanujan in the "Lost Notebook" [213], although the version stated in [20, Eq. (3.3)] is missing the equivalent of the $s^n t^n$ in the series on the left side.

The identities (18.7)–(18.9) may be more succinctly expressed using the function

$$G_3^*(s, t, q) := \sum_{n=-\infty}^{\infty} \frac{s^n t^n q^{n^2}}{(sq, tq; q)_n} \quad (18.13)$$

as follows

$$G_3^*(s, t, q) = \frac{(q/s, q/t; q)_\infty}{(sq, tq; q)_\infty} G_3^*(s^{-1}, t^{-1}, q), \quad (18.14)$$

$$= \frac{(q/t; q)_\infty}{(sq, q; q)_\infty} \sum_{r=-\infty}^{\infty} \frac{(t; q)_r (-s)^r q^{r(r+1)/2}}{(tq; q)_r}; \quad (18.15)$$

$$= \frac{(q/s, q/t; q)_\infty}{(stq, q/(st), q; q)_\infty} \sum_{r=-\infty}^{\infty} \frac{(1 - stq^{2r})(s, t; q)_r (st)^{2r} q^{2r^2}}{(1 - st)(sq, tq; q)_r}. \quad (18.16)$$

The identity at (18.7) (or (18.15)) was also proved by Choi [101, Theorem 4], and stated previously by Ramanujan (see [39, Entry 3.4.7]).

We next derive some new transformations for the third order mock theta functions of Ramanujan (similar results may be derived for the third order mock theta functions of Watson [259] and those of Gordon and McIntosh [131]).

Theorem 18.1. *If* $|q| < 1$, *then*

$$f(q) := \sum_{n=0}^{\infty} \frac{q^{n^2}}{(-q, -q; q)_n}$$
$$= -\sum_{n=1}^{\infty} (-1, -1; q)_n q^n + 4\frac{(-q; q)_{\infty}^2}{(q; q)_{\infty}^3} \sum_{r=-\infty}^{\infty} \frac{q^{2r^2+r}(4rq^r + 1)}{(1 + q^r)^2}. \quad (18.17)$$

$$\phi(q) := \sum_{n=0}^{\infty} \frac{q^{n^2}}{(-q^2; q^2)_n}$$
$$= -\sum_{n=1}^{\infty} (-1; q^2)_n q^n + 4\frac{(-q^2; q^2)_{\infty}}{(q; q)_{\infty}^3} \sum_{r=-\infty}^{\infty} \frac{q^{2r^2+2r}(2rq^{2r} + 1)}{(1 + q^{2r})^2}. \quad (18.18)$$

$$\nu(q) := \sum_{n=0}^{\infty} \frac{q^{n^2+n}}{(-q; q^2)_{n+1}} = -\sum_{n=0}^{\infty} (-q; q^2)_n q^n$$
$$+ 4\frac{(-q; q^2)_{\infty}}{(q; q)_{\infty}^3} \sum_{r=-\infty}^{\infty} \frac{q^{2r^2+2r}(r + 1)}{(1 + q^{2r+1})^2} - 2\frac{(-q; q^2)_{\infty}}{(q; q)_{\infty}^3}(-q^4, -q^{12}, q^{16}; q^{16})_{\infty}.$$
$$\quad (18.19)$$

$$\psi(q) := \sum_{n=1}^{\infty} \frac{q^{n^2}}{(q; q^2)_n}$$
$$= -\sum_{n=0}^{\infty} (q; q^2)_n (-1)^n + \frac{1}{2(q^2; q^2)_{\infty}^2} \sum_{r=-\infty}^{\infty} q^{2r^2+r}(4r + 1)(-1)^r. \quad (18.20)$$

Proof. For (18.17), replace z with z^2, s and t with $-z$ in (18.12), and then let $z \to 1$.

For each of the other identities, again replace z with z^2 in (18.12), replace s and t with expressions involving z, and let z tend the appropriate limit. The details are left as exercises (see Q18.20). Note that the convergence of the first series on the right side of (18.20) is in the Cesàro sense. □

As Watson pointed out in [260, Section 7], and as will be seen in section 18.2, bilateral sums that are essentially pairs of fifth order mock theta functions are expressible as theta functions, or combinations of infinite q-products. It seems less well known that the bilateral series associated with two of Ramanujan's third order mock theta functions are also expressible as infinite products. A similar expression is given for Watson's [259] third order mock theta function $\nu(q)$.

Theorem 18.2. *If $|q| < 1$, then*

$$\phi(q) + \sum_{r=1}^{\infty}(-1;q^2)_r q^r = \sum_{n=-\infty}^{\infty}\frac{q^{n^2}}{(-q^2;q^2)_n} = \frac{(-q,-q,q^2;q^2)_\infty}{(q,-q^2;q^2)_\infty}; \quad (18.21)$$

$$\nu(q) + \sum_{r=0}^{\infty}(-q;q^2)_r q^r = \sum_{r=-\infty}^{\infty}\frac{q^{n^2+n}}{(-q;q^2)_{n+1}} = 2(-q^2,-q^2;q^2)_\infty(q^4;q^4)_\infty; \quad (18.22)$$

$$\psi(q) + \sum_{r=0}^{\infty}(q;q^2)_r(-1)^r = \sum_{n=-\infty}^{\infty}\frac{q^{n^2}}{(q;q^2)_n} = \frac{(-q,-q,q^2;q^2)_\infty}{2(q,-q^2;q^2)_\infty}. \quad (18.23)$$

Proof. From (8.5) (replace q with q^2, set $b = -z/t$, $a = -z$, and let $c \to \infty$),

$$\sum_{r=-\infty}^{\infty}\frac{(-z/t;q)_r t^r q^{r(r+1)/2}}{(tq;q)_r} = \frac{(-t^2q^2/z,-zq,-q/z,q^2;q^2)_\infty}{(tq,-tq/z;q)_\infty},$$

and from (18.11) (with $s = -t$),

$$\sum_{n=-\infty}^{\infty}\frac{z^n q^{n^2}}{(t^2q^2;q^2)_n} = \frac{(-tq/z;q)_\infty}{(-tq,-t^2q/z;q)_\infty}\sum_{r=-\infty}^{\infty}\frac{(-z/t;q)_r(t)^r q^{r(r+1)/2}}{(tq;q)_r}.$$

Together, these equations imply that

$$\sum_{n=-\infty}^{\infty}\frac{z^n q^{n^2}}{(t^2q^2;q^2)_n} = \frac{(-zq,-q/z,q^2;q^2)_\infty}{(t^2q^2,-t^2q/z;q^2)_\infty}. \quad (18.24)$$

The identity at (18.21) is now immediate upon setting $z = 1$ and $t^2 = -1$, and that at (18.23) results similarly upon setting $z = 1$ and $t^2 = 1/q$. The identity at (18.22) follows upon setting $z = q$, $t^2 = -q$, multiplying the resulting product by $1/(1+q)$, and finally performing some elementary q-product manipulations. $\qquad\square$

Note that the convergence of the sum added to $\psi(q)$ on the left side of (18.23) is in the Cesàro sense.

The results in Theorem 18.2 are of interest in connection with [117, Theorem 1.1], in which the authors make explicit for the third order mock theta function $f(q)$ condition (0) above, namely that for every root of unity ζ, there is a θ-function $\theta_\zeta(q)$ such that the difference $f(q) - \theta_\zeta(q)$ is bounded as $q \to \zeta$ radially (they also prove a much more general result, their Theorem 1.2). Specifically, they proved the following.

Theorem 18.3. *(Folsom, Ono and Rhoades [117]) If ζ is a primitive even-order $2k$ root of unity, then, as q approaches ζ radially within the unit disk, we have that*

$$\lim_{q \to \zeta}(f(q) - (-1)^k b(q)) = -4\sum_{n=0}^{k-1}(1+\zeta)^2(1+\zeta^2)^2\ldots(1+\zeta^n)^2\zeta^{n+1}. \quad (18.25)$$

Here

$$b(q) = \frac{(q;q)_\infty}{(-q;q)_\infty^2}.$$

The infinite product representation of $b(q)$ was not stated in [117], but was stated by Rhoades in [215].

The identities in Theorem 18.2 lead immediately to the results in the following corollary, upon rearranging those identities and letting q tend radially to the specified root of unity from within the unit circle, since the other series accompanying each the mock theta functions in the bilateral sums terminates. The proofs are straightforward, so are omitted.

Corollary 18.1. *(i) If ζ is a primitive even-order $4k$ root of unity, then, as q approaches ζ radially within the unit disk, we have that*

$$\lim_{q \to \zeta}\left(\phi(q) - \frac{(q^2,-q,-q;q^2)_\infty}{(-q^2,q;q^2)_\infty}\right) = -2\sum_{n=0}^{k-1}(1+\zeta^2)(1+\zeta^4)\ldots(1+\zeta^{2n})\zeta^{n+1}. \quad (18.26)$$

(ii) If ζ is a primitive even-order $4k+2$ root of unity, then, as q approaches ζ radially within the unit disk, we have that

$$\lim_{q \to \zeta}\left(\nu(q) - 2(-q^2;q^2)_\infty^2(q^4;q^4)_\infty\right) = -\sum_{n=0}^{k}(1+\zeta)(1+\zeta^3)\ldots(1+\zeta^{2n-1})\zeta^n. \quad (18.27)$$

(iii) If ζ is a primitive odd-order $2k+1$ root of unity, then, as q approaches ζ radially within the unit disk, we have that

$$\lim_{q \to \zeta} \left(\psi(q) - \frac{(q^2, -q, -q; q^2)_\infty}{2(-q^2, q; q^2)_\infty} \right) = -\sum_{n=0}^{k} (1-\zeta)(1-\zeta^3) \dots (1-\zeta^{2n-1})(-1)^n. \tag{18.28}$$

Remark: The radial limit results for various mock theta functions stated in this chapter, and others besides, were proved by the authors in [65], using the same, or similar, methods as those employed here in this chapter.

The next results were proved by Watson [259, page 64] and used to prove some of the linear relations between the third order mock theta functions stated by Ramanujan.

Corollary 18.2. *The function $G_3(s, 1/s, q)$ at (18.2) and the mock theta functions $f(q)$, $\phi(q)$ and $\chi(q)$ have the following expressions as Lambert series:*

$$\sum_{n=0}^{\infty} \frac{q^{n^2}}{(sq, q/s; q)_n} = \frac{(1-s)}{(q; q)_\infty} \sum_{n=-\infty}^{\infty} \frac{(-1)^n q^{(3n^2+n)/2}}{1 - sq^n} \tag{18.29}$$

$$f(q)(q; q)_\infty = 1 + 4 \sum_{n=1}^{\infty} \frac{(-1)^n q^{(3n^2+n)/2}}{1 + q^n} = 2 \sum_{n=-\infty}^{\infty} \frac{(-1)^n q^{(3n^2+n)/2}}{1 + q^n}, \tag{18.30}$$

$$\phi(q)(q; q)_\infty = 1 + 2 \sum_{n=1}^{\infty} \frac{(-1)^n (1 + q^n) q^{(3n^2+n)/2}}{1 + q^{2n}}$$

$$= \sum_{n=-\infty}^{\infty} \frac{(-1)^n (1 + q^n) q^{(3n^2+n)/2}}{1 + q^{2n}}, \tag{18.31}$$

$$\chi(q)(q; q)_\infty = 1 + \sum_{n=1}^{\infty} \frac{(-1)^n (1 + q^n) q^{(3n^2+n)/2}}{1 - q^n + q^{2n}}$$

$$= \frac{1}{2} \sum_{n=-\infty}^{\infty} \frac{(-1)^n (1 + q^n) q^{(3n^2+n)/2}}{1 - q^n + q^{2n}}. \tag{18.32}$$

Proof. In (18.6), let $t = 1/s$ so that the left side of (18.6) becomes the left side of (18.29), and the identity follows upon noting that the right side becomes

$$\frac{1}{(q;q)_\infty}\left(1+\sum_{n=1}^{\infty}\frac{(1+q^n)(1-s)(1-1/s)(-1)^n q^{(3n^2+n)/2}}{(1-sq^n)(1-q^n/s)}\right)$$

$$=\frac{1}{(q;q)_\infty}\left(1+(1-s)\sum_{n=1}^{\infty}(-1)^n q^{(3n^2+n)/2}\left[\frac{1}{1-sq^r}-\frac{1/s}{1-q^r/s}\right]\right).$$

For (18.30), let $s=-1$ in (18.29). The proofs of (18.31) and (18.32) are left as exercises (see Q18.4).   □

We next prove a number of identities that express linear combinations of third order mock theta functions as infinite products, identities which were stated by Ramanujan in the "Lost Notebook" and proved by Watson [259].

Theorem 18.4. *The following identities hold:*

$$2\phi(-q)-f(q)=f(q)+4\psi(-q)=\frac{(q;q)_\infty}{(-q,-q;q)_\infty};\tag{18.33}$$

$$4\chi(q)-f(q)=\frac{3}{(q;q)_\infty}\frac{(q^3;q^3)_\infty^2}{(-q^3;q^3)_\infty^2}\tag{18.34}$$

$$2\rho(q)+\omega(q)=\frac{3}{(q^2;q^2)_\infty}\frac{(q^6;q^6)_\infty^2}{(q^3;q^6)_\infty^2}\tag{18.35}$$

$$\nu(-q)-q\omega(q^2)=\frac{(q^4;q^4)_\infty}{(q^2;q^4)_\infty^2}.\tag{18.36}$$

Proof. For the proof of the identities at (18.33) we follow Berndt [71, **Entry 1.3.1**]. The replacements $b=q$ and $c=-q$ in (4.20) gives, after some simplification and replacing j with $j-1$ in the second sum on the right, that

$$f(q)=\phi(-q)-2\psi(-q),$$

which implies the first equality at (18.33). For the second equality, set $b=-q$ and $c=q$ in (4.20) to get

$$\sum_{k=0}^{\infty}\frac{q^{k^2}}{(q,q;q)_k}=\frac{(-q;q)_\infty^2}{(q;q)_\infty^2}(\phi(-q)+2\psi(-q)),$$

and the result follows after using Euler's identity (4.9) to replace the sum on the left side.

For the next two identities we follow Watson [259]. From (18.30) and (18.31) it follows after some simple manipulation that

$$4\chi(q) - f(q) = \frac{3}{(q;q)_\infty}\left(1 + 4\sum_{n=1}^{\infty}\frac{(-1)^n q^{3(n^2+n)/2}}{1+q^{3n}}\right),$$

and (18.34) follows from the identity at (18.142).

The proof of (18.35) is similar, and is left as an exercise (see Q18.5 and Q18.6).

For (18.36), replace q with q^2 in (4.22) and then set $b = q^2$, $c = q^3$ and multiply both sides by $1/(1-q)$. The result follows after some simplification, and again using Euler's identity (4.9), with q replaced with q^4 (this is the proof given in [71, **Entry 1.3.8**], which is more straightforward than the proof of Watson [259]). $\qquad\square$

18.2   The Fifth Order Mock Theta Functions

Ramanujan's fifth order mock theta functions are the following:

$$f_0(q) = \sum_{n=0}^{\infty}\frac{q^{n^2}}{(-q;q)_n}, \qquad\qquad f_1(q) = \sum_{n=0}^{\infty}\frac{q^{n(n+1)}}{(-q;q)_n},$$

$$F_0(q) = \sum_{n=0}^{\infty}\frac{q^{2n^2}}{(q;q^2)_n}, \qquad\qquad F_1(q) = \sum_{n=0}^{\infty}\frac{q^{2n(n+1)}}{(q;q^2)_{n+1}},$$

$$\phi_0(q) = \sum_{n=0}^{\infty}q^{n^2}(-q;q^2)_n, \qquad\qquad \phi_1(q) = \sum_{n=0}^{\infty}q^{(n+1)^2}(-q;q^2)_n,$$

$$\psi_0(q) = \sum_{n=0}^{\infty}q^{(n+1)(n+2)/2}(-q;q)_n, \qquad \psi_1(q) = \sum_{n=0}^{\infty}q^{n(n+1)/2}(-q;q)_n,$$

$$\chi_0(q) = \sum_{n=0}^{\infty}\frac{q^n(q;q)_n}{(q;q)_{2n}}, \qquad\qquad \chi_1(q) = \sum_{n=0}^{\infty}\frac{q^n(q;q)_n}{(q;q)_{2n+1}}.$$

Watson [260] proved a number of relations for the fifth order mock theta functions as intermediate steps to the proof of a number of identities stated by Ramanujan in his last letter to Hardy (and which were also to be found in the last notebook). However, in proving these, we follow the method of Berndt [71], as the proofs follow as simple consequences of the general identities found in Q4.14 and Q4.15 (which were also proved by Berndt [71]).

From space considerations, we do not state and prove all of these identities, and the reader who is interested in knowing more should investigate the papers of Watson [260] and Berndt [71].

Proposition 18.2. *The following relations hold:*

$$f_0(q) = \phi_0(-q^2) + q\frac{(q^4, q^{16}, q^{20}; q^{20})_\infty}{(q^2; q^4)_\infty} - F_0(q^2) + 1, \quad (18.37)$$

$$\frac{(q^2, q^3, q^5; q^5)_\infty}{(-q; q)_\infty} = \phi_0(-q^2) - q\frac{(q^4, q^{16}, q^{20}; q^{20})_\infty}{(q^2; q^4)_\infty} + F_0(q^2) - 1, \quad (18.38)$$

$$\psi_0(q) = q\frac{(q^4, q^{16}, q^{20}; q^{20})_\infty}{(q^2; q^4)_\infty} + F_0(q^2) - 1, \quad (18.39)$$

$$f_1(q) = \frac{(q^8, q^{12}, q^{20}; q^{20})_\infty}{(q^2; q^4)_\infty} - qF_1(q^2) - \frac{\phi_1(-q^2)}{q}, \quad (18.40)$$

$$\frac{(q, q^4, q^5; q^5)_\infty}{(-q; q)_\infty} = \frac{(q^8, q^{12}, q^{20}; q^{20})_\infty}{(q^2; q^4)_\infty} - qF_1(q^2) + \frac{\phi_1(-q^2)}{q}, \quad (18.41)$$

$$\psi_1(q) = \frac{(q^8, q^{12}, q^{20}; q^{20})_\infty}{(q^2; q^4)_\infty} + qF_1(q^2). \quad (18.42)$$

Proof. For (18.37), set $b = q$ and $s = 0$ in (4.26), and use (10.18) to sum one of the resulting series.

The proof of (18.38) is similar, and follows from setting $b = -q$ and $s = 0$ in (4.26), and using both (10.17) and (10.18) to sum two of the resulting series.

For (18.39), set $s = 1$ and $z = q^2$ in (4.24), and let $a \to 0$, with (10.18) once again being used to sum one of the resulting series.

The proof of (18.40) is like that of (18.37), except $s = 1$ (and $b = q$) and uses (10.17) instead of (10.18).

The identity at (18.41) follows from (4.26), with $s = 1$ and $b = -q$, and also employing both (10.17) and (10.18).

Finally, for (18.42) set $s = 0$ and $z = q^2$ in (4.24), and let $a \to 0$, this time with (10.17) being used to sum one series.   □

The identities stated by Ramanujan, alluded to just before the proposition above, express linear combinations of the fifth order mock theta functions in terms of infinite products. This is achieved simply by either rearranging a single equation (for example, the expression for $\psi_1(q) - qF_1(q^2)$ resulting from (18.42), labelled **Entry 2.4.4** in [71]), or combining two equations (for example, the expression for $f_1(q) + 2qF_1(q^2)$ that results from combining (18.40) and (18.41) (with q replaced with $-q$).

We will not list these identities (the interested reader should consult the papers of Berndt [71] or Watson [260]). Perhaps slightly more interesting is the fact that certain combinations of *pairs* of mock theta functions of order five may be expressed as single bilateral series, and using the identities in Proposition 18.2, in terms of theta products (as was described by Watson in section 7 of [260]).

Corollary 18.3. *The following identities hold.*

$$\sum_{n=-\infty}^{\infty} \frac{q^{r^2}}{(-q;q)_r} = f_0(q) + 2\psi_0(q) = 4q\frac{(q^4,q^{16},q^{20};q^{20})_\infty}{(q^2;q^4)_\infty} + \frac{(q^2,q^3,q^5;q^5)_\infty}{(-q;q)_\infty}.$$
$$(18.43)$$

$$\sum_{n=-\infty}^{\infty} \frac{q^{r^2+r}}{(-q;q)_r} = f_1(q) + 2\psi_1(q) = 4\frac{(q^8,q^{12},q^{20};q^{20})_\infty}{(q^2;q^4)_\infty} - \frac{(q,q^4,q^5;q^5)_\infty}{(-q;q)_\infty}.$$
$$(18.44)$$

$$\sum_{n=-\infty}^{\infty} \frac{q^{4r^2}}{(q^2;q^4)_r}$$
$$= F_0(q^2) + \phi_0(-q^2) - 1 = q\frac{(q^4,q^{16},q^{20};q^{20})_\infty}{(q^2;q^4)_\infty} + \frac{(q^2,q^3,q^5;q^5)_\infty}{(-q;q)_\infty}. \quad (18.45)$$

$$\sum_{n=-\infty}^{\infty} \frac{q^{4r^2+4r}}{(q^2;q^4)_{r+1}} = F_1(q^2) - \frac{\phi_1(-q^2)}{q^2} = \frac{(q^8,q^{12},q^{20};q^{20})_\infty}{q(q^2;q^4)_\infty} - \frac{(q,q^4,q^5;q^5)_\infty}{q(-q;q)_\infty}.$$
$$(18.46)$$

Proof. The first equality in each case follows directly after applying (2.5) to the terms of negative index in the series. The second equalities in (18.43) and (18.44) and follow, respectively, from the following combinations of the identities in Proposition 18.2:

$$(18.37) + 2(18.39) - (18.37),$$
$$(18.40) + 2(18.42) + (18.41).$$

The second equalities at (18.45) and (18.46) are simply restatements of (18.38) and (18.41). ☐

These summation formulae may be rearranged and used to give explicit radial limits for the difference of certain fifth order mock theta functions and certain corresponding theta functions, as q tends to certain roots of unity from within the unit circle, as was shown by Bajpai, Kimport, Liang, Ma and Ricci in [65], using methods similar to those employed below (for ease of notation, the statements for $F_0(q)$ and $F_1(q)$ are written in terms of q^2 instead of q).

Corollary 18.4. *(i) If ζ is a primitive even-order $2k$ root of unity, then, as q approaches ζ radially within the unit disk, we have that*

$$\lim_{q\to\zeta}\left(f_0(q) - \left[4q\frac{(q^4,q^{16},q^{20};q^{20})_\infty}{(q^2;q^4)_\infty} + \frac{(q^2,q^3,q^5;q^5)_\infty}{(-q;q)_\infty}\right]\right)$$
$$= -2\sum_{n=0}^{k-1}(1+\zeta)(1+\zeta^2)\dots(1+\zeta^n)\zeta^{(n+1)(n+2)/2}. \quad (18.47)$$

(ii) If ζ is a primitive even-order $2k$ root of unity, then, as q approaches ζ radially within the unit disk, we have that

$$\lim_{q\to\zeta}\left(f_1(q) - \left[4\frac{(q^8,q^{12},q^{20};q^{20})_\infty}{(q^2;q^4)_\infty} - \frac{(q,q^4,q^5;q^5)_\infty}{(-q;q)_\infty}\right]\right)$$
$$= -2\sum_{n=0}^{k-1}(1+\zeta)(1+\zeta^2)\dots(1+\zeta^n)\zeta^{n(n+1)/2}. \quad (18.48)$$

(iii) If ζ is a primitive even-order $4k+2$ root of unity, then, as q approaches ζ radially within the unit disk, we have that

$$\lim_{q\to\zeta}\left(F_0(q^2) - \left[q\frac{(q^4,q^{16},q^{20};q^{20})_\infty}{(q^2;q^4)_\infty} + \frac{(q^2,q^3,q^5;q^5)_\infty}{(-q;q)_\infty}\right]\right)$$
$$= -\sum_{n=1}^{k}(1-\zeta^2)(1-\zeta^6)\dots(1-\zeta^{4n-2})(-1)^n\zeta^{2n^2}. \quad (18.49)$$

(iv) If ζ is a primitive odd-order $4k+2$ root of unity, then, as q approaches ζ radially within the unit disk, we have that

$$\lim_{q\to\zeta}\left(F_1(q^2) - \left[\frac{(q^8,q^{12},q^{20};q^{20})_\infty}{q(q^2;q^4)_\infty} - \frac{(q,q^4,q^5;q^5)_\infty}{q(-q;q)_\infty}\right]\right)$$
$$= -\sum_{n=0}^{k}(1-\zeta^2)(1-\zeta^6)\dots(1-\zeta^{4n-2})(-1)^n\zeta^{2n^2+4n}. \quad (18.50)$$

There are no known transformations for mock theta functions of the fifth order similar to those for mock theta functions of the third order that are implied by (18.4)–(18.6). However, there are bilateral transformations that may be applied to the bilateral series in Corollary 18.3.

Here we consider the series

$$G_5(w, y, q) = \sum_{n=0}^{\infty} \frac{w^n q^{n^2}}{(y; q)_n}, \qquad G_5^*(w, y, q) = \sum_{n=-\infty}^{\infty} \frac{w^n q^{n^2}}{(y; q)_n}. \qquad (18.51)$$

Identities for these functions are not so plentiful as those for $G_3(s, t, q)$ and $G_3^*(s, t, q)$, but two such are given in the next theorem.

Theorem 18.5. *Let $G_5^*(w, y, q)$ be as defined at (18.51). Then*

$$G_5^*(w, y, q) = \frac{(y/w; q)_\infty}{(y; q)_\infty} \sum_{r=-\infty}^{\infty} (wq/y; q)_r (-y)^r q^{r(r-1)/2}, \qquad (18.52)$$

$$= \frac{(y/w; q)_\infty}{(wq, q/w; q)_\infty} \sum_{r=-\infty}^{\infty} \frac{(1 - wq^{2r})(wq/y; q)_r (-yw^2)^r q^{(5r^2 - 3r)/2}}{(1 - w)(y; q)_r}. \qquad (18.53)$$

Proof. In (7.20) (or (7.21)), replace z with z/ac, then let $a, c \to \infty$ and $d \to 0$. Then replace z with wq and b with y, and (18.52) follows.

For (18.52), let $d, e, f \to \infty$ in (8.9), and then set $a = w$ and $c = wq/y$. □

Remark: For $G_5^*(w, y, q)$ to represent a sum of fifth order mock theta function, it is necessary to have $w = 1$ or $w = q$, and in those cases (18.52) does not provide any non-trivial results (for $w = 1$ the right side is just the series in reverse order).

The identity at (18.53) could be used to derive new expressions for the sums of fifth order mock theta functions found in Corollary 18.3. However, we instead use it to derive four identities reminiscent of the identities in Q6.6 and Q6.7.

Corollary 18.5. *The following identities hold for $|q| < 1$:*

$$\sum_{r=-\infty}^{\infty} (10r + 1)q^{(5r^2 + r)/2}$$

$$= \left(\frac{4q(q^4, q^{16}, q^{20}; q^{20})_\infty}{(q^2; q^4)_\infty} + \frac{(q^2, q^3, q^5; q^5)_\infty}{(-q; q)_\infty} \right) \frac{(q; q)_\infty^2}{(-q; q)_\infty}, \qquad (18.54)$$

$$\sum_{r=-\infty}^{\infty} (10r+3)q^{(5r^2+3r)/2}$$

$$= \left(\frac{4(q^8,q^{12},q^{20};q^{20})_\infty}{(q^2;q^4)_\infty} - \frac{(q,q^4,q^5;q^5)_\infty}{(-q;q)_\infty} \right) \frac{(q;q)_\infty^2}{(-q;q)_\infty}, \quad (18.55)$$

$$\sum_{r=-\infty}^{\infty} (5r+1)(-1)^r q^{10r^2+4r}$$

$$= \left(\frac{q(q^4,q^{16},q^{20};q^{20})_\infty}{(q^2;q^4)_\infty} + \frac{(q^2,q^3,q^5;q^5)_\infty}{(-q;q)_\infty} \right) \frac{(q^4;q^4)_\infty^2}{(q^2;q^4)_\infty}, \quad (18.56)$$

$$\sum_{r=-\infty}^{\infty} (5r+2)(-1)^r q^{10r^2+8r}$$

$$= \left(\frac{(q^8,q^{12},q^{20};q^{20})_\infty}{(q^2;q^4)_\infty} - \frac{(q,q^4,q^5;q^5)_\infty}{(-q;q)_\infty} \right) \frac{(q^4;q^4)_\infty^2}{(q^2;q^4)_\infty}. \quad (18.57)$$

Proof. For (18.54), in (18.53) replace w with w^2, set $y = -wq$ and simplify the resulting right side to get

$$\sum_{n=-\infty}^{\infty} \frac{w^{2n}q^{n^2}}{(-wq;q)_n} = \frac{(-q/w;q)_\infty}{(w^2q,q/w^2;q)_\infty} \sum_{r=-\infty}^{\infty} \frac{(1-wq^r)w^{5r}q^{(5r^2-r)/2}}{(1-w)}$$

$$= \frac{(-q/w;q)_\infty}{(w^2q,q^2/w;q)_\infty}$$

$$\times \left(1 + \sum_{r=1}^{\infty} \frac{(1-wq^r)w^{5r}q^{(5r^2-r)/2} + (1-wq^{-r})w^{-5r}q^{(5r^2+r)/2}}{1-w} \right)$$

$$= \frac{(-q/w;q)_\infty}{(w^2q,q^2/w;q)_\infty}$$

$$\times \left(1 + \sum_{r=1}^{\infty} \frac{w^{-5r}q^{(5r^2-r)/2}}{1-w} \left((1-wq^r)w^{10r} + (1-wq^{-r})q^r \right) \right)$$

$$= \frac{(-q/w;q)_\infty}{(w^2q,q^2/w;q)_\infty}$$

$$\times \left(1 + \sum_{r=1}^{\infty} w^{-5r}q^{(5r^2-r)/2} \left(-w\frac{1-w^{10r-1}}{1-w} + q^r\frac{1-w^{10r+1}}{1-w} \right) \right).$$

Now let $w \to 1$, noting that the left side above tends to the left side of (18.43), and hence to the right side of (18.43). After using L'Hospital's rule on the terms in the last series on the right side, this series becomes

$$1 + \sum_{r=1}^{\infty} q^{(5r^2-r)/2}\left(-(10r-1) + q^r(10r+1)\right)$$

$$= \sum_{r=-\infty}^{\infty} 10rq^{(5r^2+r)/2} + \sum_{r=-\infty}^{\infty} q^{(5r^2-r)/2}.$$

The result now follows.

To obtain (18.55), in (18.53) replace w with w^2q, set $y = -wq$ and simplify the resulting right side to get

$$\sum_{n=-\infty}^{\infty} \frac{w^{2n}q^{n^2+n}}{(-wq;q)_n} = \frac{(-1/w;q)_\infty}{(w^2q^2, 1/w^2;q)_\infty} \sum_{r=-\infty}^{\infty} \frac{(1-w^2q^{2r+1})w^{5r}q^{(5r^2+3r)/2}}{(1-w^2q)}$$

$$= \frac{(-1/w;q)_\infty(-w^2)}{(w^2q, q/w^2;q)_\infty} \times$$

$$\sum_{r=0}^{\infty} \frac{(1-w^2q^{2r+1})w^{5r}q^{(5r^2+3r)/2}}{(1-w^2)} + \sum_{r=0}^{\infty} \frac{(1-w^2q^{-2r-1})w^{-5r-5}q^{(5r^2+7r+2)/2}}{(1-w^2)}$$

$$= \frac{(-1/w;q)_\infty(-w^2)}{(w^2q, q/w^2;q)_\infty} \times$$

$$\sum_{r=0}^{\infty} w^{5r}q^{(5r^2+3r)/2}\left(\frac{(1-w^{-10r-3})}{(1-w^2)} - w^2q^{2r+1}\frac{(1-w^{-10r-7})}{(1-w^2)}\right),$$

where the second series in the second right side came from taking the terms of negative index in the series on the first right side, and replacing r with $-r-1$. The identity at (18.55) now follows as previously upon letting $w \to 1$, this time noting that the left side tends to (18.44).

For (18.56) and (18.57), in (18.53) replace (w, y, q) with (w^2, wq^2, q^4) and (w^2q^4, wq^6, q^4), respectively (in the case of (18.57), after making the replacements in (18.53), multiply both sides by $1/(1-wq^2)$). The details are left as exercises (see Q18.9). □

18.2.1 The Mock Theta Conjectures

Ramanujan also stated (as usual, without proof), a collection of ten identities for the fifth order mock theta functions which were considerably more difficult to prove. These identities fall naturally into two sets of five, with all of the identities in each set being equivalent (this equivalence being proved by Andrews and Garvan [46]). The identities (18.58) ((18.58)–(18.62) are equivalent) and (18.63) ((18.63)–(18.67) are equivalent) are known as *the mock theta conjectures*. The mock theta conjectures were proven by Hickerson [141], and Folsom [116] later gave a proof using Maass forms.

The proof of these conjectures is beyond the scope of the present book, but we list them in the hope that doing so may encourage the reader to seek more elementary proofs than those mentioned in the previous paragraph, as it seems likely that Ramanujan's original proofs were different from those of Hickerson [141] and Folsom [116].

Recall that $g_3(x, q) = g(x, q)$ is the function defined at (18.1).

Theorem 18.6. *The following identities hold.*

$$f_0(q) = -2q^2\, g_3(q^2, q^{10}) + \frac{(q^5; q^5)_\infty (q^5; q^{10})_\infty}{(q, q^4; q^5)_\infty}, \tag{18.58}$$

$$F_0(q) = 1 + q\, g_3(q, q^5) - \frac{q(q^{10}; q^{10})_\infty}{(q^4, q^5, q^6; q^{10})_\infty}, \tag{18.59}$$

$$\phi_0(-q) = -q\, g_3(q, q^5) + \frac{(-q^2, -q^3, q^5; q^5)_\infty}{(q^2, q^8; q^{10})_\infty}, \tag{18.60}$$

$$\psi_0(q) = q^2\, g_3(q^2, q^{10}) + q\frac{(q, q^9, q^{10}; q^{10})_\infty}{(q^2, q^3; q^5)_\infty}, \tag{18.61}$$

$$\chi_0(q) = 2 + 3q\, g_3(q, q^5) - \frac{(q^2, q^3, q^5; q^5)_\infty}{(q, q, q^4, q^4; q^5)_\infty}, \tag{18.62}$$

$$f_1(q) = -2q^3\, g_3(q^4, q^{10}) + \frac{(q^5; q^5)_\infty (q^5; q^{10})_\infty}{(q^2, q^3; q^5)_\infty}, \tag{18.63}$$

$$F_1(q) = q\, g_3(q^2, q^5) + \frac{(q^{10}; q^{10})_\infty}{(q^2, q^5, q^8; q^{10})_\infty}, \tag{18.64}$$

$$\phi_1(-q) = q^2\, g_3(q^2, q^5) - q\frac{(-q, -q^4, q^5; q^5)_\infty}{(q^4, q^6; q^{10})_\infty}, \tag{18.65}$$

$$\psi_1(q) = q^3\, g_3(q^4, q^{10}) + \frac{(q^3, q^7, q^{10}; q^{10})_\infty}{(q, q^4; q^5)_\infty}, \tag{18.66}$$

$$\chi_1(q) = 3q\, g_3(q^2, q^5) + \frac{(q, q^4, q^5; q^5)_\infty}{(q^2, q^2, q^3, q^3; q^5)_\infty}. \tag{18.67}$$

18.3   The Sixth Order Mock Theta Functions

The mock theta functions which have deemed to be of the sixth order are the following:

$$\phi(q) = \sum_{n=0}^{\infty} \frac{(-1)^n q^{n^2} (q; q^2)_n}{(-q; q)_{2n}}, \qquad \psi(q) = \sum_{n=0}^{\infty} \frac{(-1)^n q^{(n+1)^2} (q; q^2)_n}{(-q; q)_{2n+1}},$$

$$\rho(q) = \sum_{n=0}^{\infty} \frac{q^{n(n+1)/2} (-q; q)_n}{(q; q^2)_{n+1}}, \qquad \sigma(q) = \sum_{n=0}^{\infty} \frac{q^{(n+1)(n+2)/2} (-q; q)_n}{(q; q^2)_{n+1}},$$

$$\lambda(q) = \sum_{n=0}^{\infty} \frac{(-1)^n q^n (q; q^2)_n}{(-q; q)_n}, \qquad \mu(q) = \sum_{n=0}^{\infty} \frac{(-1)^n (q; q^2)_n}{(-q; q)_n},$$

$$\phi_-(q) = \sum_{n=1}^{\infty} \frac{q^n (-q; q)_{2n-1}}{(q; q^2)_n}, \qquad \psi_-(q) = \sum_{n=1}^{\infty} \frac{q^n (-q; q)_{2n-2}}{(q; q^2)_n},$$

$$\beta(q) = \sum_{n=0}^{\infty} \frac{q^{3n^2+3n+1}}{(q; q^3)_{n+1} (q^2; q^3)_{n+1}}, \qquad \gamma(q) = \sum_{n=0}^{\infty} \frac{q^{n^2} (q; q)_n}{(q^3; q^3)_n}.$$

As remarked in the paragraph following the definition of $g(x, q)$ at (18.1), Gordon and McIntosh indicate in [132] that $\beta(q)$ and $\gamma(q)$ should probably be reclassified as being of the third order. Note that the series stated for $\mu(q)$ does not converge, but that the sequence of even-indexed partial sums and the sequence of odd-indexed partial sums do converge, and $\mu(q)$ is defined to be the average of these two values.

Ramanujan stated a number of identities for the sixth order mock theta functions in the Lost Notebook [213], and these were proved by Andrews and Hickerson [47]. A number of similar identities were proved by Berndt and Chan [72]. The proofs in these papers were quite involved, employing both Bailey pairs and the constant term method. Here we will give the proofs of Lovejoy [186], for four of the identities proved by Andrews and Hickerson [47], Lovejoy's proofs probably being more straightforward. Before coming to these identities, some preliminary results are necessary.

Lemma 18.1. *The following identities hold.*

$$\mu(q) = \frac{1}{2}\phi(q^2) - \sigma(-q), \tag{18.68}$$

$$(q; q^2)_\infty^3 (-q, -q^2, q^3; q^3)_\infty = \phi(q^2) + 2\sigma(-q), \tag{18.69}$$

$$\lambda(q) = \rho(-q) + q^{-1}\psi(q^2), \tag{18.70}$$

$$(q; q^2)_\infty^3 (-q^3, -q^3, q^3; q^3)_\infty = \rho(-q) - q^{-1}\psi(q^2), \tag{18.71}$$

Proof. In (4.30), set $a = 1$ and let $x \to q$ radially towards the origin, so that $|x| > |q|$ or $|-q/x| < 1$, so that the series on the left of (4.30) converges for each such x and has value thus equal to half the sum of the limit of the even-indexed partial sums and the limit of the odd-indexed partial sums. In the limit as $x \to q$ then the left side of (4.30) converges to $\mu(q)$ according to the definition above, and when the right side is simplified we are led to (18.68).

Next, note that by setting $d = -e = \sqrt{q}$ and letting $c \to \infty$ and $a \to 1$ in (8.8), there results the identity

$$2 \lim_{a \to 1} \sum_{j=0}^{\infty} \frac{(q; q^2)_j (-a)^j}{(q; q)_j} = \frac{(q; q^2)_\infty (-q, -q^2, q^3; q^3)_\infty}{(q^2; q^2)_\infty}.$$

The identity at (18.69) now follows from setting $a = -1$ in (4.30) and similarly letting $x \to q$ so that $|-q/x| < 1$, and using the above limit.

The identity at (18.70) follows directly upon setting $a = 1$ and $x = q$ in (4.34).

Finally, set $d = -e = \sqrt{q}, a = q$ in (8.8), and let $c \to \infty$ and to get

$$\sum_{j=0}^{\infty} \frac{(q; q^2)_j (-q)^j}{(q; q)_j} = \frac{(q; q^2)_\infty (-q^3, -q^3, q^3; q^3)_\infty}{(q^2; q^2)_\infty},$$

and (18.71) follows upon setting $a = -1$ and $x = q$ in (4.34), after using the above summation formula on one of the resulting series.   □

Theorem 18.7. *The following identities hold for $|q| < 1$.*

$$q^{-1} \psi(q^2) + \rho(q) = (-q; q^2)_\infty^2 (-q, -q^5, q^6; q^6)_\infty, \qquad (18.72)$$

$$\phi(q^2) + 2\sigma(q) = (-q; q^2)_\infty^2 (-q^3, -q^3, q^6; q^6)_\infty, \qquad (18.73)$$

$$2\phi(q^2) - 2\mu(-q) = (-q; q^2)_\infty^2 (-q^3, -q^3, q^6; q^6)_\infty, \qquad (18.74)$$

$$2q^{-1} \psi(q^2) + \lambda(-q) = (-q; q^2)_\infty^2 (-q, -q^5, q^6; q^6)_\infty. \qquad (18.75)$$

Proof. These identities all follow directly by elementary manipulations from the identities in Lemma 18.1.   □

As the authors remark in [132], just as the mock theta functions of odd order are related to the function $g_3(x, q)$ at (18.1), so are those of even order related to

$$g_2(x, q) := \sum_{k=0}^{\infty} \frac{q^{k(k+1)/2}(-q; q)_k}{(x, q/x; q)_{k+1}}. \qquad (18.76)$$

One example of this relationship is its appearance in one of the sixth order mock theta conjectures ([132, Eq. (5.10), first equation]):

$$\phi(q^4) = \frac{(q^2;q^2)_\infty^3\,(q^3;q^3)_\infty^2\,(q^{12};q^{12})_\infty^3}{(q;q)_\infty^2\,(q^6;q^6)_\infty^3\,(q^8;q^8)_\infty\,(q^{24};q^{24})_\infty} - 2qg_2(q,q^6). \qquad (18.77)$$

This conjecture was stated without proof in both [132] and [190], and as with the fifth order mock theta conjectures, we do not give a proof here.

In [190, Eq. (2.7)], McIntosh proves the following transformation.

Theorem 18.8. *If $|q| < 1$ and $q^n x \neq 1$ for any $n \in \mathbb{Z}$, then*

$$g_2(x,q) = \frac{(-q;q)_\infty}{(q;q)_\infty} \sum_{n=-\infty}^{\infty} \frac{(-1)^n q^{n(n+1)}}{1 - xq^n}. \qquad (18.78)$$

Proof. In (5.12) let $d, e \to \infty$ and set $a = q$, $b = x$, $c = q/x$, $f = -q$ and then multiply both sides of the resulting identity by $(1-q)/((1-x)(1-q/x))$ to get

$$\sum_{n=0}^{\infty} \frac{(1 - q^{2n+1})(-1)^n q^{n(n+1)}}{(1 - xq^n)(1 - q^{n+1}/x)} = \frac{(q;q)_\infty}{(-q;q)_\infty} \sum_{k=0}^{\infty} \frac{q^{k(k+1)/2}(-q;q)_k}{(x, q/x; q)_{k+1}}. \qquad (18.79)$$

Now set

$$\frac{(1 - q^{2n+1})}{(1 - xq^n)(1 - q^{n+1}/x)} = \frac{1}{(1 - xq^n)} - \frac{1}{(1 - xq^{-(n+1)})},$$

split the series on the left of (18.79) into two series, replace n with $-n - 1$ in the second series (so that the summation runs from $-\infty$ to -1), and the result follows. $\qquad \square$

We next give some transformation for a general bilateral series, which as will be seen shortly, permits the derivation of several identities for pairs of sixth order mock theta functions.

Theorem 18.9. *(i) If $|q|, |bd/azq| < 1$, then*

$$G_6(a,b,d,z,q) := \sum_{n=-\infty}^{\infty} \frac{(a;q^2)_r z^r q^{r^2}}{(b,d;q^2)_r} = \frac{(-zq, -qb/az, -qd/az; q^2)_\infty}{(b, d, q^2/a; q^2)_\infty}$$

$$\times \sum_{n=-\infty}^{\infty} \frac{(-azq/b, -azq/d; q^2)_r}{(-zq; q^2)_r} \left(\frac{-bd}{azq}\right)^r. \qquad (18.80)$$

(ii) If $|q|, |bd/azq|, |b/a| < 1$, then

$$\sum_{n=-\infty}^{\infty} \frac{(a;q^2)_r z^r q^{r^2}}{(b,d;q^2)_r} = \frac{(b/a, -qb/az; q^2)_\infty}{(b, -bd/azq; q^2)_\infty}$$

$$\times \sum_{n=-\infty}^{\infty} \frac{(-azq/b, a; q^2)_r}{(d; q^2)_r} \left(\frac{b}{a}\right)^r. \quad (18.81)$$

(iii) If $|q|, |bd/azq| < 1$, then

$$\sum_{n=-\infty}^{\infty} \frac{(a;q^2)_r z^r q^{r^2}}{(b,d;q^2)_r} = \frac{(-bq/az, -dq/az, -qz; q^2)_\infty}{(-bd/aqz, -q^3/az, -aqz; q^2)_\infty}$$

$$\times \sum_{n=-\infty}^{\infty} \frac{(1 + azq^{4r-1})(a, -azq/b, -azq/d; q^2)_r (bdz)^r q^{3r^2-4r}}{(1 + az/q)(b, d, -zq; q^2)_r}. \quad (18.82)$$

Proof. For (18.80) and (18.81), replace q with q^2, z with $-zq/c$ and then let $c \to \infty$ in (7.20) and (7.21), respectively.

The identity (18.82) is a consequence of replacing q with q^2 in (8.9), and then replacing c with aq^2/b and d with aq^2/d, letting $f \to \infty$, and replacing, in turn, a with $-ze/q$ and finally e with a. $\square$

The sums of various pairs of sixth order mock theta functions may be expressed in terms of $G_6(a, b, d, z, q)$, and the above theorem allows the derivation of some alternative expressions for these sums.

Corollary 18.6. *The following identities hold for $|q| < 1$.*

$$4\sigma(q) + 2\mu(q) = \frac{(q;q^2)_\infty}{(q;q)_\infty^2} \sum_{r=-\infty}^{\infty} (-1)^r (6r+1) q^{r(3r+1)/2}, \quad (18.83)$$

$$\phi(q) + 2\phi_-(q) = \frac{(-q;q)_\infty}{(q^2;q^4)_\infty} \quad (18.84)$$
$$\times [2(-q^2;q^4)_\infty^2(-q^6, -q^6, q^{12}; q^{12})_\infty - (q^2;q^4)_\infty^2(q^6, q^6, q^{12}; q^{12})_\infty],$$

$$\psi(q) + 2\psi_-(q) = \frac{3q(-q;q)_\infty(q^6, q^6, q^6; q^6)_\infty}{(q^2;q^2)_\infty^2} \quad (18.85)$$

$$2\rho(q) + \lambda(q) = \frac{3(q;q^2)_\infty(q^3, q^3, q^3; q^3)_\infty}{(q;q)_\infty^2} \quad (18.86)$$

Proof. In (18.82) replace z with $-zq^3$ and set $a = zq^2$, $b = zq^3$ and $d = -zq^3$ to get

$$\sum_{r=-\infty}^{\infty} \frac{(zq^2; q^2)_r(-z)^r q^{r^2+3r}}{(zq^3, -zq^3; q^2)_r} = \frac{(1/qz, -1/qz, zq^4; q^2)_\infty}{(-1, 1/q^2 z^2, q^6 z^2; q^2)_\infty}$$

$$\times \sum_{r=-\infty}^{\infty} \frac{q^{3r^2+5r} z^{3r} \left(1 - z^2 q^{4r+4}\right) \left(q^2 z; q^2\right)_r}{\left(1 - q^4 z^2\right)\left(q^4 z; q^2\right)_r},$$

$$= \frac{(q^2/z^2; q^4)_\infty (zq^4; q^2)_\infty z}{2(1 - 1/z)(-q^2, q^2/z^2, q^6 z^2; q^2)_\infty} \sum_{r=-\infty}^{\infty} \frac{q^{r(3r+5)} z^{3r} \left(1 + zq^{2r+2}\right)}{(1 + q^2 z)(1 + z)}.$$

Now multiply both sides by $q^2/(1 - z^2 q^2)$ and let $z \to -1$, noting that the left side tends to $\sigma(q^2) + \mu(q^2)/2$ using the definitions above and (2.5). On the right side replace r with $r - 1$ and rewrite the resulting series as

$$\sum_{r=-\infty}^{\infty} \frac{(1 + zq^{2r})z^{3r} q^{3r^2-r}}{1+z} = \sum_{r=-\infty}^{\infty} \frac{z^{3r} q^{3r^2-r} + z^{3r+1} q^{3r^2+r}}{1+z}$$

$$= \sum_{r=-\infty}^{\infty} \frac{z^{3r} q^{3r^2-r} + z^{-3r+1} q^{3r^2-r}}{1+z} = \sum_{r=-\infty}^{\infty} z^{3r} q^{3r^2-r} \frac{1 + z^{-6r+1}}{1+z},$$

where the second equality follows from reversing the order of summation for the second terms in the sum. Now let $z \to -1$ to arrive at

$$\sigma(q^2) + \frac{\mu(q^2)}{2} = \frac{(q^2; q^4)_\infty}{4(q^2; q^2)_\infty^2} \sum_{r=-\infty}^{\infty} (6r + 1)(-1)^r q^{3r^2+r}.$$

The identity at (18.83) now follows upon multiplying this last identity by 4 and replacing q with $q^{1/2}$.

Note that the expression for $4\sigma(q) + 2\mu(q)$ deriving from (18.83) and (18.73) together with (18.74) imply that

$$\sum_{r=-\infty}^{\infty} (6r + 1)(-1)^r q^{(3r^2+r)/2} = \frac{(q; q)_\infty^2}{(q; q^2)_\infty}$$

$$\times \left[2(-q; q^2)_\infty^2 (-q^3, -q^3, q^6; q^6)_\infty - (q; q^2)_\infty^2 (q^3, q^3, q^6; q^6)_\infty\right]. \quad (18.87)$$

For (18.84), set $a = -zq$, $b = zq$ and $d = zq^2$ in (18.82) to get, after simplifying the right side,

$$\sum_{r=-\infty}^{\infty} \frac{(-zq; q^2)_r z^r q^{r^2}}{(zq, zq^2; q^2)_r} = \frac{(q/z, q^2/z, -zq; q^2)_\infty}{(q, q^2/z^2, q^2 z^2; q^2)_\infty} \sum_{r=-\infty}^{\infty} \frac{(1 + zq^{2r})z^{3r} q^{3r^2-r}}{1+z}.$$

Let $z \to -1$ on the left side to get, once again using the definitions above and (2.5),

$$\phi(q) + 2\phi_-(q) = \frac{(-q;q)_\infty}{(q^2;q^2)_\infty^2} \sum_{r=-\infty}^{\infty} (6r+1)(-1)^r q^{3r^2+r}.$$

An application of (18.87), with q replaced with q^2, gives the result.

Similarly, for (18.85), replace z with zq^2, a with $-zq$, b with zq^2 and d with zq^3 in (18.82) to get, after once again simplifying the right side, that

$$\sum_{r=-\infty}^{\infty} \frac{(-zq;q^2)_r z^r q^{r^2+2r}}{(zq^2,zq^3;q^2)_r} = \frac{(1/z,q/z,-zq^3;q^2)_\infty}{(q,1/z^2,q^4z^2;q^2)_\infty} \sum_{r=-\infty}^{\infty} z^{3r}q^{3r^2+3r}$$

$$= \frac{(1/z,q/z,-zq^3;q^2)_\infty(-q^6z^3,-1/z^3,q^6;q^6)_\infty}{(q,1/z^2,q^4z^2;q^2)_\infty},$$

where the Jacobi triple product identity (6.1) has been used at the last step. The result now follows after multiplying both sides by $q/(1+q)$ and letting $z \to -1$ as before.

We leave the details of the proof of (18.86) as an exercise. Briefly, replace z with zq, a with $-zq^2$, b with zq^3 and d with $-zq^3$ in (18.82), simplify and sum the right side using the Jacobi triple product identity (6.1), let $z \to 1$, multiply both sides by $2/(1-q^2)$, and finally replace q with $q^{1/2}$. $\square$

Remark: Choi [101, p. 370] also gave expressions for each of the sums of sixth order mock theta functions in Corollary 18.6, but with different combinations of theta functions on the right sides. Yet another version of (18.84) was stated by Ramanujan [213, p. 6 and p. 16] (see also [102, p. 1740]). Different proofs of (18.85) and (18.86) were given by Choi and Kim [102, Theorem 1.4, p. 1742].

The identities in Corollary 18.6 may be used to describe the asymptotic behavior of each of the two sixth order mock theta functions on the left side of each identity, at particular classes of roots of unity. For example, (18.83) may be used in conjunction with (18.87), to make condition (0) in the interpretation by Andrews and Hickerson of a mock theta function explicit for both $\sigma(q)$ at primitive roots of unity of odd order, and for $\mu(q)$ at primitive roots of unity of even order. We state the result for just one of each pair of mock theta functions, and leave the result for the other mock theta function of each pair to the reader.

Corollary 18.7. *(i) If ζ is a primitive odd-order $2k+1$ root of unity, then, as q approaches ζ radially within the unit disk, we have that*

$$\lim_{q \to \zeta} \left(\sigma(q) - \frac{1}{4} \left[2(-q;q^2)^2_\infty (-q^3,-q^3,q^6;q^6)_\infty - (q;q^2)^2_\infty (q^3,q^3,q^6;q^6)_\infty \right] \right)$$

$$= -\frac{1}{2} \sum_{n=0}^{k} \frac{(1-\zeta)(1-\zeta^3)\dots(1-\zeta^{2n-1})}{(1+\zeta)(1+\zeta^2)\dots(1+\zeta^n)} (-1)^n. \quad (18.88)$$

(ii) If ζ is a primitive even-order $2k$ root of unity, then, as q approaches ζ radially within the unit disk, we have that

$$\lim_{q \to \zeta} \left(\phi(q) - \frac{(-q;q)_\infty}{(q^2;q^4)_\infty} \right.$$

$$\left. \times \left[2(-q^2;q^4)^2_\infty (-q^6,-q^6,q^{12};q^{12})_\infty - (q^2;q^4)^2_\infty (q^6,q^6,q^{12};q^{12})_\infty \right] \right)$$

$$= -2 \sum_{n=1}^{k} \frac{(1+\zeta)(1+\zeta^2)\dots(1+\zeta^{2n-1})}{(1-\zeta)(1-\zeta^3)\dots(1-\zeta^{2n-1})} \zeta^n. \quad (18.89)$$

(iii) If ζ is a primitive even-order $2k$ root of unity, then, as q approaches ζ radially within the unit disk, we have that

$$\lim_{q \to \zeta} \left(\psi(q) - \frac{3q(-q;q)_\infty (q^6,q^6,q^6;q^6)_\infty}{(q^2;q^2)^2_\infty} \right)$$

$$= -2 \sum_{n=1}^{k} \frac{(1+\zeta)(1+\zeta^2)\dots(1+\zeta^{2n-2})}{(1-\zeta)(1-\zeta^3)\dots(1-\zeta^{2n-1})} \zeta^n. \quad (18.90)$$

(iv) If ζ is a primitive odd-order $2k+1$ root of unity, then, as q approaches ζ radially within the unit disk, we have that

$$\lim_{q \to \zeta} \left(\rho(q) - \frac{3(q;q^2)_\infty (q^3,q^3,q^3;q^3)_\infty}{2(q;q)^2_\infty} \right)$$

$$= -\frac{1}{2} \sum_{n=0}^{k} \frac{(1-\zeta)(1-\zeta^3)\dots(1-\zeta^{2n-1})}{(1+\zeta)(1+\zeta^2)\dots(1+\zeta^n)} (-\zeta)^n. \quad (18.91)$$

Note that combining the identities (18.72), (18.75) and (18.86) implies that

$$\frac{3(q;q^2)_\infty (q^3,q^3,q^3;q^3)_\infty}{(q;q)^2_\infty}$$

$$= 2(-q;q^2)^2_\infty (-q,-q^5,q^6;q^6)_\infty + (q;q^2)^2_\infty (q,q^5,q^6;q^6)_\infty. \quad (18.92)$$

In [72], Berndt and Chan gave representations of each of the functions $\rho(q)-2q^{-1}\psi_{-1}(q^2)$, $\sigma(q)-\phi_-(q^2)$, $4\phi_-(q^2)+2\mu(q)$ and $4q^{-1}\psi_{-1}(q^2)+\lambda(q)$ as infinite products. By combining the identities in Theorem 18.7 with those in Corollary 18.6 is it also possible to express each of these functions in terms of infinite products, but with two terms instead of one. For example, [72, Eq. (1.8)] is the identity

$$\sigma(q) - \phi_-(q^2) = q(-q^2;q^2)_\infty^2(-q^6,-q^6,q^6;q^6)_\infty, \tag{18.93}$$

while the corresponding identity implied by Theorem 18.7 and Corollary 18.6

$$\sigma(q) - \phi_-(q^2) = \frac{1}{2}(-q;q^2)_\infty^2(-q^3,-q^3,q^6;q^6)_\infty - \frac{1}{2}\frac{(-q^2;q^2)_\infty}{(q^4;q^8)_\infty}$$
$$\times \left[2(-q^4;q^8)_\infty^2(-q^{12},-q^{12},q^{24};q^{24})_\infty - (q^4;q^8)_\infty^2(q^{12},q^{12},q^{24};q^{24})_\infty\right]. \tag{18.94}$$

The reader may enjoy the challenge of proving the equality of the right sides of (18.93) and (18.94) directly (see Q18.13).

18.4   The Eighth Order Mock Theta Functions

In [130], Gordon and McIntosh introduce the following eight mock theta functions of order eight:

$$S_0(q) = \sum_{n=0}^\infty \frac{q^{n^2}(-q;q^2)_n}{(-q^2;q^2)_n}, \qquad S_1(q) = \sum_{n=0}^\infty \frac{q^{n(n+2)}(-q;q^2)_n}{(-q^2;q^2)_n}, \tag{18.95}$$

$$T_0(q) = \sum_{n=0}^\infty \frac{q^{(n+1)(n+2)}(-q^2;q^2)_n}{(-q;q^2)_{n+1}}, \quad T_1(q) = \sum_{n=0}^\infty \frac{q^{n(n+1)}(-q^2;q^2)_n}{(-q;q^2)_{n+1}}, \tag{18.96}$$

$$U_0(q) = \sum_{n=0}^\infty \frac{q^{n^2}(-q;q^2)_n}{(-q^4;q^4)_n}, \qquad U_1(q) = \sum_{n=0}^\infty \frac{q^{(n+1)^2}(-q;q^2)_n}{(-q^2;q^4)_{n+1}}, \tag{18.97}$$

$$V_0(q) = -1 + 2\sum_{n=0}^\infty \frac{q^{n^2}(-q;q^2)_n}{(q;q^2)_n}, \qquad V_1(q) = \sum_{n=0}^\infty \frac{q^{(n+1)^2}(-q;q^2)_n}{(q;q^2)_{n+1}}. \tag{18.98}$$

We will prove the following alternative expressions for $V_0(q)$ and $V_1(q)$ below.

$$V_0(q) = -1 + 2\sum_{n=0}^\infty \frac{q^{2n^2}(-q^2;q^4)_n}{(q;q^2)_{2n+1}}, \quad V_1(q) = \sum_{n=0}^\infty \frac{q^{2n^2+2n+1}(-q^4;q^4)_n}{(q;q^2)_{2n+2}}. \tag{18.99}$$

To accomplish this, we will prove the following more general transformation, from which the identities at (18.99) will follow as special cases.

Lemma 18.2. *If* $|q| < 1$, *and* $z \neq q^{-2k}$ *for any integer* $k \geq 0$, *then*

$$\sum_{n=0}^{\infty} \frac{(-q;q^2)_n q^{n(n-1)} z^n}{(z;q^2)_n} = \sum_{n=0}^{\infty} \frac{(-zq;q^4)_n q^{n(2n-1)} z^n}{(z;q^2)_{2n+1}}. \qquad (18.100)$$

Proof. The proof we will give is a slight variation of the proof given by the authors in [130], which was communicated to them by George Andrews. Let $L(z)$ and $R(z)$ denote the left- and right sides of (18.100), and define

$$F(z) := \sum_{n=0}^{\infty} \frac{(-q;q^2)_n q^{n(n-1)} z^n}{(z;q^2)_{n+1}},$$

and note that

$$F(z) = \sum_{n=0}^{\infty} \frac{(-q;q^2)_n q^{n(n-1)} z^n (1 - zq^{2n} + zq^{2n})}{(z;q^2)_{n+1}} = L(z) + \frac{z}{1-z} L(zq^2). \qquad (18.101)$$

It also follows that

$$\begin{aligned}
F(z) &= \frac{1}{1-z} + \sum_{n=0}^{\infty} \frac{(-q;q^2)_{n+1} q^{n(n+1)} z^{n+1}}{(z;q^2)_{n+2}} \\
&= \frac{1}{1-z} + \frac{z}{(1-z)(1-zq^2)} \sum_{n=0}^{\infty} \frac{(-q;q^2)_n (1 + q^{2n+1}) q^{n(n-1)} q^{2n} z^n}{(zq^4;q^2)_n} \\
&= \frac{1}{1-z} + \frac{z}{1-z} \sum_{n=0}^{\infty} \frac{(-q;q^2)_n q^{n(n-1)} q^{2n} z^n}{(zq^2;q^2)_{n+1}} \\
&\quad + \frac{zq}{(1-z)(1-zq^2)} \sum_{n=0}^{\infty} \frac{(-q;q^2)_n q^{n(n-1)} (zq^4)^n}{(zq^4;q^2)_n} \\
&= \frac{1}{1-z} + \frac{z}{1-z} F(zq^2) + \frac{zq}{(1-z)(1-zq^2)} L(zq^4). \qquad (18.102)
\end{aligned}$$

When (18.101) is used to eliminate $F(z)$ and $F(zq^2)$ from (18.102), there results the identity

$$L(z) = \frac{1}{1-z} + \frac{zq(1+zq)}{(1-z)(1-zq^2)} L(q^4). \qquad (18.103)$$

If this transformation is iterated $N + 1$ times, it follows that

$$L(z) = \sum_{n=0}^{N} \frac{(-zq; q^4)_n q^{n(2n-1)} z^n}{(z; q^2)_{2n+1}}$$
$$+ \frac{(-zq; q^4)_{N+1} q^{(N+1)(2(N+1)-1)} z^{N+1}}{(z; q^2)_{2(N+1)}} L(zq^{4(N+1)}). \quad (18.104)$$

Since $L(0) = 1$, letting $N \to \infty$ gives that $L(z) = R(z)$. $\qquad \square$

The identities at (18.99) now follow upon letting $z = q$ and $z = q^3$ in (18.100).

The identities at (18.99) are now used in the proof of two of the following identities, all of which are stated and proved in [130] (we omit a summation formula for $V_0(q) + V_0(-q)$ stated in [130], as its proof requires modular transformations which are not proved here).

Theorem 18.10. *(Gordon and McIntosh [130]) If* $|q| < 1$, *then*

$$U_0(q) = 2 \frac{(-q; q^2)_\infty}{(q^2; q^2)_\infty} \sum_{n=-\infty}^{\infty} \frac{(-1)^n q^{n(2n+1)}}{1 + q^{4n}}, \quad (18.105)$$

$$U_1(q) = \frac{(-q; q^2)_\infty}{(q^2; q^2)_\infty} \sum_{n=-\infty}^{\infty} \frac{(-1)^n q^{(n+1)(2n+1)}}{1 + q^{4n+2}}, \quad (18.106)$$

$$V_0(q) = -1 + 2 \frac{(-q^2; q^4)_\infty}{(q^4; q^4)_\infty} \sum_{n=-\infty}^{\infty} \frac{(-1)^n q^{2n(2n+1)}}{1 - q^{4n+1}}, \quad (18.107)$$

$$V_1(q) = \frac{(-q^4; q^4)_\infty}{(q^4; q^4)_\infty} \sum_{n=-\infty}^{\infty} \frac{(-1)^n q^{(2n+1)^2}}{1 - q^{4n+1}}, \quad (18.108)$$

$$U_0(q) = S_0(q^2) + q S_1(q^2), \quad (18.109)$$

$$U_1(q) = T_0(q^2) + q T_1(q^2), \quad (18.110)$$

$$U_0(q) + 2U_1(q) = (-q; q^2)_\infty^3 (q^2; q^2)_\infty (q^2; q^4)_\infty, \quad (18.111)$$

$$V_1(q) - V_1(-q) = 2q(-q^2; q^2)_\infty (-q^4; q^4)_\infty^2 (q^8; q^8)_\infty. \quad (18.112)$$

Proof. In (5.12), replace q with q^2 and let $d, e \to \infty$ to get

$$\sum_{n=0}^{\infty} \frac{(1 - aq^{4n})(a, b, c, f; q^2)_n}{(1 - a)(aq^2/b, aq^2/c, aq^2/f, q^2; q^2)_n} \left(\frac{a^2}{bcf} \right)^n q^{2n(n+1)}$$
$$= \frac{(aq^2; q^2)_\infty}{(aq^2/f; q^2)_\infty} \sum_{n=0}^{\infty} \frac{(aq^2/bc, f; q^2)_n}{(aq^2/b, aq^2/c, q^2; q^2)_n} \left(\frac{-a}{f} \right)^n q^{n(n+1)}. \quad (18.113)$$

Let $a \to 1$, set $b = i$, $c = -i$ and $f = -q$ in (18.113) and simplify to get

$$1 + 2 \sum_{n=1}^{\infty} \frac{(1 + q^{2n})(-1)^n q^{2n^2+n}}{1 + q^{4n}} = \frac{(q^2; q^2)_\infty}{(-q; q^2)_\infty} \sum_{n=0}^{\infty} \frac{(-q; q^2)_n q^{n^2}}{(-q^4; q^4)_n}.$$

The identity at (18.105) now follows, since

$$1 + 2 \sum_{n=1}^{\infty} \frac{(1 + q^{2n})(-1)^n q^{2n^2+n}}{1 + q^{4n}}$$

$$= 1 + 2 \sum_{n=1}^{\infty} \frac{(-1)^n q^{2n^2+n}}{1 + q^{4n}} + 2 \sum_{n=1}^{\infty} \frac{(-1)^n q^{2n^2-n}}{1 + q^{-4n}} = 2 \sum_{n=-\infty}^{\infty} \frac{(-1)^n q^{2n^2+n}}{1 + q^{4n}}.$$

The substitutions $a = q^2$, $b = iq$, $c = -iq$, $f = -q$ in (18.113) followed by similar elementary manipulations lead to (18.106).

For (18.107), replace q with q^2 in (18.113), then make the replacements $a \to 1$, $b = q$, $c = 1/q$ and $f = -q^2$. Multiply both sides of the resulting identity by $1/(1-q)$, use the fact that

$$\frac{(1 + q^{4n})(1 - 1/q)}{(1 - q^{4n+1})(1 - q^{4n-1})} = \frac{1}{1 - q^{4n+1}} - \frac{q^{-1}}{1 - q^{4n-1}}$$

to split the sum on the left side into two sums, shift the index of summation on the second sum and (18.107) follows from (18.99).

The proof of (18.108) is similar to that of (18.107). Replace q with q^2 in (18.113), then make the replacements $a = q^4$, $b = q$, $c = q^3$ and $f = -q^4$. Multiply both sides of the resulting identity by $q(1 - q^4)/((1 - q)(1 - q^3))$, this time using the fact that

$$\frac{1 - q^{8n+4}}{(1 - q^{4n+1})(1 - q^{4n+3})} = \frac{1}{1 - q^{4n+1}} + \frac{q^{4n+3}}{1 - q^{4n+3}}$$

to split the sum on the left side into two sums. Once again shift the index of summation on the second sum and the result once more follows from (18.99).

The identity at (18.109) also follows from (18.113), after setting $a = q$, $b = iq$, $c = -iq$ and $f = q^2$, simplifying the resulting series, multiplying both sides by $1 - q$, separating the left side into two sums, and finally replacing q with $-q$.

For (18.110), in (18.113) set $a = q^3$, $b = iq^2$, $c = -iq^2$ and $f = q^2$, again simplify the resulting expression, multiply both sides by $q/(1+q^2)$, separate the left side into two sums, and finally once again replace q with $-q$.

From (18.105) and (18.106),

$$
U_0(-q) + 2U_1(-q) = 2\frac{(q;q^2)_\infty}{(q^2;q^2)_\infty}\left[\sum_{n=-\infty}^{\infty}\frac{q^{n(2n+1)}}{1+q^{4n}} - \sum_{n=-\infty}^{\infty}\frac{q^{(n+1)(2n+1)}}{1+q^{4n+2}}\right]
$$

$$
= 2\frac{(q;q^2)_\infty}{(q^2;q^2)_\infty}\left[\sum_{n=-\infty}^{\infty}\frac{q^{2n(2n+1)/2}}{1+q^{2(2n)}} - \sum_{n=-\infty}^{\infty}\frac{q^{(2n+1)(2n+2)/2}}{1+q^{2(2n+1)}}\right]
$$

$$
= \frac{(q;q^2)_\infty}{(q^2;q^2)_\infty}\sum_{n=-\infty}^{\infty}\frac{2(-1)^n q^{n(n+1)/2}}{1+q^{2n}}
$$

$$
= \frac{(q;q^2)_\infty}{(q^2;q^2)_\infty}\frac{(q,q;q)_\infty}{(-q^2;q^2)_\infty},
$$

where the last equality follows from (8.10) with $b = i$, $c = -i$ and $d \to \infty$. The summation formula at (18.111) follows after replacing q with $-q$ and performing some elementary manipulations.

The proof of (18.112) is left as an exercise — see Q18.16. (Note that the statement of this identity in [130, Eq. (1.11)] incorrectly has $(-q^2;q^2)_\infty^3$ instead of $(-q^2;q^2)_\infty$, while the authors state the correct form in [131, page 221].)   □

Next, we consider bilateral sums related to the eighth order mock theta functions (the eight order equivalent of Theorem 18.9).

Theorem 18.11. *(i) If* $|q| < 1$, *then*

$$
G_8(a,b,d,z,q) := \sum_{n=-\infty}^{\infty}\frac{(a;q^2)_r z^r q^{r^2}}{(b,q^2)_r} = \frac{(-zq,-qb/az;q^2)_\infty}{(b,q^2/a;q^2)_\infty}
$$

$$
\times \sum_{n=-\infty}^{\infty}\frac{(-azq/b;q^2)_r}{(-zq;q^2)_r}(-b)^r q^{r^2-r}. \quad (18.114)
$$

(ii) If $|q|, |b/a| < 1$, *then*

$$
\sum_{n=-\infty}^{\infty}\frac{(a;q^2)_r z^r q^{r^2}}{(b;q^2)_r} = \frac{(b/a,-qb/az;q^2)_\infty}{(b;q^2)_\infty} \times \sum_{n=-\infty}^{\infty}(-azq/b,a;q^2)_r\left(\frac{b}{a}\right)^r.
$$

$$
(18.115)
$$

(iii) If $|q|, |bd/azq| < 1$, then

$$\sum_{n=-\infty}^{\infty} \frac{(a;q^2)_r z^r q^{r^2}}{(b;q^2)_r} = \frac{(-bq/az, -qz; q^2)_\infty}{(-q^3/az, -aqz; q^2)_\infty}$$

$$\times \sum_{n=-\infty}^{\infty} \frac{(1+azq^{4r-1})(a, -azq/b; q^2)_r \left(baz^2\right)^r q^{4r^2-4r}}{(1+az/q)(b, -zq; q^2)_r}. \qquad (18.116)$$

Proof. Let $d \to 0$ in (18.80), (18.81) and (18.82), respectively. □

The reason we state these summation formulae is that, as with mock theta functions of other orders, certain sums of eighth order mock theta functions may be written as single bilateral series, and the above transformations lead to new identities for these sums. It is a straightforward consequence of the definitions and (2.5) that

$$S_0(q) + 2T_0(q) = \sum_{r=-\infty}^{\infty} \frac{(-q; q^2)_r q^{r^2}}{(-q^2; q^2)_r}, \qquad (18.117)$$

$$S_1(q) + 2T_1(q) = \sum_{r=-\infty}^{\infty} \frac{(-q; q^2)_r q^{r^2+2r}}{(-q^2; q^2)_r}. \qquad (18.118)$$

In fact, following the method of the authors in [130], it will be shown that each of these sums has an expression in terms of infinite products.

Theorem 18.12. *If $|q| < 1$, then*

$$S_0(q^2) + 2T_0(q^2) = \frac{(q^2; q^2)_\infty \left[(q; q^2)_\infty^3 + (-q; q^2)_\infty^3\right]}{2(-q^2; q^2)_\infty}, \qquad (18.119)$$

$$S_1(q^2) + 2T_1(q^2) = \frac{(q^2; q^2)_\infty \left[(-q; q^2)_\infty^3 - (q; q^2)_\infty^3\right]}{2q(-q^2; q^2)_\infty}. \qquad (18.120)$$

Proof. Let $R_0(q)$ and $R_1(q)$ denote the series on the right side of (18.117) and (18.118), respectively. Next, in (8.1) replace q with q^2, set $a = q$, $b = iq$ and $c = -iq$, and then let $d, e \to \infty$. This leads to

$$\frac{R_0(q^2) - qR_1(q^2)}{1-q} = \sum_{r=-\infty}^{\infty} \frac{(1-q^{4r+1})(-q^2; q^4)_r q^{2r^2}}{(1-q)(-q^4; q^4)_r}$$

$$= \frac{(q^3, q, q^2, q; q^2)_\infty}{(iq^2, -iq^2, iq, -iq; q^2)_\infty} = \frac{(q^2; q^2)_\infty (q; q^2)_\infty^3}{(1-q)(-q^2; q^2)_\infty}.$$

Multiply through by $1 - q$ and then replace q with $-q$ to get an expression for $R_0(q^2) + qR_1(q^2)$. The pair of equations may then be solved for, in turn, $R_0(q^2)$ and $R_1(q^2)$, to give the results.   □

The following identities are a consequence of combining the results in Theorems 18.11 and 18.12.

Corollary 18.8. *If* $|q| < 1$, *then*

$$\sum_{r=-\infty}^{\infty} \frac{q^{4r^2}}{(-q^2; q^2)_{2r}} = \frac{(q^2; q^2)_{\infty}}{2} \left[(q; q^2)_{\infty}^3 + (-q; q^2)_{\infty}^3 \right], \qquad (18.121)$$

$$\sum_{r=-\infty}^{\infty} \frac{q^{4r^2+4r}}{(-q^2; q^2)_{2r+1}} = \frac{(q^2; q^2)_{\infty}}{2q} \left[(-q; q^2)_{\infty}^3 - (q; q^2)_{\infty}^3 \right], \qquad (18.122)$$

$$\sum_{r=-\infty}^{\infty} (8r + 1)q^{8r^2+2r} = \frac{(q^2; q^2)_{\infty}^3}{2} \left[(q; q^2)_{\infty}^3 + (-q; q^2)_{\infty}^3 \right], \qquad (18.123)$$

$$\sum_{r=-\infty}^{\infty} (8r + 3)q^{8r^2+6r} = \frac{(q^2; q^2)_{\infty}^3}{2q} \left[(-q; q^2)_{\infty}^3 - (q; q^2)_{\infty}^3 \right]. \qquad (18.124)$$

Proof. The proofs are left as exercises — see Q18.17.   □

Note that if the series on the left side of (18.121) and (18.122) are restricted to non-negative terms, the series of Andrews [32, Eqs. (1.14), (1.15)] are obtained (with q replaced with q^2).

The identities in Theorem 18.12 also contain implications for the limiting behaviour of each of the four eighth order mock theta functions that appear in these identities, as q tends to certain classes of roots of unity from within the unit circle. We state these for $S_0(q)$ and $S_1(q)$, as those for $T_0(q)$ and $T_1(q)$ are equally easily derived. To avoid fractional exponents, we state the results for q^2 instead of q.

Corollary 18.9. *(i) If* ζ *is a primitive even-order 8k root of unity, then, as* q *approaches* ζ *radially within the unit disk, we have that*

$$\lim_{q \to \zeta} \left(S_0(q^2) - \frac{(q^2; q^2)_{\infty} \left[(q; q^2)_{\infty}^3 + (-q; q^2)_{\infty}^3 \right]}{2(-q^2; q^2)_{\infty}} \right)$$

$$= -2 \sum_{n=0}^{k-1} \frac{(1 + \zeta^4)(1 + \zeta^8) \dots (1 + \zeta^{4n})}{(1 + \zeta^2)(1 + \zeta^6) \dots (1 + \zeta^{4n+2})} \zeta^{2n^2+6n+4}. \quad (18.125)$$

(ii) If ζ is a primitive even-order $8k$ root of unity, then, as q approaches ζ radially within the unit disk, we have that

$$\lim_{q \to \zeta} \left(S_1(q^2) - \frac{(q^2; q^2)_\infty \left[(-q; q^2)_\infty^3 - (q; q^2)_\infty^3 \right]}{2q(-q^2; q^2)_\infty} \right)$$

$$= -2 \sum_{n=0}^{k-1} \frac{(1 + \zeta^4)(1 + \zeta^8) \dots (1 + \zeta^{4n})}{(1 + \zeta^2)(1 + \zeta^6) \dots (1 + \zeta^{4n+2})} \zeta^{2n^2 + 2n}. \quad (18.126)$$

18.5 The Second Order Mock Theta Functions

The second order mock theta functions are

$$A(q) = \sum_{n=0}^{\infty} \frac{q^{(n+1)^2}(-q; q^2)_n}{(q; q^2)_{n+1}^2}, \quad B(q) = \sum_{n=0}^{\infty} \frac{q^{n(n+1)}(-q^2; q^2)_n}{(q; q^2)_{n+1}^2}, \quad (18.127)$$

$$\mu(q) = \sum_{n=0}^{\infty} \frac{(-1)^n q^{n^2}(q; q^2)_n}{(-q^2; q^2)_n^2}.$$

The first two were stated by McIntosh [189], and the third was given by Ramanujan in the Lost Notebook [213]. We first prove alternative expressions for $A(q)$ and $B(q)$.

Lemma 18.3. *(McIntosh [189]) The following identities hold:*

$$A(q) = \sum_{n=0}^{\infty} \frac{q^{n+1}(-q^2; q^2)_n}{(q; q^2)_{n+1}}, \quad (18.128)$$

$$B(q) = \sum_{n=0}^{\infty} \frac{q^n(-q; q^2)_n}{(q; q^2)_{n+1}}. \quad (18.129)$$

Proof. In (4.17), replace q with q^2 and set $a = q^2$, $b = -q^2/\tau$, $c = -q^2/z$ and $d = e = q^3$ to get

$$\sum_{r=0}^{\infty} \frac{(-q^2/\tau, -q^2/z; q^2)_r z^r \tau^r}{(q^3, q^3; q^2)_r} = \frac{(q, z\tau q^2; q^2)_\infty}{(q^3, z\tau; q^2)_\infty} \sum_{r=0}^{\infty} \frac{(-\tau q, -zq; q^2)_r q^r}{(q^3, \tau z q^2; q^2)_r}.$$

Now multiply both sides by $1/(1 - q)^2$ and let $\tau \to 0$ to get

$$\sum_{r=0}^{\infty} \frac{(-q^2/z; q^2)_r z^r q^{r(r+1)}}{(q; q^2)_{r+1}^2} = \sum_{r=0}^{\infty} \frac{(-zq; q^2)_r q^r}{(q; q^2)_{r+1}}. \quad (18.130)$$

The identities at (18.128) and (18.129) now follow upon letting $z = q$ and $z = 1$, respectively. $\square$

Remark: The identity at (18.130) also follows from (9.12), so that all of the identities for the Fine function $F(a, b; t)$ at (9.2) may be used to derive identities for $A(q)$ and $B(q)$.

Theorem 18.13. *(McIntosh [189]) Let $U_0(q)$, $U_1(q)$, $V_0(q)$ and $V_1(q)$ be the eighth order mock theta functions defined at (18.97) and (18.98). Then*

$$U_0(q) - 2U_1(q) = \mu(q). \tag{18.131}$$

$$V_0(q) - V_0(-q) = 4qB(q^2), \tag{18.132}$$

$$V_1(q) + V_1(-q) = 2A(q^2). \tag{18.133}$$

Proof. From (18.105) and (18.106),

$$U_0(-q) - 2U_1(-q) = 2\frac{(q; q^2)_\infty}{(q^2; q^2)_\infty} \left[\sum_{n=-\infty}^{\infty} \frac{q^{2n(2n+1)/2}}{1 + q^{2(2n)}} + \sum_{n=-\infty}^{\infty} \frac{q^{(2n+1)(2n+2)/2}}{1 + q^{2(2n+1)}} \right]$$

$$= 2\frac{(q; q^2)_\infty}{(q^2; q^2)_\infty} \sum_{n=-\infty}^{\infty} \frac{q^{n(n+1)/2}}{1 + q^{2n}}$$

$$= \frac{(q; q^2)_\infty}{(q^2; q^2)_\infty} \left[1 + \sum_{n=1}^{\infty} \frac{2q^{n(n+1)/2}(1 + q^n)}{1 + q^{2n}} \right]$$

$$= 2 \sum_{n=0}^{\infty} \frac{(q; q^2)_n(-1)^n}{(-q^2; q^2)_n}$$

$$= \sum_{n=0}^{\infty} \frac{q^{n^2}(-q; q^2)_n}{(-q^2; q^2)_n^2} = \mu(-q),$$

completing the proof of (18.131), upon replacing q with $-q$. The fourth equality follows from (5.26) with $b = i$, $c = -i$, $d = q^{1/2}$, $e = -q^{1/2}$ and $a \to 1$, with convergence of the next-to-last series being in the Cesàro sense (see the discussion of the convergence of the sixth order mock theta function $\mu(q)$ on page 305). The next to last equality follows from (4.17), with q replaced with q^2, $a = q^2$, $c = -q$, $d = e = -q^2$ and $b \to \infty$.

For (18.132), from (18.107),

$$V_0(iq) - V_0(-iq) = 2\frac{(q^2; q^4)_\infty}{(q^4; q^4)_\infty} \left[\sum_{n=-\infty}^{\infty} \frac{q^{2n(2n+1)}}{1 - iq^{4n+1}} - \sum_{n=-\infty}^{\infty} \frac{q^{2n(2n+1)}}{1 + iq^{4n+1}} \right]$$

$$= 4iq\frac{(q^2; q^4)_\infty}{(q^4; q^4)_\infty} \sum_{n=-\infty}^{\infty} \frac{q^{4n^2+6n}}{1 + q^{8n+2}}$$

$$= 4iq \frac{(q^2;q^4)_\infty}{(q^4;q^4)_\infty} \left[\sum_{n=0}^{\infty} \frac{q^{4n^2+6n}}{1+q^{8n+2}} + \sum_{n=0}^{\infty} \frac{q^{4n^2+10n+4}}{1+q^{8n+6}} \right]$$

$$= 4iq \frac{(q^2;q^4)_\infty}{(q^4;q^4)_\infty} \sum_{n=0}^{\infty} \frac{q^{n^2+3n}}{1+q^{4n+2}}$$

$$= 4iq \frac{(q^2;q^4)_\infty}{(q^4;q^4)_\infty} \frac{(q^4,q^2;q^2)_\infty}{(iq^3,-iq^3;q^4)_\infty} \frac{1}{1+q^2}$$
$$\times \sum_{n=0}^{\infty} \frac{(iq,-iq,-q^2;q^2)_n}{(q^3,-q^3,q^2;q^2)_n} q^{2n}$$

$$= \frac{4iq}{1+q^2} \sum_{n=0}^{\infty} \frac{(q,-q;q^2)_n}{(iq^3,-iq^3;q^2)_n} (-q^2)^n = 4iq B((iq)^2).$$

The proof of (18.132) is complete upon replacing q with $-iq$. The second sum in the third equality arises from re-indexing with negative n in the previous sum. The fourth equality follows from (5.26) with q replaced with q^2, $a = q^2$, $b = -c = q$ and $d = -e = iq$. The next to last equality follows from (4.18), with q replaced with q^2, $a = -q^2$, $b = iq$, $c = -iq$, $d = -e = q^3$, and the final equality comes from (18.129) above.

Similarly, for (18.133), from (18.108),

$$V_1(q) + V_1(q) = \frac{(-q^4;q^4)_\infty}{(q^4;q^4)_\infty} \left[\sum_{n=-\infty}^{\infty} \frac{(-1)^n q^{(2n+1)^2}}{1-q^{4n+1}} - \sum_{n=-\infty}^{\infty} \frac{(-1)^n q^{(2n+1)^2}}{1+q^{4n+1}} \right]$$

$$= 2q^2 \frac{(-q^4;q^4)_\infty}{(q^4;q^4)_\infty} \sum_{n=-\infty}^{\infty} \frac{(-1)^n q^{4n^2+8n}}{1-q^{8n+2}}$$

$$= 2q^2 \frac{(-q^4;q^4)_\infty}{(q^4;q^4)_\infty} (1+q^2) \sum_{n=0}^{\infty} \frac{(-1)^n q^{4n^2+8n}(1-q^{8n+4})}{(1-q^{8n+2})(1-q^{8n+6})}$$

$$= 2q^2 \frac{(-q^4;q^4)_\infty}{(q^4;q^4)_\infty} \frac{(1+q^2)(1-q^8)}{(1+q^4)(1-q^2)(1-q^6)}$$
$$\times \sum_{n=0}^{\infty} \frac{1-q^{16n+8}}{1-q^8} \frac{(q^8,-q^4,q^2,q^6,-q^8;q^8)_n (-1)^n q^{4n^2+8n}}{(-q^{12},q^{14},q^{10},-q^8,q^8;q^8)_n}$$

$$= 2q^2 \frac{(-q^4;q^4)_\infty}{(q^4;q^4)_\infty} \frac{(q^8,q^8;q^8)_\infty}{(q^2,q^6;q^8)_\infty} \frac{(1+q^2)}{(1+q^4)}$$
$$\times \sum_{n=0}^{\infty} \frac{(q^2,q^4,q^6;q^8)_n q^{8n}}{(-q^{12},-q^8,q^8;q^8)_n}$$

$$= 2q^2 \frac{(-q^4; q^4)_\infty}{(q^4; q^4)_\infty} \frac{(q^8, q^8; q^8)_\infty}{(q^2, q^6; q^8)_\infty} \frac{(1 + q^2)}{(1 + q^4)}$$

$$\times \sum_{n=0}^{\infty} \frac{(q^4; q^8)_n (q^2; q^4)_{2n} q^{8n}}{(q^8; q^8)_n (-q^8; q^4)_{2n}}$$

$$= \frac{2q^2}{1 - q^2} \sum_{n=0}^{\infty} \frac{(-q^6; q^4)_n (q^8; q^8)_n}{(q^4; q^4)_n (q^{12}; q^8)_n} q^{2n} = 2A(q^2).$$

The third equality arises from re-indexing terms with negative n in the previous sum to get a second series and combining the two series. The fourth equality is just the third equality written in a more complicated fashion. The fifth equality follows from (5.26) with q replaced with q^8, $a = q^8$, $b = -q^4$, $c = -q^8$ and $d = q^2$ and $e = q^6$. The next to last equality follows from (4.2), with q replaced with q^4 and $h = 2$, $a = q^4$, $b = q^2$, $c = -q^8$ and $z = q^8$. The final equality comes from (18.128) above, after simplifying the right side.   □

18.6   The Mock Theta Functions of Other Orders — Some Brief Comments

The transformations which were specialized to derive various identities for the second-, third-, fifth-, sixth- and eighth order mock theta functions may not be applied to produce identities for either the seventh- or tenth order mock theta functions, and identities for these mock theta functions are more difficult to prove. However, just to give some flavour of their nature, we briefly discuss them, and state (without proof) some examples of the identities they satisfy.

18.6.1   The Tenth Order Mock Theta Functions

Ramanujan's tenth order mock theta functions, which were written down by him in the Lost Notebook [213], are the following:

$$\phi(q) = \sum_{n=0}^{\infty} \frac{q^{n(n+1)/2}}{(q; q^2)_{n+1}}, \qquad \Psi(q) = \sum_{n=0}^{\infty} \frac{q^{(n+1)(n+2)/2}}{(q; q^2)_{n+1}}, \qquad (18.134)$$

$$X(q) = \sum_{n=0}^{\infty} \frac{(-1)^n q^{n^2}}{(-q; q)_{2n}}, \qquad \chi(q) = \sum_{n=0}^{\infty} \frac{(-1)^n q^{(n+1)^2}}{(-q; q)_{2n+1}}.$$

In [213, p. 9], Ramanujan stated eight identities for the these functions, and these were proved by Choi [97–100]. The proofs were long and beyond the scope of this text, but we state one of the identities (in the form proved by Choi [97, Eq. (1.5)]) to provide an indication of their complexity:

$$q^2\phi(q^9) - \frac{\Psi(\omega q) - \Psi(\omega^2 q)}{\omega - \omega^2} = -q\frac{(q;q)_\infty(q,q^5,q^6;q^6)_\infty(q^3,q^{12},q^{15};q^{15})_\infty}{(q^3;q^3)_\infty^2},$$

$$(18.135)$$

where ω is a primitive cube root of unity.

18.6.2 The Seventh Order Mock Theta Functions

The seventh order mock theta functions are the following:

$$\mathcal{F}_0(q) = \sum_{n=0}^{\infty} \frac{q^{n^2}(q;q)_n}{(q;q)_{2n}}, \qquad (18.136)$$

$$\mathcal{F}_1(q) = \sum_{n=0}^{\infty} \frac{q^{(n+1)^2}(q;q)_n}{(q;q)_{2n+1}}, \qquad (18.137)$$

$$\mathcal{F}_2(q) = \sum_{n=0}^{\infty} \frac{q^{n(n+1)}(q;q)_n}{(q;q)_{2n+1}}. \qquad (18.138)$$

As Andrews remarked in [25] "These three functions are perhaps the most mysterious in all of Ramanujan's work." Nothing substantial had been proved about them before this paper [25] of Andrews, in which he derived Hecke type series for both the fifth and seventh order mock theta functions. Such expansions were derived elsewhere for mock theta functions of other orders, including those of sixth order by Andrews and Hickerson [47] and those of second order and eighth order by Srivastava in [242] and [241] respectively. As an example, Andrews [25, Eq. (7.22)] found the following representation (corrected):

$$\mathcal{F}_0(q) = \frac{1}{(q;q)_\infty}\left[\sum_{n=0}^{\infty}\sum_{j=-n}^{n} q^{7n^2+n-j^2}(1-q^{12n+6})\right.$$

$$\left. - 2q^2\sum_{n=0}^{\infty}\sum_{j=0}^{n} q^{7n^2+8n-j^2-j}(1-q^{12n+12})\right]. \quad (18.139)$$

Hickerson [142] stated "seventh order mock theta conjectures" for $\mathcal{F}_0(q)$, $\mathcal{F}_1(q)$ and $\mathcal{F}_2(q)$. His identity for $\mathcal{F}_0(q)$ is

$$\mathcal{F}_0(q) - 2 = 2q\,g(q,q^7) - \frac{(q^3,q^4,q^7;q^7)_\infty^2}{(q;q)_\infty}, \qquad (18.140)$$

where $g(x, q)$ is the universal order mock theta function defined at (18.1).

<h2 style="text-align:center">Exercises</h2>

18.1 By considering certain summation formulae for third order mock theta functions stated earlier in the chapter, or otherwise, prove the identity

$$\sum_{n=-\infty}^{\infty} \frac{q^{n^2}}{(-q^2; q^2)_n} = 2 \sum_{n=-\infty}^{\infty} \frac{q^{n^2}}{(q; q^2)_n}, \tag{18.141}$$

where convergence of the sum on the right for negative n is in the Cesàro sense.

18.2 By using the series on the right side of (2.28) in conjunction with (18.6), prove that the third order mock theta function $\psi(q)$ satisfies

$$\psi(q) = \frac{1}{(q^4; q^4)_\infty} \sum_{n=-\infty}^{\infty} \frac{(-1)^n q^{6n^2+6n+1}}{1 - q^{4n+1}}. \tag{18.142}$$

18.3 By using (8.5) or otherwise, prove that

$$\frac{(q; q)_\infty}{(-q; q)_\infty^2} = 1 + 4 \sum_{n=1}^{\infty} \frac{(-1)^n q^{(n^2+n)/2}}{1 + q^n}. \tag{18.143}$$

18.4 Prove the identities at (18.31) and (18.32).

18.5 By writing the third order mock theta functions $\omega(q)$, $\nu(q)$ and $\rho(q)$ in terms of the function $G_3(s, t, q)$, and then employing (18.6), prove the identities

$$\omega(q)(q^2; q^2)_\infty = \sum_{n=0}^{\infty} (-1)^n q^{3(n^2+n)} \frac{1 + q^{2n+1}}{1 - q^{2n+1}} = \sum_{n=-\infty}^{\infty} \frac{(-1)^n q^{3(n^2+n)}}{1 - q^{2n+1}}, \tag{18.144}$$

$$\nu(q)(q; q)_\infty = \sum_{n=0}^{\infty} (-1)^n q^{3(n^2+n)/2} \frac{1 - q^{2n+1}}{1 + q^{2n+1}} = \sum_{n=-\infty}^{\infty} \frac{(-1)^n q^{3(n^2+n)/2}}{1 + q^{2n+1}}, \tag{18.145}$$

$$\rho(q)(q^2; q^2)_\infty = \sum_{n=0}^{\infty} (-1)^n q^{3(n^2+n)} \frac{1 - q^{4n+2}}{1 + q^{2n+1} + q^{4n+2}}$$

$$= \sum_{n=-\infty}^{\infty} \frac{(-1)^n q^{3(n^2+n)}}{1 + q^{2n+1} + q^{4n+2}}. \tag{18.146}$$

18.6 (i) By using (5.7) or otherwise, prove the identity

$$\sum_{n=0}^{\infty} \frac{1+q^{2n+1}}{1-q^{2n+1}} (-1)^n q^{n^2+n} = \frac{(q^2;q^2)_\infty^2}{(q;q^2)_\infty^2}. \tag{18.147}$$

(ii) By using the above identity or otherwise, prove the identity at (18.35).

18.7 (i) Deduce from (9.20) that

$$\sum_{n=1}^{\infty} (-1;q^2)_n q^n = 2 \sum_{n=1}^{\infty} \frac{q^{n^2}}{(q;q^2)_n}. \tag{18.148}$$

(ii) (Fine, [115, p. 60]) Hence deduce from the $_1\psi_1$ summation formula (7.1) that (here $\phi(q)$ and $\psi(q)$ are the third order mock theta functions)

$$\phi(q) + 2\psi(q) = (q^2;q^2)_\infty (-q;q^2)_\infty^3. \tag{18.149}$$

18.8 By using (3.6) or otherwise, prove that the third order mock theta function $\chi_0(q)$ satisfies

$$\chi_0(q) = 1 + \sum_{m=0}^{\infty} \frac{(q;q)_m q^{2m+1}}{(q;q)_{2m+1}}. \tag{18.150}$$

18.9 Complete the proof of the identities at (18.56) and (18.57).

18.10 By employing the identities in Corollary 18.3 or otherwise, prove the identities

$$\frac{2q(-q^2;q^2)_\infty^2}{(-q^4,-q^6;q^{10})_\infty} = \frac{(-q;q^2)_\infty^2}{(-q,-q^9;q^{10})_\infty} - \frac{(q;q^2)_\infty^2}{(q,q^9;q^{10})_\infty}, \tag{18.151}$$

$$\frac{2(-q^2;q^2)_\infty^2}{(-q^2,-q^8;q^{10})_\infty} = \frac{(-q;q^2)_\infty^2}{(-q^3,-q^7;q^{10})_\infty} + \frac{(q;q^2)_\infty^2}{(q^3,q^7;q^{10})_\infty}. \tag{18.152}$$

18.11 (i) By employing either the identity at (8.9) or (8.12), or otherwise, prove that

$$\sum_{n=-\infty}^{\infty} \frac{(-1)^n q^{n(n+1)/2}}{1-xq^n} = \frac{(q,q;q)_\infty}{(x,q/x;q)_\infty}. \tag{18.153}$$

(ii) Hence or otherwise prove that

$$g_2(x,q) + g_2(-x,q) = \frac{2(q^2;q^2)_\infty}{(q,q,x^2,q^2/x^2;q^2)_\infty}. \tag{18.154}$$

18.12 Prove that

$$xg_2(x,q) + x^{-1}g_2(x^{-1},q) = -1. \tag{18.155}$$

18.13 Prove the equality of the right sides of (18.93) and (18.94) without appealing to the mock theta function identities in this chapter.

18.14 Let $\lambda(q)$, $\mu(q)$, $\rho(q)$ and $\sigma(q)$ be the sixth order mock theta functions defined earlier. Prove that

(i) $\lambda(-q) + \lambda(q) = \rho(-q) + \rho(q)$
$$= (-q;q^2)^2_\infty(-q,-q^5,q^6;q^6)_\infty + (q;q^2)^2_\infty(q,q^5,q^6;q^6)_\infty, \tag{18.156}$$

(ii) $2\mu(q) - 2\mu(q) = 2\sigma(q) - 2\sigma(-q)$
$$= (-q;q^2)^2_\infty(-q^3,-q^3,q^6;q^6)_\infty - (q;q^2)^2_\infty(q^3,q^3,q^6;q^6)_\infty. \tag{18.157}$$

18.15 Prove the identities at (18.109) and (18.110) directly from the transformation at (18.113).

18.16 By using (18.153) or otherwise, prove the identity at (18.112).

18.17 By combining the results in Theorems 18.11 and 18.12 and the formulae at (18.117) and (18.118), or otherwise, prove the identities in Corollary 18.8.

18.18 By using the $_1\psi_1$ summation (7.1) or otherwise, prove that

$$U_0(q) - 2U_1(q) + 4A(q) = (-q;q^2)^5_\infty(q^2;q^2)_\infty, \tag{18.158}$$

where $U_0(q)$ and $U_1(q)$ are the eighth order mock theta functions defined above, and $A(q)$ is the second order mock theta functions.

18.19 (i) By using the transformation at (9.12) or otherwise, prove that the second order mock theta function $\mu(q)$ satisfies

$$\mu(q) = 2\sum_{n=0}^{\infty} \frac{(-q;q^2)_n(-1)^n}{(-q^2;q^2)_n}, \tag{18.159}$$

where convergence of the series on the right side is once again in the Cesàro sense (see the discussion of the convergence of the sixth order mock theta function $\mu(q)$ on page 305).

(ii) By applying various transformations of Heine from Chapter 4 to the right side of (18.159), or otherwise, prove the following alternative expressions for $\mu(q)$:

$$\mu(q) = \frac{(q;q)_\infty}{(-q^2,-q^2;q^2)_\infty} \sum_{n=0}^{\infty} \frac{(-1,-1;q^2)_n q^{2n}}{(q;q)_{2n}}, \tag{18.160}$$

$$= \frac{(-q;q^2)_\infty}{(-q^2;q^2)_\infty} \sum_{n=0}^{\infty} \frac{(-1,q;q^2)_n (-q)^n}{(q^4;q^4)_n}, \tag{18.161}$$

$$= \frac{(q,q;q^2)_\infty}{(-q^2,-q^2;q^2)_\infty} \sum_{n=0}^{\infty} \frac{(-q,-q;q^2)_n q^n}{(q;q)_{2n}}. \tag{18.162}$$

Srivastava [243] proved (18.159) and (18.161). Alternative expressions may similarly be derived for the second order mock theta functions $A(q)$ and $B(q)$, and some of these were also derived by Srivastava [243].

18.20 Prove the transformations at (18.18), (18.19) and (18.20).

18.21 By considering various identities for the third order mock theta functions earlier in the chapter, or otherwise, prove the identities

$$4 \sum_{r=-\infty}^{\infty} \frac{q^{r^2+r}(2rq^r+1)}{(1+q^r)^2} = (q;q^2)_\infty^4 (q;q)_\infty^4, \tag{18.163}$$

$$2 \sum_{r=-\infty}^{\infty} \frac{q^{2r^2+2r}(r+1)}{(1+q^{2r+1})^2} = \frac{(-q^2;q^2)_\infty^2 (q;q)_\infty^3 (q^4;q^4)_\infty}{(-q;q^2)_\infty}$$

$$+ (-q^4, -q^{12}, q^{16}; q^{16})_\infty. \tag{18.164}$$

Appendix I: Frequently Used Theorems

In this appendix, we include three important theorems whose implications are used fairly often to prove results about basic hypergeometric series. They are included for the benefit of those who may not have encountered them yet in a course on real- or complex analysis.

The first of these is useful if it is desired to interchange limit and sum when dealing with various infinite series.

Theorem I.1. *(Tannery's Theorem [247]) For each $n \in \mathbb{N}$, suppose that $\{f_k(n)\}_{k=1}^{\infty}$ is such that $\sum_{k=1}^{\infty} f_k(n)$ converges. Suppose further that for each $k \in \mathbb{N}$, $\lim_{n \to \infty} f_k(n) = f_k$, that there exists constants M_k, $k \geq 1$ such that $|f_k(n)| \leq M_k$ for all k, $n \geq 1$, and that $\sum_{k=1}^{\infty} M_k$ converges. Then*

$$\lim_{n \to \infty} \sum_{k=1}^{\infty} f_k(n) = \sum_{k=1}^{\infty} f_k. \tag{I.1}$$

Proof. Let $\epsilon > 0$. Choose an integer m such that $\sum_{k=m+1}^{\infty} M_k < \epsilon/3$. Since $|f_k(n)| \leq M_k$ for all k, $n \geq 1$, letting $n \to \infty$ gives that $|f_k| < M_k, \forall k \geq 1$, and thus that

$$|f_k(n) - f_k| \leq |f_k(n)| + |f_k| \leq M_k + M_k = 2M_k.$$

Hence

$$\left| \sum_{k=1}^{\infty} f_k(n) - \sum_{k=1}^{\infty} f_k \right| = \left| \sum_{k=1}^{m} (f_k(n) - f_k) + \sum_{k=m+1}^{\infty} (f_k(n) - f_k) \right|$$

$$\leq \sum_{k=1}^{m} |f_k(n) - f_k| + \sum_{k=m+1}^{\infty} |f_k(n) - f_k|$$

$$\leq \sum_{k=1}^{m} |f_k(n) - f_k| + \sum_{k=m+1}^{\infty} 2M_k$$

$$< \sum_{k=1}^{m} |f_k(n) - f_k| + \frac{2\epsilon}{3}.$$

Next, since $\lim_{n \to \infty} f_k(n) = f_k$, there exists an integer N, such that

$$|f_k(n) - f_k| < \frac{\epsilon}{3m}, \quad 1 \leq k \leq m, \quad \forall n > N.$$

Thus, if $n > N$, then

$$\left| \sum_{k=1}^{\infty} f_k(n) - \sum_{k=1}^{\infty} f_k \right| < \sum_{k=1}^{m} \frac{\epsilon}{3m} + \frac{2\epsilon}{3} = \epsilon,$$

completing the proof. $\qquad \square$

Remark: The identity at (I.1) may be rewritten as

$$\lim_{n \to \infty} \sum_{k=1}^{\infty} f_k(n) = \sum_{k=1}^{\infty} \lim_{n \to \infty} f_k(n). \tag{I.2}$$

Theorem I.2. *(Identity Theorem for complex functions)*
   Let f and g be holomorphic functions defined on an open connected set $D \subseteq \mathbb{C}$. If

$$S := \{z \in D | f(z) = g(z)\}$$

has an accumulation point, then $f(z) = g(z)$, for all $z \in D$.

Proof. Let c be an accumulation point in S. By uniqueness of Taylor series, it is sufficient to show that $f^{(k)}(c) = g^{(k)}(c)$ for all $k \geq 0$. Suppose this is not the case, and let m be the least positive integer such that $f^{(m)}(c) \neq g^{(m)}(c)$.

Since f and g are holomorphic in D, it follows that in some open set around c,

$$(f - g)(z) = \sum_{k=m}^{\infty} \frac{(f - g)^{(k)}(c)(z - c)^k}{k!}$$

$$= (z - c)^m \sum_{k=m}^{\infty} \frac{(f - g)^{(k)}(c)(z - c)^{k-m}}{k!}$$

$$=: (z - c)^m h(z).$$

With $h(c) = (f - g)^{(m)}(c)/m! \neq 0$, by continuity $h(z) \neq 0$ in some open disc B about c. Hence $(f - g)(z) \neq 0$, or $f(z) \neq g(z)$, in $B \setminus \{c\}$, contradicting the fact that c is an accumulation point of S.

Thus no such integer m exists, $f(z) = g(z)$ on D, and the theorem is proved. $\qquad\square$

The next theorem has applications, as will be indicated, for the separate convergence of numerators and denominators of q-continued fractions.

Theorem I.3. *(Abel's Theorem [1] for power series)*[1]
Let

$$\mathcal{A}(z) := \sum_{k=0}^{\infty} A_k z^k$$

be a series with complex coefficients A_k that is convergent for $|z| < 1$. If $\sum_{k=0}^{\infty} A_k$ converges, then

$$\lim_{z \to 1} \mathcal{A}(z) = \sum_{k=0}^{\infty} A_k, \qquad (I.3)$$

where $z \to 1$ in any Stolz sector, that is, a region of the open unit disc that is also in the region defined by

$$|1 - z| \leq M(1 - |z|), \qquad some\ M > 0.$$

Proof. Suppose

$$\sum_{k=0}^{\infty} A_k = s.$$

---

[1]The proof given here is based on that at
https://en.wikipedia.org/wiki/Abel's_theorem.

By replacing a_0 with $a_0 - s$, it may be assumed that $\sum_{k=0}^{\infty} A_k = 0$. Define

$$s_k := \sum_{j=0}^{k} A_j,$$

so that $A_k = s_k - s_{k-1}$, and $\lim_{k \to \infty} s_k = 0$. Then

$$\mathcal{A}(z) = \sum_{k=0}^{\infty} (s_k - s_{k-1}) z^k = (1 - z) \sum_{k=0}^{\infty} s_k z^k.$$

Fix $M > 0$ and $\epsilon > 0$. Choose n large enough so that

$$|s_k| < \frac{\epsilon}{2M}, \quad \forall\, k \geq n.$$

Hence, for any z in the Stolz sector described above,

$$\left| (1 - z) \sum_{k=n}^{\infty} s_k z^k \right| \leq |1 - z| \sum_{k=n}^{\infty} |s_k| |z|^k$$

$$< \frac{|1 - z|\epsilon}{2M} \sum_{k=n}^{\infty} |z|^k = \frac{|1 - z|\epsilon |z|^n}{2M(1 - |z|)} < \frac{\epsilon}{2}.$$

On the other hand, since n is independent of z, there is a $\delta > 0$ such that if $|1 - z| < \delta$,

$$\left| (1 - z) \sum_{k=0}^{n-1} s_k z^k \right| < \frac{\epsilon}{2}.$$

Thus, $|\mathcal{A}(z)| < \epsilon$ when $|1 - z| < \delta$ with z in the stated Stolz sector, and the result is proved. $\qquad\square$

The following corollary is sometimes useful.

Corollary I.1. *Suppose that $\{B_k\}$ is a sequence such that $\lim_{k \to \infty} B_k = B \in \mathbb{C}$. Suppose also that*

$$\sum_{k=0}^{\infty} B_k t^k$$

converges for $-1 < t < 1$. Then

$$\lim_{t \to 1-} (1 - t) \sum_{k=0}^{\infty} B_k t^k = B. \tag{I.4}$$

Proof. Write

$$(1 - t) \sum_{k=0}^{\infty} B_k t^k = B_0 + \sum_{k=1}^{\infty} (B_k - B_{k-1}) t^k$$

and apply Theorem I.1 with $A_0 = B_0$ and $A_k = B_k - B_{k-1}$ for $k \geq 1$. $\qquad \square$

As an example of an application of this corollary, suppose that it can be determined that the sequence $\{A_n\}$ of numerators in the continued fraction in Theorem 16.10 converges separately in the case $b = 1$, without knowing an exact formula for $\lim_{n \to \infty} A_n$. Then this limit may be determined directly from the expression for $F(t)$ at (16.42), by setting $b = 1$, multiplying by $1 - t$, and then letting $t \to 1-$.

Appendix II: WP-Bailey Chains

If $(\alpha_n(a,k), \beta_n(a,k))$ is a WP-Bailey pair with respect to a and k (in other words, they satisfy (12.3)), then so is the pair $(\alpha'_n(a,k), \beta'_n(a,k))$, where $(\alpha'_n(a,k), \beta'_n(a,k))$ is defined by any of the following pairs of relations.

The WP-Bailey chains of Andrews (Andrews [31]):

$$\alpha'_n(a,k) = \frac{(\rho_1, \rho_2; q)_n}{(aq/\rho_1, aq/\rho_2; q)_n} \left(\frac{k}{c}\right)^n \alpha_n(a,c),$$

$$\beta'_n(a,k) = \frac{(k\rho_1/a, k\rho_2/a; q)_n}{(aq/\rho_1, aq/\rho_2; q)_n}$$

$$\times \sum_{j=0}^n \frac{(1-cq^{2j})(\rho_1,\rho_2;q)_j(k/c;q)_{n-j}(k;q)_{n+j}}{(1-c)(k\rho_1/a, k\rho_2/a;q)_j(q;q)_{n-j}(qc;q)_{n+j}} \left(\frac{k}{c}\right)^j \beta_j(a,c). \quad \text{(II.1)}$$

$$\alpha'_n(a,k) = \frac{(qa^2/k)_{2n}}{(k)_{2n}} \left(\frac{k^2}{qa^2}\right)^n \alpha_n\left(a, \frac{qa^2}{k}\right),$$

$$\beta'_n(a,k) = \sum_{j=0}^n \frac{(k^2/qa^2)_{n-j}}{(q)_{n-j}} \left(\frac{k^2}{qa^2}\right)^j \beta_j\left(a, \frac{qa^2}{k}\right). \quad \text{(II.2)}$$

The WP-Bailey chains of Warnaar ([256]):

$$\alpha'_n(a,k) = \frac{1-\sigma k^{1/2}}{1-\sigma k^{1/2}q^n} \frac{1+\sigma a/k^{1/2}q^n}{1+\sigma a/k^{1/2}} \frac{(a^2/k;q)_{2n}}{(k;q)_{2n}} \left(\frac{k^2}{a^2}\right)^n \alpha_n\left(a, \frac{a^2}{k}\right),$$

$$\beta'_n(a,k) = \frac{1-\sigma k^{1/2}}{1-\sigma k^{1/2}q^n} \sum_{j=0}^n \frac{1+\sigma a/k^{1/2}q^j}{1+\sigma a/k^{1/2}} \frac{(k^2/a^2;q)_{n-j}}{(q;q)_{n-j}} \left(\frac{k^2}{a^2}\right)^j \beta_j\left(a, \frac{a^2}{k}\right),$$

$$\text{where } \sigma \in \{1, -1\} \quad \text{(II.3)}$$

333

$$\alpha'_n(a, k, q^2) = \alpha_n\left(\sqrt{a}, \frac{k}{q\sqrt{a}}, q\right),$$

$$\beta'_n(a, k, q^2) = \frac{\left(\frac{-k}{\sqrt{a}}; q\right)_{2n}}{(-q\sqrt{a}; q)_{2n}} \sum_{j=0}^n \frac{\left(1 - \frac{kq^{2j}}{q\sqrt{a}}\right)\left(\frac{aq^2}{k}; q^2\right)_{n-j}}{\left(1 - \frac{k}{q\sqrt{a}}\right)(q^2; q^2)_{n-j}} \frac{(k; q^2)_{n+j}}{\left(\frac{k^2}{a}; q^2\right)_{n+j}}$$

$$\times \left(\frac{k}{aq}\right)^{n-j} \beta_j\left(\sqrt{a}, \frac{k}{q\sqrt{a}}, q\right). \quad \text{(II.4)}$$

$$\alpha'_n(a, k, q^2) = q^{-n}\frac{1 + \sqrt{a}q^{2n}}{1 + \sqrt{a}}\alpha_n\left(\sqrt{a}, \frac{k}{\sqrt{a}}, q\right),$$

$$\beta'_n(a, k, q^2) = q^{-n}\frac{\left(\frac{-kq}{\sqrt{a}}; q\right)_{2n}}{(-\sqrt{a}; q)_{2n}} \sum_{j=0}^n \frac{\left(1 - \frac{kq^{2j}}{\sqrt{a}}\right)\left(\frac{a}{k}; q^2\right)_{n-j}(k; q^2)_{n+j}}{\left(1 - \frac{k}{\sqrt{a}}\right)(q^2; q^2)_{n-j}\left(\frac{k^2q^2}{a}; q^2\right)_{n+j}}$$

$$\times \left(\frac{k}{a}\right)^{n-j} \beta_j\left(\sqrt{a}, \frac{k}{\sqrt{a}}, q\right). \quad \text{(II.5)}$$

$$\alpha'_{2n}(a, k, q) = \alpha_n\left(a, \frac{k^2}{a}, q^2\right), \qquad \alpha'_{2n+1}(a, k, q) = 0$$

$$\beta'_n(a, k, q) = \frac{\left(\frac{k^2q}{a}; q^2\right)_n}{(aq; q^2)_n} \sum_{j=0}^{\lfloor n/2 \rfloor} \frac{\left(1 - \frac{k^2q^{4j}}{a}\right)\left(\frac{a}{k}; q\right)_{n-2j}(k; q)_{n+2j}}{\left(1 - \frac{k^2}{a}\right)(q; q)_{n-2j}\left(\frac{k^2q}{a}; q\right)_{n+2j}}$$

$$\times \left(\frac{-k}{a}\right)^{n-2j} \beta_j\left(a, \frac{k^2}{a}, q^2\right). \quad \text{(II.6)}$$

The WP-Bailey chain of Liu and Ma ([181]):

$$\alpha'_n(a, k) = \frac{(a^2q/k; q^2)_n}{(kq; q^2)_n}\left(\frac{-k}{a}\right)^n \alpha_n\left(a, \frac{a^2}{k}\right),$$

$$\beta'_n(a, k) = \sum_{j=0}^{n/2} \frac{\left(1 - \frac{a^2q^{2n-4j}}{k}\right)(k; q^2)_{n-j}\left(\frac{k^2}{a^2}; q^2\right)_j}{\left(1 - \frac{a^2}{k}\right)\left(\frac{a^2q^2}{k}; q^2\right)_{n-j}(q^2; q^2)_j}\left(\frac{-k}{a}\right)^{n-2j} \beta_{n-2j}\left(a, \frac{a^2}{k}\right).$$

$$\text{(II.7)}$$

The WP-Bailey chains of Mc Laughlin and Zimmer ([201]):

$$\alpha'_n(a,k,q) = \frac{\left(\frac{a^2}{k};q\right)_{2n}}{(kq;q)_{2n}} \left(\frac{k^2q}{a^2}\right)^n \alpha_n\left(a,\frac{a^2}{kq},q\right),$$

$$\beta'_n(a,k,q) = \frac{1-k}{1-kq^{2n}} \sum_{j=0}^{n} \frac{\left(1-\frac{a^2q^{2j}}{kq}\right)}{\left(1-\frac{a^2}{kq}\right)} \frac{\left(\frac{k^2q}{a^2};q\right)_{n-j}}{(q;q)_{n-j}} \left(\frac{k^2q}{a^2}\right)^j \beta_j\left(a,\frac{a^2}{kq},q\right).$$

$$\tag{II.8}$$

$$\alpha'_n(a,k,q) = \frac{1+a}{1+aq^{2n}} q^n \alpha_n\left(a^2,\frac{ak}{q},q^2\right),$$

$$\beta'_n(a,k,q) = \sum_{j=0}^{n} \frac{\left(1-\frac{ak}{q}q^{4j}\right)}{\left(1-\frac{ak}{q}\right)} \frac{(-a;q)_{2j}}{(-k;q)_{2j}} \frac{\left(\frac{qk}{a};q^2\right)_{n-j}}{(q^2;q^2)_{n-j}} \frac{(k^2;q^2)_{n+j}}{(akq;q^2)_{n+j}}$$

$$\times q^j \beta_j\left(a^2,\frac{ak}{q},q^2\right).$$

$$\tag{II.9}$$

$$\alpha'_n(a,k,q^2) = \alpha_n\left(\sqrt{a},\frac{k}{\sqrt{a}},q\right),$$

$$\beta'_n(a,k,q^2) = \frac{\left(\frac{-k}{\sqrt{a}};q\right)_{2n}}{(-q\sqrt{a};q)_{2n}} \sum_{j=0}^{n} \frac{\left(1-\frac{k^2q^{4j}}{a}\right)}{\left(1-\frac{k^2}{a}\right)} \frac{\left(\frac{a}{k};q^2\right)_{n-j}}{(q^2;q^2)_{n-j}} \frac{(k;q^2)_{n+j}}{\left(\frac{k^2q^2}{a};q^2\right)_{n+j}}$$

$$\times \left(\frac{kq}{a}\right)^{n-j} \beta_j\left(\sqrt{a},\frac{k}{\sqrt{a}},q\right).$$

$$\tag{II.10}$$

Appendix III: WP-Bailey pairs

The unit WP-Bailey pair:

$$\alpha_n(a,k) = \frac{(1-aq^{2n})(a,\frac{a}{k};q)_n}{(1-a)(q,kq;q)_n}\left(\frac{k}{a}\right)^n \qquad \beta_n(a,k) = \delta_{n,0}. \qquad \text{(III.0)}$$

The WP-Bailey pairs (III.1)–(III.10) arise from inserting the unit WP-Bailey pair in the corresponding WP-Bailey chains in Appendix II, but some may have been found previously by other methods.

$$\alpha_n(a,k) = \frac{(1-aq^{2n})\left(a,\rho_1,\rho_2,\frac{a^2q}{k\rho_1\rho_2};q\right)_n}{(1-a)\left(\frac{aq}{\rho_1},\frac{aq}{\rho_2},\frac{k\rho_1\rho_2}{a},q;q\right)_n}\left(\frac{k}{a}\right)^n,$$

$$\beta_n(a,k) = \frac{\left(k,\frac{k\rho_1}{a},\frac{k\rho_2}{a},\frac{aq}{\rho_1\rho_2};q\right)_n}{\left(\frac{aq}{\rho_1},\frac{aq}{\rho_2},\frac{k\rho_1\rho_2}{a},q;q\right)_n}. \qquad \text{(III.1)}$$

$$\alpha_n(a,k) = \frac{(1-aq^{2n})\left(a,\frac{k}{aq};q\right)_n\left(\frac{a^2q}{k};q\right)_{2n}}{(1-a)\left(\frac{a^2q^2}{k},q;q\right)_n(k;q)_{2n}}\left(\frac{k}{a}\right)^n,$$

$$\beta_n(a,k) = \frac{\left(\frac{k^2}{a^2q};q\right)_n}{(q;q)_n}. \qquad \text{(III.2)}$$

$$\alpha_n(a,k) = \frac{(1-aq^{2n})\left(1-\sigma_2\sqrt{k}\right)\left(1+\sigma_2\frac{a}{\sqrt{k}}q^n\right)\left(a,\frac{k}{a};q\right)_n\left(\frac{a^2}{k};q\right)_{2n}}{(1-a)\left(1-\sigma_2\sqrt{k}q^n\right)\left(1+\sigma_2\frac{a}{\sqrt{k}}\right)\left(\frac{a^2q}{k},q;q\right)_n(k;q)_{2n}}\left(\frac{k}{a}\right)^n,$$

336

$$\beta_n(a,k) = \frac{\left(1 - \sigma_2\sqrt{k}\right)\left(\frac{k^2}{a^2};q\right)_n}{\left(1 - \sigma_2\sqrt{k}q^n\right)(q;q)_n}, \qquad \text{where } \sigma_2 \in \{-1,1\}. \tag{III.3}$$

$$\alpha_n(a,k) = \frac{(1 - \sqrt{a}q^n)\left(\frac{a\sqrt{q}}{k}, \sqrt{a}; \sqrt{q}\right)_n}{(1 - \sqrt{a})\left(\frac{k}{\sqrt{a}}, \sqrt{q}; \sqrt{q}\right)_n}\left(\frac{k}{a\sqrt{q}}\right)^n,$$

$$\beta_n(a,k) = \frac{\left(k, -\frac{k}{\sqrt{a}}, -\frac{k\sqrt{q}}{\sqrt{a}}, \frac{aq}{k}; q\right)_n}{\left(-\sqrt{aq}, -\sqrt{aq}, \frac{k^2}{a}, q; q\right)_n}\left(\frac{k}{a\sqrt{q}}\right)^n. \tag{III.4}$$

$$\alpha_n(a,k) = \frac{\left(1 - aq^{2n}\right)\left(\sqrt{a}, \frac{a}{k}; \sqrt{q}\right)_n}{(1 - a)\left(\frac{k\sqrt{q}}{\sqrt{a}}, \sqrt{q}; \sqrt{q}\right)_n}\left(\frac{k}{a\sqrt{q}}\right)^n,$$

$$\beta_n(a,k) = \frac{\left(k, \frac{a}{k}, -\frac{k\sqrt{q}}{\sqrt{a}}, -\frac{kq}{\sqrt{a}}; q\right)_n}{\left(-\sqrt{a}, -\sqrt{aq}, \frac{k^2q}{a}, q; q\right)_n}\left(\frac{k}{a\sqrt{q}}\right)^n. \tag{III.5}$$

$$\alpha_{2n}(a,k) = \frac{\left(1 - aq^{4n}\right)\left(a, \frac{a^2}{k^2}; q^2\right)_n}{(1 - a)\left(\frac{k^2q^2}{a}, q^2; q^2\right)_n}\left(\frac{k^2}{a^2}\right)^n, \qquad \alpha_{2n+1}(a,k) = 0,$$

$$\beta_n(a,k) = \frac{\left(k, \frac{a}{k}; q\right)_n\left(\frac{k^2q}{a}; q^2\right)_n}{\left(\frac{k^2q}{a}, q; q\right)_n(aq; q^2)_n}\left(-\frac{k}{a}\right)^n. \tag{III.6}$$

$$\alpha_n(a,k) = \frac{(-1)^n\left(1 - aq^{2n}\right)\left(a, \frac{k}{a}; q\right)_n\left(\frac{a^2q}{k}; q^2\right)_n}{(1 - a)\left(\frac{a^2q}{k}, q; q\right)_n(kq; q^2)_n},$$

$$\beta_{2n}(a,k) = \frac{\left(k, \frac{k^2}{a^2}; q^2\right)_n}{\left(\frac{a^2q^2}{k}, q^2; q^2\right)_n}, \qquad \beta_{2n+1}(a,k) = 0. \tag{III.7}$$

$$\alpha_n(a,k) = \frac{\left(1 - aq^{2n}\right)\left(a, \frac{kq}{a}; q\right)_n\left(\frac{a^2}{k}; q\right)_{2n}}{(1 - a)\left(\frac{a^2}{k}, q; q\right)_n(kq; q)_{2n}}\left(\frac{k}{a}\right)^n,$$

$$\beta_n(a,k) = \frac{(1-k)\left(\frac{k^2q}{a^2};q\right)_n}{(1-kq^{2n})(q;q)_n}. \tag{III.8}$$

$$\alpha_n(a,k) = \frac{(1-aq^{2n})\left(a^2,\frac{aq}{k};q^2\right)_n}{(1-a)(akq,q^2;q^2)_n}\left(\frac{k}{a}\right)^n,$$

$$\beta_n(a,k) = \frac{\left(k^2,\frac{kq}{a};q^2\right)_n}{(akq,q^2;q^2)_n}. \tag{III.9}$$

$$\alpha_n(a,k) = \frac{(1-\sqrt{a}q^n)\left(\sqrt{a},\frac{a}{k};\sqrt{q}\right)_n}{(1-\sqrt{a})\left(\frac{k\sqrt{q}}{\sqrt{a}},\sqrt{q};\sqrt{q}\right)_n}\left(\frac{k}{a}\right)^n,$$

$$\beta_n(a,k) = \frac{\left(k,\frac{a}{k},-\frac{k}{\sqrt{a}},-\frac{k\sqrt{q}}{\sqrt{a}};q\right)_n}{\left(-\sqrt{aq},-\sqrt{a}q,\frac{k^2q}{a},q;q\right)_n}\left(\frac{k\sqrt{q}}{a}\right)^n. \tag{III.10}$$

The WP-Bailey pairs (III.11)–(III.16) arise through applying Corollary 13.1 to the pairs (III.1)–(III.10) (only six new pairs arise in this manner), but, as with the pairs (III.1)–(III.10), some may also have been found previously by other methods.

$$\alpha_n(a,k) = \frac{(1-aq^{2n})\left(\frac{a^2}{k^2q};q\right)_n}{(1-a)(q;q)_n}\left(\frac{k}{a}\right)^n,$$

$$\beta_n(a,k) = \frac{\left(k,\frac{a}{kq};q\right)_n\left(\frac{k^2q}{a};q\right)_{2n}}{\left(\frac{k^2q^2}{a},q;q\right)_n(a;q)_{2n}}. \tag{III.11}$$

$$\alpha_n(a,k) = \frac{(1+\sigma_2\sqrt{a}q^n)\left(\frac{a^2}{k^2};q\right)_n}{(1+\sigma_2\sqrt{a})(q;q)_n}\left(\frac{k}{a}\right)^n,$$

$$\beta_n(a,k) = \frac{(1-\sigma_2\sqrt{a})\left(1+\sigma_2\frac{k}{\sqrt{a}}q^n\right)\left(k,\frac{a}{k};q\right)_n\left(\frac{k^2}{a};q\right)_{2n}}{(1-\sigma_2\sqrt{a}q^n)\left(1+\sigma_2\frac{k}{\sqrt{a}}\right)\left(\frac{k^2q}{a},q;q\right)_n(a;q)_{2n}},$$

where $\sigma_2\in\{-1,1\}$. \hfill (III.12)

$$\alpha_n(a,k) = \frac{(1-aq^{2n})\left(a,-\frac{a}{\sqrt{k}},-\frac{a\sqrt{q}}{\sqrt{k}},\frac{kq}{a};q\right)_n}{(1-a)\left(-\sqrt{kq},-\sqrt{k}q,\frac{a^2}{k},q;q\right)_n}\left(\frac{1}{\sqrt{q}}\right)^n,$$

$$\beta_n(a,k) = \frac{\left(1+\sqrt{k}\right)\left(\sqrt{k}, \frac{k\sqrt{q}}{a}; \sqrt{q}\right)_n}{\left(1+\sqrt{k}q^n\right)\left(\frac{a}{\sqrt{k}}, \sqrt{q}; \sqrt{q}\right)_n}\left(\frac{1}{\sqrt{q}}\right)^n. \tag{III.13}$$

$$\alpha_n(a,k) = \frac{\left(1-aq^{2n}\right)\left(a, \frac{k}{a}, -\frac{a\sqrt{q}}{\sqrt{k}}, -\frac{aq}{\sqrt{k}}; q\right)_n}{\left(1-a\right)\left(-\sqrt{k}, -\sqrt{k}q, \frac{a^2q}{k}, q; q\right)_n}\left(\frac{1}{\sqrt{q}}\right)^n,$$

$$\beta_n(a,k) = \frac{\left(\sqrt{k}, \frac{k}{a}; \sqrt{q}\right)_n}{\left(\frac{a\sqrt{q}}{\sqrt{k}}, \sqrt{q}; \sqrt{q}\right)_n}\left(\frac{1}{\sqrt{q}}\right)^n. \tag{III.14}$$

$$\alpha_n(a,k) = \frac{\left(\frac{a^2q}{k^2}; q\right)_n}{(q;q)_n}\left(\frac{k}{a}\right)^n,$$

$$\beta_n(a,k) = \frac{\left(\frac{aq}{k}, k; q\right)_n\left(\frac{k^2}{a}; q\right)_{2n}}{\left(\frac{k^2}{a}, q; q\right)_n(aq; q)_{2n}}. \tag{III.15}$$

$$\alpha_n(a,k) = \frac{\left(1-aq^{2n}\right)\left(a, -\frac{a}{\sqrt{k}}, \frac{k}{a}, -\frac{a\sqrt{q}}{\sqrt{k}}; q\right)_n}{\left(1-a\right)\left(-\sqrt{kq}, -\sqrt{k}q, \frac{a^2q}{k}, q; q\right)_n}\left(\frac{1}{\sqrt{q}}\right)^n,$$

$$\beta_n(a,k) = \frac{\left(1+\sqrt{k}\right)\left(\sqrt{k}, \frac{k}{a}; \sqrt{q}\right)_n}{\left(1+\sqrt{k}q^n\right)\left(\frac{a\sqrt{q}}{\sqrt{k}}, \sqrt{q}; \sqrt{q}\right)_n}. \tag{III.16}$$

Appendix IV: Bailey Chains

If $(\alpha_n(a,q), \beta_n(a,q))$ is a Bailey pair with respect to a and q (in other words, they satisfy (10.5)), then so is the pair $(\alpha'_n(a,q), \beta'_n(a,q))$, where $(\alpha'_n(a,q), \beta'_n(a,q))$ is defined by any of the following pairs of relations.

The Bailey chain of Andrews ([24, 51])

$$\alpha'_n(a,q) = \frac{(\rho_1, \rho_2; q)_n}{(aq/\rho_1, aq/\rho_2; q)_n} \left(\frac{aq}{\rho_1\rho_2}\right)^n \alpha_n(a,q),$$

$$\beta'_n(a,q) = \frac{1}{(aq/\rho_1, aq/\rho_2; q)_n} \sum_{j=0}^{n} \frac{(\rho_1, \rho_2; q)_j (aq/\rho_1\rho_2; q)_{n-j}}{(q; q)_{n-j}} \left(\frac{aq}{\rho_1\rho_2}\right)^j \beta_j(a,q).$$

$$\text{(IV.1)}$$

The Bailey chains of Bressoud, Ismail and Stanton ([85])

$$\alpha'_n(a,q) = \frac{(-B; q)_n}{(-aq/B; q)_n} B^{-n} q^{-\binom{n}{2}} \alpha_n(a^2, q^2),$$

$$\beta'_n(a,q) = \sum_{k=0}^{n} \frac{(-aq; q)_{2k} (B^2; q^2)_k (q^{-k}/B, Bq^{k+1}; q)_{n-k}}{(-aq/B, B; q)_n (q^2; q^2)_{n-k}} B^{-k} q^{-\binom{k}{2}} \beta_k(a^2, q^2).$$

$$\text{(IV.2)}$$

$$\alpha'_n(a^4, q^4) = \frac{(-Bq; q^2)_n}{(-qa^2/B; q^2)_n} a^{2n} B^{-n} q^{n^2} \alpha_n(a^2, q^2),$$

$$\beta'_n(a^4, q^4) = \sum_{k=0}^{n} \frac{(qa^2/B; q^2)_{2n-k} (-Bq; q^2)_k a^{2k} B^{-k} q^{k^2}}{(-q^2 a^2; q^2)_{2n} (a^4 q^2/B^2; q^4)_n (q^4; q^4)_{n-k}} \beta_k(a^2, q^2). \quad \text{(IV.3)}$$

$$\alpha'_n(a^3, q^3) = a^n q^{n^2} \alpha_n(a,q),$$

$$\beta'_n(a^3, q^3) = \frac{1}{(a^3 q^3; q^3)_{2n}} \sum_{k=0}^{n} \frac{(aq; q)_{3n-k} a^k q^{k^2}}{(q^3; q^3)_{n-k}} \beta_k(a,q). \quad \text{(IV.4)}$$

$$\alpha_n'(a,q) = a^{-n}q^{-n^2}\alpha_n(a^3,q^3),$$

$$\beta_n'(a,q) = \frac{1}{(aq;q)_{2n}} \sum_{k=0}^{n} \frac{(aq^{2n+1};q^{-1})_{3k}(a^3q^3;q^3)_{2(n-k)}}{(q^3;q^3)_k}$$

$$\times (-1)^k q^{3\binom{k}{2}-n^2} a^{-n} \beta_{n-k}(a^3,q^3). \tag{IV.5}$$

$$\alpha_n'(a,q) = \frac{(-B;q)_n}{(-aq/B;q)_n} \frac{1+a}{1+aq^{2n}} B^{-n}q^{n-\binom{n}{2}}\alpha_n(a^2,q^2),$$

$$\beta_n'(a,q) = \sum_{k=0}^{n} \frac{(-a;q)_{2k}(B^2;q^2)_k(q^{-k+1}/B,Bq^k;q)_{n-k}}{(-aq/B,B;q)_n(q^2;q^2)_{n-k}} B^{-k}q^{k-\binom{k}{2}}\beta_k(a^2,q^2).$$

$$\tag{IV.6}$$

The Bailey chains of Berkovich and Warnaar ([68])

$$\alpha_{2n}'(a;q) = (-1)^n b^n q^{n^2} \frac{(aq/b;q^2)_n}{(bq;q^2)_n}\alpha_n(a;q^2), \qquad \alpha_{2n+1}'(a;q) = 0,$$

$$\beta_n'(a;q) = \frac{(b;q^2)_n}{(q,b;q)_n(aq;q^2)_n} \sum_{r=0}^{\lfloor n/2 \rfloor} \frac{(aq/b;q^2)_r(q^{-n};q)_{2r}q^{2r}}{(q^{2-2n}/b;q^2)_r}\beta_r(a;q^2). \tag{IV.7}$$

$$\alpha_{2n}'(a;q) = \alpha_n(a^2;q^4), \qquad \alpha_{2n+1}'(a;q) = 0,$$

$$\beta_n'(a;q) = \frac{q^n(-q^{-1};q^2)_n}{(q^2,aq;q^2)_n} \sum_{r=0}^{\lfloor n/2 \rfloor} \frac{(-aq^2,q^{-2n};q^2)_{2r}q^{4r}}{(-q^{3-2n};q^2)_{2r}}\beta_r(a^2;q^4). \tag{IV.8}$$

$$\alpha_{2n}'(a;q) = q^{2n}\frac{1+a}{1+aq^{4n}}\alpha_n(a^2;q^4), \qquad \alpha_{2n+1}'(a;q) = 0,$$

$$\beta_n'(a;q) = \frac{(-q;q^2)_n}{(q^2,aq;q^2)_n} \sum_{r=0}^{\lfloor n/2 \rfloor} \frac{(-a,q^{-2n};q^2)_{2r}q^{4r}}{(-q^{1-2n};q^2)_{2r}}\beta_r(a^2;q^4). \tag{IV.9}$$

$$\alpha_{3n}'(a;q) = a^n q^{3n^2}\alpha_n(a;q^3), \qquad \alpha_{3n\pm1}'(a;q) = 0,$$

$$\beta_n'(a;q) = \frac{(a;q^3)_n}{(q;q)_n(a;q)_{2n}} \sum_{r=0}^{\lfloor n/3 \rfloor} \frac{(q^{-n};q)_{3r}q^{3r}}{(q^{3-3n}/a;q^3)_r}\beta_r(a;q^3). \tag{IV.10}$$

$$\alpha'_{2n+1}(a;q) = (-1)^n b^n q^{n^2} \frac{(aq^3/b;q^2)_n}{(bq;q^2)_n} \alpha_n(aq^2;q^2), \qquad \alpha'_{2n}(a;q) = 0,$$

$$\beta'_{n+1}(a;q) = \frac{(b;q^2)_n}{(q,b;q)_n(aq^3;q^2)_n} \sum_{r=0}^{\lfloor n/2 \rfloor} \frac{(aq^3/b;q^2)_r(q^{-n};q)_{2r}q^{2r}}{(aq;q)_2(q^{2-2n}/b;q^2)_r} \beta_r(aq^2;q^2).$$

$$\text{(IV.11)}$$

$$\alpha'_n(a;q^2) = (-1)^n b^{-n} q^{-n^2} \frac{(bq;q^2)_n}{(aq/b;q^2)_n} \alpha_{2n}(a;q),$$

$$\beta'_n(a;q^2) = \frac{(1/b;q^2)_n}{(q;q)_{2n}(aq/b;q^2)_n} \sum_{r=0}^{2n} \frac{(aq;q^2)_r(bq,q^{-2n};q)_r q^r}{(bq^{2-2n};q^2)_r} \beta_r(a;q). \quad \text{(IV.12)}$$

$$\alpha'_n(a^2;q^4) = \alpha_{2n}(a;q),$$

$$\beta'_n(a^2;q^4) = \frac{(-q;q^2)_{2n}}{(q^2,-aq^2;q^2)_{2n}} \sum_{r=0}^{2n} \frac{(aq,q^{-4n};q^2)_r q^r}{(-q^{1-4n};q^2)_r} \beta_r(a;q). \qquad \text{(IV.13)}$$

$$\alpha'_n(a^2;q^4) = q^{-2n} \frac{1+aq^{4n}}{1+a} \alpha_{2n}(a;q),$$

$$\beta'_n(a^2;q^4) = \frac{(-q^{-1};q^2)_{2n}}{(q^2,-a;q^2)_{2n}} \sum_{r=0}^{2n} \frac{(aq,q^{-4n};q^2)_r q^{2r}}{(-q^{3-4n};q^2)_r} \beta_r(a;q). \qquad \text{(IV.14)}$$

$$\alpha'_n(a;q^3) = a^{-n} q^{-3n^2} \alpha_{3n}(a;q),$$

$$\beta'_n(a;q^3) = \frac{(1/a;q^3)_n}{(q;q)_{3n}} \sum_{r=0}^{3n} \frac{(aq;q)_{2r}(q^{-3n};q)_r q^r}{(aq^{3-3n};q^3)_r} \beta_r(a;q). \qquad \text{(IV.15)}$$

Appendix V: Bailey Pairs

This list will not contain every Bailey pair listed by Slater [236, 237], but instead will contain general Bailey pairs from which the specific Bailey pairs listed by Slater are derived. The first pair was stated by Slater [236, Eq. (4.1)], while the following collections of mod 3, mod 2 and mod 4 pairs (with respect to the stated values of a) are from [197] and were derived using Slater's methods (Slater could easily have derived these more general pairs, but she was primarily interested in Bailey pairs that led to Rogers–Ramanujan type identities).

$$\alpha_n = \frac{(1 - aq^{2n})(a, b, c; q)_n}{(1 - a)(aq/b, aq/c, q; q)} \left(\frac{-a}{bc}\right)^n q^{(n^2+n)/2}, \qquad (V.1.1)$$

$$\beta_n = \frac{(aq/bc; q)_n}{(aq/b, aq/c, q; q)_n}, \quad \text{with respect to } a = a.$$

Note that all the Bailey pairs in Slater's **B**, **F** and **H** tables, as well as pairs **E(3)**, **E(6)** and **E(7)** (see [236, page 468]), are derived from the Bailey pair at (V.1.1).

Mod 3 Bailey Pairs

$$\alpha_{3r} = \frac{1 - aq^{6r}}{1 - a} \frac{(a; q^3)_r}{(q^3; q^3)_r} q^{\frac{9r^2-3r}{2}} (-a)^r, \ \alpha_{3r\pm1} = 0, \qquad (V.3.1)$$

$$\beta_n = \frac{(a; q^3)_n}{(a; q)_{2n}(q; q)_n}, \quad \text{with respect to } a = a.$$

$$\alpha_{3r} = \frac{(aq; q^3)_r}{(q^3; q^3)_r} q^{\frac{9r^2-r}{2}} (-a)^r, \ \alpha_{3r+1} = \frac{(aq; q^3)_r}{(q^3; q^3)_r} q^{\frac{9r^2+11r}{2}+1} (-a)^{r+1}, \quad (V.3.2)$$

$$\alpha_{3r-1} = 0, \ \beta_n = \frac{(aq; q^3)_n}{(aq; q)_{2n}(q; q)_n}, \quad \text{with respect to } a = a.$$

343

$$\alpha_{3r} = \frac{(aq;q^3)_r}{(q^3;q^3)_r} q^{\frac{9r^2+5r}{2}}(-a)^r, \quad \alpha_{3r+1} = -\frac{(aq;q^3)_r}{(q^3;q^3)_r} q^{\frac{9r^2+5r}{2}}(-a)^r, \quad \text{(V.3.3)}$$

$$\alpha_{3r-1} = 0, \quad \beta_n = \frac{(aq;q^3)_n q^n}{(aq;q)_{2n}(q;q)_n}, \quad \text{with respect to } a = a.$$

$$\alpha_{3r} = (-1)^r \left(\frac{q^{9r^2/2+r/2}(e;q^3)_r}{(q^4/e;q^3)_r e^r} + \frac{q^{9r^2/2+7r/2}(e/q;q^3)_r}{(q^3/e;q^3)_r e^r} \right), \quad \text{(V.3.4)}$$

$$\alpha_{3r+1} = \frac{(-1)^{r+1}q^{9r^2/2+13r/2+1}(e;q^3)_r}{(q^4/e;q^3)_r e^r},$$

$$\alpha_{3r-1} = \frac{(-1)^{r+1}q^{9r^2/2-5r/2+1}(e/q;q^3)_r}{(q^3/e;q^3)_r e^r},$$

$$\beta_n = \frac{(q^2/e;q^3)_n}{(q;q)_{2n}(q^2/e;q)_n}, \quad \text{with respect to } a = 1.$$

$$\alpha_{3r} = (-1)^r \left(\frac{q^{9r^2/2+7r/2}(e;q^3)_r}{(q^4/e;q^3)_r e^r} + \frac{q^{9r^2/2+r/2}(e/q;q^3)_r}{(q^3/e;q^3)_r e^r} \right), \quad \text{(V.3.5)}$$

$$\alpha_{3r+1} = \frac{(-1)^{r+1}q^{9r^2/2+7r/2}(e;q^3)_r}{(q^4/e;q^3)_r e^r}, \quad \alpha_{3r-1} = \frac{(-1)^{r+1}q^{9r^2/2+r/2}(e/q;q^3)_r}{(q^3/e;q^3)_r e^r},$$

$$\beta_n = \frac{q^n(q^2/e;q^3)_n}{(q;q)_{2n}(q^2/e;q)_n}, \quad \text{with respect to } a = 1.$$

$$\alpha_{3r} = (-1)^r \frac{1-q^{6r+1}}{1-q} q^{\frac{9r^2+r}{2}} \frac{(e;q^3)_r}{(q^4/e;q^3)_r \, e^r}, \quad \alpha_{3r+1} = 0, \quad \text{(V.3.6)}$$

$$\alpha_{3r-1} = (-1)^{r+1}\frac{1-q^{6r-1}}{1-q} q^{\frac{9r^2-5r}{2}+1} \frac{(e/q;q^3)_r}{(q^3/e;q^3)_r \, e^r},$$

$$\beta_n = \frac{(q^2/e;q^3)_n}{(q;q)_{2n}(q^2/e;q)_n}, \quad \text{with respect to } a = q.$$

$$\alpha_{3r} = \frac{(-1)^r q^{9r^2/2+5r/2}(e;q^3)_r}{(q^5/e;q^3)_r \, e^r}, \quad \text{(V.3.7)}$$

$$\alpha_{3r-1} = \frac{(-1)^r q^{9r^2/2+5r/2}\left(\frac{e}{q^2};q^3\right)_r}{(q^3/e;q^3)_r e^r},$$

$$\alpha_{3r+1} = \frac{(-1)^r q^{9r^2/2+11r/2+3}(e/q^2;q^3)_{r+1}}{(q^3/e;q^3)_{r+1}e^{r+1}} + \frac{(-1)^{r+1}q^{9r^2/2+17r/2+2}(e;q^3)_r}{(q^5/e;q^3)_r e^r},$$

$$\beta_n = \frac{(q^4/e;q^3)_n}{(q^2;q)_{2n}(q^3/e;q)_n}, \quad \text{with respect to } a = q.$$

$$\alpha_{3r} = \frac{(-1)^r q^{9r^2/2+11r/2}(e;q^3)_r}{(q^5/e;q^3)_r e^r}, \tag{V.3.8}$$

$$\alpha_{3r-1} = \frac{(-1)^r q^{9r^2/2-r/2}(e/q^2;q^3)_r}{(q^3/e;q^3)_r e^r},$$

$$\alpha_{3r+1} = \frac{(-1)^r q^{9r^2/2+17r/2+4}(e/q^2;q^3)_{r+1}}{(q^3/e;q^3)_{r+1}e^{r+1}} + \frac{(-1)^{r+1}q^{9r^2/2+11r/2}(e;q^3)_r}{(q^5/e;q^3)_r e^r},$$

$$\beta_n = \frac{q^n(q^4/e;q^3)_n}{(q^2;q)_{2n}(q^3/e;q)_n}, \quad \text{with respect to } a = q.$$

$$\alpha_{3r} = (-1)^r \frac{1-q^{6r+2}}{1-q^2} q^{\frac{9r^2+5r}{2}} \frac{(e;q^3)_r}{(q^5/e;q^3)_r e^r}, \quad \alpha_{3r-1} = 0, \tag{V.3.8}$$

$$\alpha_{3r+1} = (-1)^r \frac{1-q^{6r+4}}{1-q^2} q^{\frac{9r^2+11r}{2}+3} \frac{(e/q^2;q^3)_{r+1}}{(q^3/e;q^3)_{r+1} e^{r+1}},$$

$$\beta_n = \frac{(q^4/e;q^3)_n}{(q^2;q)_{2n}(q^3/e;q)_n} \quad \text{with respect to } a = q^2.$$

All eight pairs in Slater's **A** table ([236, page 463]) and all six on her **J** list ([237, pp. 148–149]) are derived from the five Bailey pairs (V.3.1), (V.3.4), (V.3.5), (V.3.7), (V.3.8) above, for particular values of e.

Mod 2 Bailey Pairs

$$\alpha_{2r} = \frac{1-aq^{4r}}{1-a} \frac{(a,d;q^2)_r}{(aq^2/d,q^2;q^2)_r} \frac{a^r q^{2r^2}}{d^r}, \quad \alpha_{2r-1} = 0, \tag{V.2.1}$$

$$\beta_n = \frac{(aq/d;q^2)_n}{(aq;q^2)_n(aq/d,q;q)_n}, \quad \text{with respect to } a = a.$$

$$\alpha_{2r} = \frac{(aq,d;q^2)_r q^{2r^2+r}a^r}{(aq^3/d,q^2;q^2)_r d^r}, \quad \alpha_{2r+1} = -\frac{(aq,d;q^2)_r q^{2r^2+5r+1}a^{r+1}}{(aq^3/d,q^2;q^2)_r d^r}, \tag{V.2.2}$$

$$\beta_n = \frac{(aq^2/d; q^2)_n}{(aq^2; q^2)_n(aq^2/d, q; q)_n} \quad \text{with respect to } a = a.$$

$$\alpha_{2r} = \frac{(aq, d; q^2)_r q^{2r^2+3r} a^r}{(aq^3/d, q^2; q^2)_r d^r}, \quad \alpha_{2r+1} = -\frac{(aq, d; q^2)_r q^{2r^2+3r} a^r}{(aq^3/d, q^2; q^2)_r d^r}, \quad \text{(V.2.3)}$$

$$\beta_n = \frac{(aq^2/d; q^2)_n q^n}{(aq^2; q^2)_n(aq^2/d, q; q)_n}, \quad \text{with respect to } a = a.$$

$$\alpha_{2r} = \frac{(1 - q^{4r+1})q^{2r^2+2r}(d; q^2)_r(e; q^2)_r}{(1 - q)(q^3/d; q^2)_r(q^3/e; q^2)_r d^r e^r}, \quad \text{(V.2.4)}$$

$$\alpha_{2r-1} = -\frac{(1 - q^{4r-1})q^{2r^2+1}(d/q; q^2)_r(e/q; q^2)_r}{(1 - q)(q^2/d; q^2)_r(q^2/e; q^2)_r d^r e^r},$$

$$\beta_n = \frac{(q^3/de; q^2)_n}{(q^2; q^2)_n(q^2/d; q)_n(q^2/e; q)_n}, \quad \text{with respect to } a = q.$$

$$\alpha_{2r} = \frac{q^{2r^2+4r}(d/q; q^2)_r(e/q; q^2)_r}{(q^2/d; q^2)_r(q^2/e; q^2)_r d^r e^r} + \frac{q^{2r^2+2r}(d; q^2)_r(e; q^2)_r}{(q^3/d; q^2)_r(q^3/e; q^2)_r d^r e^r}, \quad \text{(V.2.5)}$$

$$\alpha_{2r+1} = -\frac{q^{2r^2+4r+3}(d/q; q^2)_{r+1}(e/q; q^2)_{r+1}}{(q^2/d; q^2)_{r+1}(q^2/e; q^2)_{r+1} d^{r+1} e^{r+1}} - \frac{q^{2r^2+6r+1}(d; q^2)_r(e; q^2)_r}{(q^3/d; q^2)_r(q^3/e; q^2)_r d^r e^r},$$

$$\beta_n = \frac{(q^3/de; q^2)_n}{(q^2; q^2)_n(q^2/d; q)_n(q^2/e; q)_n}, \quad \text{with respect to } a = 1.$$

$$\alpha_{2r} = \frac{q^{2r^2+2r}(d/q; q^2)_r(e/q; q^2)_r}{(q^2/d; q^2)_r(q^2/e; q^2)_r d^r e^r} + \frac{q^{2r^2+4r}(d; q^2)_r(e; q^2)_r}{(q^3/d; q^2)_r(q^3/e; q^2)_r d^r e^r}, \quad \text{(V.2.6)}$$

$$\alpha_{2r+1} = -\frac{q^{2r^2+6r+4}(d/q; q^2)_{r+1}(e/q; q^2)_{r+1}}{(q^2/d; q^2)_{r+1}(q^2/e; q^2)_{r+1} d^{r+1} e^{r+1}} - \frac{q^{2r^2+4r}(d; q^2)_r(e; q^2)_r}{(q^3/d; q^2)_r(q^3/e; q^2)_r d^r e^r},$$

$$\beta_n = \frac{q^n(q^3/de; q^2)_n}{(q^2; q^2)_n(q^2/d; q)_n(q^2/e; q)_n}, \quad \text{with respect to } a = 1.$$

The Bailey pairs in Slater's **G-**, **C-** and **I** tables, as well as pairs **E(1)**, **E(2)**, **E(4)** and **E(5)** (see [236, pages 469 and 470]), are derived from the six mod 2 Bailey pairs above.

Mod 4 Bailey Pairs

The next six Bailey pairs (with $\alpha_0 = \beta_0 = 1$ in each case) are to be found in Slater's **K** table [236, page 471].

$$\alpha_{4r} = q^{8r^2-2r} + q^{8r^2+2r}, \ \alpha_{4r+1} = -q^{8r^2+6r+1}, \ \alpha_{4r-1} = -q^{8r^2-6r+1}, \ \ (\text{V.4.1})$$

$$\alpha_{4r-2} = 0, \qquad \beta_n = \frac{(-q^2;q^2)_{n-1}}{(q;q)_{2n}} \ \text{with respect to } a = 1.$$

$$\alpha_{4r} = q^{8r^2+2r} + q^{8r^2-2r}, \ \alpha_{4r+1} = -q^{8r^2+2r}, \ \ \alpha_{4r-1} = -q^{8r^2-2r} \ \ (\text{V.4.2})$$

$$\alpha_{4r-2} = 0, \qquad \beta_n = \frac{q^n \, (-q^2;q^2)_{n-1}}{(q;q)_{2n}} \ \text{with respect to } a = 1.$$

$$\alpha_{4r} = q^{8r^2}, \ \ \alpha_{4r+1} = -q^{8r^2+8r+2}, \ \ \alpha_{4r-1} = q^{8r^2}, \qquad\qquad (\text{V.4.3})$$

$$\alpha_{4r-2} = -q^{8r^2-8r+2}, \qquad \beta_n = \frac{(-q;q^2)_n}{(q^2;q^2)_{2n}} \ \text{with respect to } a = q.$$

$$\alpha_{4r} = q^{8r^2+4r}, \ \ \alpha_{4r+1} = -q^{8r^2+4r}, \ \ \alpha_{4r-1} = q^{8r^2-4r}, \qquad (\text{V.4.4})$$

$$\alpha_{4r-2} = -q^{8r^2-4r}, \qquad \beta_n = \frac{q^n(-q;q^2)_n}{(q^2;q^2)_{2n}} \ \text{with respect to } a = q.$$

$$\alpha_{4r} = q^{8r^2+2r}, \ \ \alpha_{4r+1} = -q^{8r^2+10r+3} - q^{8r^2+6r+1}, \ \ \alpha_{4r-1} = 0, \ \ (\text{V.4.5})$$

$$\alpha_{4r-2} = q^{8r^2-2r}, \qquad \beta_n = \frac{(-q^2;q^2)_n}{(q^3;q)_{2n}} \ \text{with respect to } a = q^2.$$

$$\alpha_{4r} = q^{8r^2+6r}, \ \ \alpha_{4r+1} = -q^{8r^2+6r} - q^{8r^2+10r+2}, \ \ \alpha_{4r-1} = 0, \ \ (\text{V.4.6})$$

$$\alpha_{4r-2} = q^{8r^2-6r}, \qquad \beta_n = q^n\frac{(-q^2;q^2)_n}{(q^3;q)_{2n}} \ \text{with respect to } a = q^2.$$

Appendix VI: Mock Theta Functions

Second order:

$$A(q) = \sum_{n=0}^{\infty} \frac{q^{(n+1)^2}(-q;q^2)_n}{(q;q^2)_{n+1}^2}. \tag{VI.2.1}$$

$$B(q) = \sum_{n=0}^{\infty} \frac{q^{n(n+1)}(-q^2;q^2)_n}{(q;q^2)_{n+1}^2}. \tag{VI.2.2}$$

$$\mu(q) = \sum_{n=0}^{\infty} \frac{(-1)^n q^{n^2}(q;q^2)_n}{(-q^2;q^2)_n^2}. \tag{VI.2.3}$$

The functions $A(q)$ and $B(q)$ above were given by McIntosh [189], and $\mu(q)$ was given by Ramanujan in the Lost Notebook [213].

Third order:

$$f(q) = \sum_{n=0}^{\infty} \frac{q^{n^2}}{(-q,-q;q)_n}. \tag{VI.3.1}$$

$$\phi(q) = \sum_{n=0}^{\infty} \frac{q^{n^2}}{(-q^2;q^2)_n}. \tag{VI.3.2}$$

$$\chi(q) = \sum_{n=0}^{\infty} \frac{q^{n^2}(-q;q)_n}{(-q^3;q^3)_n}. \tag{VI.3.3}$$

$$\psi(q) = \sum_{n=1}^{\infty} \frac{q^{n^2}}{(q;q^2)_n}. \tag{VI.3.4}$$

$$\omega(q) = \sum_{n=0}^{\infty} \frac{q^{2n^2+2n}}{(q,q;q^2)_{n+1}}. \tag{VI.3.5}$$

$$\nu(q) = \sum_{n=0}^{\infty} \frac{q^{n^2+n}}{(-q;q^2)_{n+1}}. \tag{VI.3.6}$$

$$\rho(q) = \sum_{n=0}^{\infty} \frac{q^{2n^2+2n}(q;q^2)_{n+1}}{(q^3;q^6)_{n+1}}. \tag{VI.3.7}$$

$$\xi(q) = 1 + 2\sum_{n=1}^{\infty} \frac{q^{6n^2-6n}}{(q,q^5;q^6)_n}. \tag{VI.3.8}$$

$$\sigma(q) = \sum_{n=1}^{\infty} \frac{q^{3n^2-3n}}{(-q,-q^2;q^3)_n}. \tag{VI.3.9}$$

The first four third order mock theta functions listed above were stated by Ramanujan ([214, pp. 354–355], [73, pp. 220–223]), the next three by Watson [259], and the final two by Gordon and McIntosh [131].

Fifth order:

$$f_0(q) = \sum_{n=0}^{\infty} \frac{q^{n^2}}{(-q;q)_n}. \tag{VI.5.1}$$

$$F_0(q) = \sum_{n=0}^{\infty} \frac{q^{2n^2}}{(q;q^2)_n}. \tag{VI.5.2}$$

$$\phi_0(q) = \sum_{n=0}^{\infty} q^{n^2}(-q;q^2)_n. \tag{VI.5.3}$$

$$\psi_0(q) = \sum_{n=0}^{\infty} q^{(n+1)(n+2)/2}(-q;q)_n. \tag{VI.5.4}$$

$$\chi_0(q) = \sum_{n=0}^{\infty} \frac{q^n(q;q)_n}{(q;q)_{2n}}. \tag{VI.5.5}$$

$$f_1(q) = \sum_{n=0}^{\infty} \frac{q^{n(n+1)}}{(-q;q)_n}. \tag{VI.5.6}$$

$$F_1(q) = \sum_{n=0}^{\infty} \frac{q^{2n(n+1)}}{(q;q^2)_{n+1}}. \tag{VI.5.7}$$

$$\phi_1(q) = \sum_{n=0}^{\infty} q^{(n+1)^2}(-q;q^2)_n. \tag{VI.5.8}$$

$$\psi_1(q) = \sum_{n=0}^{\infty} q^{n(n+1)/2}(-q;q)_n. \tag{VI.5.9}$$

$$\chi_1(q) = \sum_{n=0}^{\infty} \frac{q^n(q;q)_n}{(q;q)_{2n+1}}. \tag{VI.5.10}$$

The ten fifth order mock theta functions were listed by Ramanujan in his letter to Hardy (see [214, pp. 354–355], [73, pp. 220–223]).

Sixth order:

$$\beta(q) = \sum_{n=0}^{\infty} \frac{q^{3n^2+3n+1}}{(q;q^3)_{n+1}(q^2;q^3)_{n+1}}. \tag{VI.6.1}$$

$$\gamma(q) = \sum_{n=0}^{\infty} \frac{q^{n^2}(q;q)_n}{(q^3;q^3)_n}. \tag{VI.6.2}$$

$$\phi(q) = \sum_{n=0}^{\infty} \frac{(-1)^n q^{n^2}(q;q^2)_n}{(-q;q)_{2n}}. \tag{VI.6.3}$$

$$\psi(q) = \sum_{n=0}^{\infty} \frac{(-1)^n q^{(n+1)^2}(q;q^2)_n}{(-q;q)_{2n+1}}. \tag{VI.6.4}$$

$$\rho(q) = \sum_{n=0}^{\infty} \frac{q^{n(n+1)/2}(-q;q)_n}{(q;q^2)_{n+1}}. \tag{VI.6.5}$$

$$\sigma(q) = \sum_{n=0}^{\infty} \frac{q^{(n+1)(n+2)/2}(-q;q)_n}{(q;q^2)_{n+1}}. \tag{VI.6.6}$$

$$\lambda(q) = \sum_{n=0}^{\infty} \frac{(-1)^n q^n(q;q^2)_n}{(-q;q)_n}. \tag{VI.6.7}$$

$$\mu(q) = \sum_{n=0}^{\infty} \frac{(-1)^n(q;q^2)_n}{(-q;q)_n}. \tag{VI.6.8}$$

$$\phi_-(q) = \sum_{n=1}^{\infty} \frac{q^n(-q;q)_{2n-1}}{(q;q^2)_n}. \tag{VI.6.9}$$

$$\psi_-(q) = \sum_{n=1}^{\infty} \frac{q^n(-q;q)_{2n-2}}{(q;q^2)_n}. \tag{VI.6.10}$$

The function $\beta(q)$ was stated by McIntosh [191], the next seven sixth order mock theta functions are to be found in Ramanujan's Lost Notebook [213], and the final two were discovered independently by Berndt and Chan [72] and McIntosh [191].

Seventh order:

$$\mathcal{F}_0(q) = \sum_{n=0}^{\infty} \frac{q^{n^2}(q;q)_n}{(q;q)_{2n}}. \tag{VI.7.1}$$

$$\mathcal{F}_1(q) = \sum_{n=0}^{\infty} \frac{q^{(n+1)^2}(q;q)_n}{(q;q)_{2n+1}}. \tag{VI.7.2}$$

$$\mathcal{F}_2(q) = \sum_{n=0}^{\infty} \frac{q^{n(n+1)}(q;q)_n}{(q;q)_{2n+1}}. \tag{VI.7.3}$$

The three seventh order mock theta functions were also included in Ramanujan's letter to Hardy (see [214, pp. 354–355], [73, pp. 220–223]).

Eighth order:

$$S_0(q) = \sum_{n=0}^{\infty} \frac{q^{n^2}(-q;q^2)_n}{(-q^2;q^2)_n}. \tag{VI.8.1}$$

$$S_1(q) = \sum_{n=0}^{\infty} \frac{q^{n(n+2)}(-q;q^2)_n}{(-q^2;q^2)_n}. \tag{VI.8.2}$$

$$T_0(q) = \sum_{n=0}^{\infty} \frac{q^{(n+1)(n+2)}(-q^2;q^2)_n}{(-q;q^2)_{n+1}}. \tag{VI.8.3}$$

$$T_1(q) = \sum_{n=0}^{\infty} \frac{q^{n(n+1)}(-q^2;q^2)_n}{(-q;q^2)_{n+1}}. \tag{VI.8.4}$$

$$U_0(q) = \sum_{n=0}^{\infty} \frac{q^{n^2}(-q;q^2)_n}{(-q^4;q^4)_n}. \tag{VI.8.5}$$

$$U_1(q) = \sum_{n=0}^{\infty} \frac{q^{(n+1)^2}(-q;q^2)_n}{(-q^2;q^4)_{n+1}}. \tag{VI.8.6}$$

$$V_0(q) = -1 + 2\sum_{n=0}^{\infty} \frac{q^{n^2}(-q;q^2)_n}{(q;q^2)_n}. \tag{VI.8.7}$$

$$V_1(q) = \sum_{n=0}^{\infty} \frac{q^{(n+1)^2}(-q;q^2)_n}{(q;q^2)_{n+1}}. \tag{VI.8.8}$$

The eighth order mock theta functions were discovered by Gordon and McIntosh [130].

Tenth order:

$$\phi(q) = \sum_{n=0}^{\infty} \frac{q^{n(n+1)/2}}{(q;q^2)_{n+1}}. \tag{VI.10.1}$$

$$\psi(q) = \sum_{n=0}^{\infty} \frac{q^{(n+1)(n+2)/2}}{(q;q^2)_{n+1}}. \tag{VI.10.2}$$

$$X(q) = \sum_{n=0}^{\infty} \frac{(-1)^n q^{n^2}}{(-q;q)_{2n}}. \tag{VI.10.3}$$

$$\chi(q) = \sum_{n=0}^{\infty} \frac{(-1)^n q^{(n+1)^2}}{(-q;q)_{2n+1}}. \tag{VI.10.4}$$

The tenth order mock theta functions were written down by Ramanujan in the Lost Notebook [213].

Appendix VII: Selected Summation Formulae

A large number of identities of Rogers–Ramanujan–Slater type are to be found in the literature, particularly in the two papers of Slater [236, 237]. A more recent survey of such identities may be found here [196]. We do not include any of those identities here (apart from a small number encountered in this present text and which are listed at the end of this appendix), instead focusing on other summation formulae encountered in the text, some containing one or more free parameters.

The q-binomial theorem (page 12):

$$\sum_{n=0}^{\infty} \frac{(a;q)_n}{(q;q)_n} z^n = \frac{(az;q)_\infty}{(z;q)_\infty}. \tag{VII.1}$$

Two special cases of the q-binomial theorem (page 13):

$$\sum_{n=0}^{N} \begin{bmatrix} N \\ n \end{bmatrix} (-z)^n q^{n(n-1)/2} = (z;q)_N, \tag{VII.2}$$

$$\sum_{n=0}^{\infty} \begin{bmatrix} N+n-1 \\ n \end{bmatrix} z^n = \frac{1}{(z;q)_N}, \quad |z| < 1,\ |q| < 1. \tag{VII.3}$$

Variants of the q-Chu–Vandermonde summation formula (pages 14, 15):

$$\sum_{k=0}^{n} \frac{(b, q^{-n};q)_k}{(q, q^{1-n}/a;q)_k} q^k = \frac{(ab;q)_n}{(a;q)_n}. \tag{VII.4}$$

$$\sum_{k=0}^{n} \frac{(b, q^{-n};q)_k}{(c, q;q)_k} q^k = \frac{(c/b;q)_n}{(c;q)_n} b^n. \tag{VII.5}$$

$$\sum_{k=0}^{n} \frac{(b, q^{-n}; q)_k}{(c, q; q)_k} \left(\frac{cq^n}{b}\right)^k = \frac{(c/b; q)_n}{(c; q)_n}. \tag{VII.6}$$

Two parity-dependent $_2\phi_1$ sums (pages 15, 16):

$$\sum_{k=0}^{n} \frac{(a, q^{-n}; q)_k}{(q, q^{1-n}/a; q)_k} \left(\frac{-q}{a}\right)^k = \begin{cases} \dfrac{(a^2, q; q^2)_{n/2}}{(a; q)_n}, & n \text{ even}, \\ 0, & n \text{ odd}. \end{cases} \tag{VII.7}$$

$$\sum_{k=0}^{n} \frac{(a; q)_{n-k}(a; q)_k}{(q; q)_{n-k}(q; q)_k}(-1)^k = \begin{cases} \frac{(a^2; q^2)_{n/2}}{(q^2; q^2)_{n/2}} & n \text{ even}, \\ 0, & n \text{ odd}. \end{cases} \tag{VII.8}$$

Four identities involving q-binomial coefficientss (pages 15, 16):

$$\sum_{k=0}^{n} \begin{bmatrix} M \\ k \end{bmatrix} \begin{bmatrix} N \\ n-k \end{bmatrix} q^{(M-k)(n-k)} = \begin{bmatrix} M+N \\ n \end{bmatrix}. \tag{VII.9}$$

$$\sum_{k=0}^{N} \begin{bmatrix} N \\ k \end{bmatrix}^2 q^{k^2} = \begin{bmatrix} 2N \\ N \end{bmatrix}. \tag{VII.10}$$

$$\sum_{m=0}^{n} \begin{bmatrix} N \\ m \end{bmatrix} \begin{bmatrix} N+n-m-1 \\ n-m \end{bmatrix} (-1)^m q^{m(m-1)/2} = 0. \tag{VII.11}$$

$$\sum_{j=0}^{n} \begin{bmatrix} n \\ j \end{bmatrix} (-1)^j q^{j(j-1)/2} = 0. \tag{VII.12}$$

The q-Pfaff–Saalschütz sum (page 18):

$$\sum_{k=0}^{n} \frac{(a, b, q^{-n}; q)_k}{(c, abq^{1-n}/c, q; q)_k} q^k = \frac{(c/a, c/b; q)_n}{(c, c/ab; q)_n}. \tag{VII.13}$$

Heine's q-Gauss sum (page 20):

$$\sum_{k=0}^{\infty} \frac{(a, b; q)_k}{(c, q; q)_k} \left(\frac{c}{ab}\right)^k = \frac{(c/a, c/b; q)_\infty}{(c, c/ab; q)_\infty}. \tag{VII.14}$$

The Bailey–Daum summation formula (page 20):

$$\sum_{k=0}^{\infty} \frac{(a, b; q)_k}{(aq/b, q; q)_k} \left(\frac{-q}{b}\right)^k = \frac{(-q; q)_\infty(aq, aq^2/b^2; q^2)_\infty}{(aq/b, -q/b; q)_\infty}. \tag{VII.15}$$

A summation formula of Alladi (page 23):

$$\sum_{n=0}^{\infty} \frac{(a;q)_n q^n}{(bq;q)_n} = \frac{(1-b)}{(b-a)} \left(\frac{(a;q)_\infty}{(b;q)_\infty} - 1 \right).$$ (VII.16)

Lebesgue's identity (page 23):

$$\sum_{n=0}^{\infty} \frac{(-aq;q)_n q^{n(n+1)/2}}{(q;q)_n} = (-q;q)_\infty (-aq^2;q^2)_\infty.$$ (VII.17)

An identity of Andrews (page 24):

$$\sum_{k=0}^{\infty} \frac{(b, q/b; q)_k c^k q^{k(k-1)/2}}{(c;q)_k (q^2;q^2)_k} = \frac{(bc, qc/b; q^2)_\infty}{(c;q)_\infty}.$$ (VII.18)

A zero sum (page 24):

$$\sum_{j=0}^{n} \frac{(-1;q)_{n-j}(-1,q)_j}{(q;q)_{n-j}(q,q)_j} (-1)^j = 0.$$ (VII.19)

Gasper's bibasic sum and a special case (pages 29, 30):

$$\sum_{k=0}^{n} \frac{(1-ap^k q^k)(1-bp^k q^{-k})}{(1-a)(1-b)} \frac{(a,b;p)_k (c, a/bc; q)_k}{(ap/c, bcp; p)_k (aq/b, q; q)_k} q^k$$
$$= \frac{(ap, bp; p)_n (cq, aq/bc; q)_n}{(ap/c, bcp; p)_n (aq/b, q; q)_n}.$$ (VII.20)

$$\sum_{k=0}^{n} \frac{(1-aq^{2k})}{(1-a)} \frac{(a, b, q^{-n}, aq^n/b; q)_k}{(aq^{n+1}, bq^{1-n}, aq/b, q; q)_k} q^k = \delta_{0,n}.$$ (VII.21)

A $_6\phi_5$ summation formula and a terminating special case (page 31):

$$\sum_{k=0}^{\infty} \frac{1-aq^{2k}}{1-a} \frac{(a,b,c,d;q)_k}{(aq/b, aq/c, aq/d, q; q)_k} \left(\frac{aq}{bcd} \right)^k$$
$$= \frac{(aq, aq/bc, aq/bd, aq/cd; q)_\infty}{(aq/b, aq/c, aq/d, aq/bcd; q)_\infty}.$$ (VII.22)

$$\sum_{k=0}^{n} \frac{1-aq^{2k}}{1-a} \frac{(a,b,c,q^{-n};q)_k}{(aq/b, aq/c, aq^{n+1}, q; q)_k} \left(\frac{aq^{n+1}}{bc} \right)^k = \frac{(aq, aq/bc; q)_n}{(aq/b, aq/c; q)_n}.$$ (VII.23)

The q-Dixon sum (special cases of (VII.22) and (VII.23)) (page 32):

$$\sum_{k=0}^{\infty} \frac{(a, -\sqrt{a}q, b, c; q)_k}{(-\sqrt{a}, aq/b, aq/c, q; q)_k} \left(\frac{\sqrt{a}q}{bc} \right)^k = \frac{(aq, aq/bc, \sqrt{a}q/b, \sqrt{a}q/c; q)_\infty}{(aq/b, aq/c, \sqrt{a}q, \sqrt{a}q/bc; q)_\infty}.$$
(VII.24)

$$\sum_{k=0}^{n} \frac{(a, -\sqrt{a}q, b, q^{-n}; q)_k}{(-\sqrt{a}, aq/b, aq^{n+1}, q; q)_k} \left(\frac{\sqrt{a}q^{n+1}}{b} \right)^k = \frac{(aq, \sqrt{a}q/b; q)_n}{(aq/b, \sqrt{a}q; q)_n}.$$
(VII.25)

Jackson's $_8\phi_7$ summation formula (page 33):

$$\sum_{k=0}^{n} \frac{1 - aq^{2k}}{1 - a} \frac{(a, b, c, d, e, q^{-n}; q)_k}{(aq/b, aq/c, aq/d, aq/e, aq^{n+1}, q; q)_k} q^k$$
$$= \frac{(aq, aq/bc, aq/bd, aq/cd; q)_n}{(aq/b, aq/c, aq/d, aq/bcd; q)_n}.$$
(VII.26)

A q-analogue of Watson's $_3F_2$ sum (page 39):

$$\sum_{k=0}^{\infty} \frac{1 - \lambda q^{2k}}{1 - \lambda} \frac{(\lambda, a, b, c, -c, \lambda q/c^2; q)_k}{(\lambda q/a, \lambda q/b, \lambda q/c, -\lambda q/c, c^2, q; q)_k} \left(-\frac{\lambda q}{ab} \right)^k$$
$$= \frac{(\lambda q, c^2/\lambda; q)_\infty (aq, bq, c^2 q/a, c^2 q/b; q^2)_\infty}{(\lambda q/a, \lambda q/b; q)_\infty (q, abq, c^2 q, c^2 q/ab; q^2)_\infty}, \text{ where } \lambda = -c(ab/q)^{1/2}.$$
(VII.27)

A q-analogue of Whipple's $_3F_2$ sum (page 40):

$$\sum_{k=0}^{\infty} \frac{1 - (-c)q^{2k}}{1 - (-c)} \frac{(-c, a, q/a, c, -d, -q/d; q)_k}{(-cq/a, -ac, -q, cq/d, cd, q; q)_k} c^k$$
$$= \frac{(-c, -cq; q)_\infty (acd, acq/d, cdq/a, cq^2/ad; q^2)_\infty}{(cd, cq/d, -ac, -cq/a; q)_\infty}.$$
(VII.28)

Two parity-dependent $_8\phi_7$ sums (page 40):

$$\sum_{k=0}^{\infty} \frac{1 - q^{2k+1}/ge}{1 - q/ge} \frac{(q/ge, -q/ge, e, q/e, q^{1+n}/g, gq^{-n}; q)_k}{(q^{1-n}/e, q/g, -q, q^2/ge^2, q^{2+n}/eg^2, q; q)_k} \left(\frac{-q}{ge} \right)^k$$

$$= \begin{cases} 0, & n \text{ odd}, \\ \dfrac{(q/eg, q^2/eg; q)_\infty (q, q^2/e^2, q^2/g^2, q^3/e^2g^2; q^2)_\infty}{(q/e, q/g, q^2/e^2g, q^2/eg^2; q)_\infty} \\ \quad \times \dfrac{(q, e^2; q^2)_{n/2}(q^2/eg^2; q)_n}{(q^2/g^2, q^3/e^2g^2; q^2)_{n/2}(e; q)_n}, & n \text{ even}. \end{cases}$$
(VII.29)

$$\sum_{k=0}^{n} \frac{1-q^{2k+1}/e}{1-q/e} \frac{(q/e,-q/e,e,q/e,q^{1+n},q^{-n};q)_k}{(q^{1-n}/e,q,-q,q^2/e^2,q^{2+n}/e,q;q)_k} \left(\frac{-q}{e}\right)^k$$

$$= \begin{cases} 0, & n \text{ odd}, \\ \dfrac{(q,e^2;q^2)_{n/2}(q^2/e;q)_n}{(q^2,q^3/e^2;q^2)_{n/2}(e;q)_n}, & n \text{ even}. \end{cases} \qquad \text{(VII.30)}$$

A finite $_6\phi_5$ sum (page 40):

$$\sum_{k=0}^{n} \frac{1-aq^{2k}}{1-a} \frac{(a,b,c,a/bc;q)_k}{(aq/c,bcq,aq/b,q;q)_k} q^k = \frac{(aq,bq,cq,aq/bc;q)_n}{(aq/c,bcq,aq/b,q;q)_n}. \qquad \text{(VII.31)}$$

A terminating q-analogue of Watson's $_3F_2$ sum due to Andrews (page 41):

$$\sum_{k=0}^{n} \frac{(c,-c,a^2q^{n+1},q^{-n};q)_k}{(aq,-aq,c^2,q;q)_k} q^k = \begin{cases} 0, & n \text{ odd}, \\ \dfrac{(q,a^2q^2/c^2;q^2)_{n/2}c^n}{(c^2q,a^2q^2;q^2)_{n/2}}, & n \text{ even}. \end{cases} \qquad \text{(VII.32)}$$

The terminating q-analogue of Whipple's $_3F_2$ sum of Andrews (page 42):

$$\sum_{k=0}^{n} \frac{(f,-f,q^{n+1},q^{-n};q)_k}{(e,f^2q/e,-q,q;q)_k} q^k = \frac{(q/e;q)_n(eq^{1-n}/f^2;q^2)_n f^{2n}}{(f^2q/e;q)_n(eq^{1-n};q^2)_n}. \qquad \text{(VII.33)}$$

Two summation formulae of Warnaar (page 42): when $e = -bdq$ or $e = d^2q/b$,

$$\sum_{k=0}^{n} \frac{(dq,b,d^2,q^{-n};q)_k q^k}{(d,e,bd^2q^{2-n}/e,q;q)_k} = \frac{1+dq^n/b}{1+d/b} \frac{(e/d^2q,e/bq;q)_n}{(e,e/bd^2q;q)_n}. \qquad \text{(VII.34)}$$

$$\sum_{k=0}^{n} \frac{1-aq^{2k}}{1-a} \frac{(a,b,aq^n/b^{1/2},-aq^n/b^{1/2},-q^{-n},q^{-n};q)_k}{(aq/b,b^{1/2}q^{1-n},-b^{1/2}q^{1-n},-aq^{n+1},aq^{n+1},q;q)_k} q^{2k}$$

$$= \frac{(-a/b;q)_{2n}}{(-aq;q)_{2n}} \frac{(a^2q^2,b;q^2)_n}{(1/b,a^2q^2/b^2;q^2)_n} \left(\frac{q}{b}\right)^n. \qquad \text{(VII.35)}$$

The Jacobi triple product identity (page 44):

$$\sum_{n=-\infty}^{\infty} (-z)^n q^{n^2} = (zq,q/z,q^2;q^2)_\infty. \qquad \text{(VII.36)}$$

Jacobi's identity (page 46):

$$\sum_{n=0}^{\infty} (-1)^n (2n+1) q^{n(n+1)/2} = (q;q)_{\infty}^3.$$ (VII.37)

An identity of Fine (page 54):

$$\sum_{n=-\infty}^{\infty} (6n+1) q^{n(3n+1)/2} = (q;q)_{\infty}^3 (q;q^2)_{\infty}^2.$$ (VII.38)

Another Fine-type identity (page 54):

$$\sum_{n=-\infty}^{\infty} (3n+1) q^{n(3n+2)} = (q;q)_{\infty}^2 \frac{(q^4;q^4)_{\infty}}{(q^2;q^4)_{\infty}}.$$ (VII.39)

Ramanujan's $_1\psi_1$ summation formula (page 56):

$$_1\psi_1(a;b;q,z) := \sum_{n=-\infty}^{\infty} \frac{(a;q)_n}{(b;q)_n} z^n = \frac{(q, b/a, az, q/az; q)_{\infty}}{(b, q/a, z, b/az; q)_{\infty}}.$$ (VII.40)

A parity-dependent bilateral sum, valid for $n \in \mathbb{Z}$ (page 64):

$$\sum_{k=-\infty}^{\infty} \frac{(q^{1-n}/b, a; q)_k}{(q^{1-n}/a, b; q)_k} \left(\frac{-b}{a}\right)^k$$

$$= \begin{cases} \dfrac{(q, b/a, -b, -q/a; q)_{\infty}}{(-q, -b/a, b, q/a; q)_{\infty}} \dfrac{(a^2, b, bq; q^2)_{n/2}}{(b^2, a, aq; q^2)_{n/2}}, & n \text{ even,} \\ 0, & n \text{ odd.} \end{cases}$$ (VII.41)

Bailey's $_6\psi_6$ summation and two special cases (pages 65, 67):

$$\sum_{n=-\infty}^{\infty} \frac{(q\sqrt{a}, -q\sqrt{a}, b, c, d, e; q)_n}{(\sqrt{a}, -\sqrt{a}, aq/b, aq/c, aq/d, aq/e; q)_n} \left(\frac{qa^2}{bcde}\right)^n$$

$$= \frac{(aq, aq/bc, aq/bd, aq/be, aq/cd, aq/ce, aq/de, q, q/a; q)_{\infty}}{(aq/b, aq/c, aq/d, aq/e, q/b, q/c, q/d, q/e, qa^2/bcde; q)_{\infty}}.$$ (VII.42)

$$\sum_{n=-\infty}^{\infty} \frac{(-q\sqrt{a}, b, c, d; q)_n}{(-\sqrt{a}, aq/b, aq/c, aq/d; q)_n} \left(\frac{qa^{3/2}}{bcd}\right)^n$$

$$= \frac{(aq, aq/bc, aq/bd, \sqrt{a}q/b, aq/cd, \sqrt{a}q/c, \sqrt{a}q/d, q, q/a; q)_{\infty}}{(aq/b, aq/c, aq/d, \sqrt{a}q, q/b, q/c, q/d, q/\sqrt{a}, qa^{3/2}/bcd; q)_{\infty}}.$$ (VII.43)

$$\sum_{n=-\infty}^{\infty} \frac{(b,c;q)_n}{(aq/b,aq/c;q)_n} \left(\frac{-qa}{bc}\right)^n$$

$$= \frac{(aq/bc;q)_\infty (aq^2/b^2, aq^2/c^2, q^2, aq, q/a; q^2)_\infty}{(aq/b, aq/c, q/b, q/c, -qa/bc; q)_\infty}. \quad \text{(VII.44)}$$

Four Lambert series summations of Andrews that follow from Bailey's $_6\psi_6$ summation formula (VII.42) (pages 68, 69):

$$1 + 4\sum_{n=1}^{\infty} \frac{(-1)^n q^{n(n+1)/2}}{1+q^n} = \left(\frac{(q;q)_\infty}{(-q;q)_\infty}\right)^2. \quad \text{(VII.45)}$$

$$1 + 8\sum_{n=1}^{\infty} \frac{(-q)^n}{(1+q^n)^2} = \left(\frac{(q;q)_\infty}{(-q;q)_\infty}\right)^4. \quad \text{(VII.46)}$$

$$1 + 16\sum_{n=1}^{\infty} \frac{-q^n + 4q^{2n} - q^{3n}}{(1+q^n)^4} = \left(\frac{(q;q)_\infty}{(-q;q)_\infty}\right)^8. \quad \text{(VII.47)}$$

$$\sum_{n=0}^{\infty} \left\{ \frac{q^{5n+1}}{(1-q^{5n+1})^2} - \frac{q^{5n+2}}{(1-q^{5n+2})^2} - \frac{q^{5n+3}}{(1-q^{5n+3})^2} + \frac{q^{5n+4}}{(1-q^{5n+4})^2} \right\}$$

$$= q\frac{(q^5;q^5)_\infty^5}{(q;q)_\infty}. \quad \text{(VII.48)}$$

Two $_3\psi_3$ summation formula and two finite sums that follow from them (page 70):

$$\sum_{n=-\infty}^{\infty} \frac{(b,c,d;q)_n}{(q/b,q/c,q/d;q)_n} \left(\frac{q}{bcd}\right)^n = \frac{(q,q/bc,q/bd,q/cd;q)_\infty}{(q/b,q/c,q/d,q/bcd;q)_\infty}. \quad \text{(VII.49)}$$

$$\sum_{n=-\infty}^{\infty} \frac{(b,c,d;q)_n}{(q^2/b,q^2/c,q^2/d;q)_n} \left(\frac{q^2}{bcd}\right)^n = \frac{(q,q^2/bc,q^2/bd,q^2/cd;q)_\infty}{(q^2/b,q^2/c,q^2/d,q^2/bcd;q)_\infty}. \quad \text{(VII.50)}$$

$$\sum_{n=-N}^{N} (-1)^n \begin{bmatrix} 2N \\ N+n \end{bmatrix}_q^3 q^{n(3n+1)/2} = \frac{(q;q)_{3N}}{(q;q)_N^3}. \quad \text{(VII.51)}$$

$$\sum_{n=-N-1}^{N} (-1)^n \begin{bmatrix} 2N+1 \\ N+n+1 \end{bmatrix}_q^3 q^{n(3n+1)/2} = \frac{(q;q)_{3N+1}}{(q;q)_N^3}. \quad \text{(VII.52)}$$

Three terminating $_{10}\phi_9$ summations (pages 119, 120, 129):

$$\sum_{j=0}^{n} \frac{\left(q\sqrt{a}, -q\sqrt{a}, iq\sqrt{a}, -iq\sqrt{a}, a, \frac{a^2}{k^2}, kq^n, -kq^n, -q^{-n}, q^{-n}; q\right)_j}{\left(\sqrt{a}, -\sqrt{a}, i\sqrt{a}, -i\sqrt{a}, \frac{k^2 q}{a}, -\frac{aq^{1-n}}{k}, \frac{aq^{1-n}}{k}, -aq^{n+1}, aq^{n+1}, q; q\right)_j} q^j$$

$$= \frac{(a^2 q^2, a^2/k^2; q^2)_n (-k^2 q/a; q)_{2n}}{(k^2/a^2, k^4 q^2/a^2; q^2)_n (-a; q)_{2n}} \left(\frac{k^2}{a^2 q}\right)^n. \quad \text{(VII.53)}$$

$$\sum_{j=0}^{n} \frac{(1 - aq^{2j})(a, k/aq, a\sqrt{q/k}, -a\sqrt{q/k}, aq/\sqrt{k}, -aq/\sqrt{k}, kq^n, q^{-n}; q)_j q^j}{(1 - a)(q^2 a^2/k, \sqrt{k}, -\sqrt{k}, \sqrt{kq}, -\sqrt{kq}, aq^{n+1}, aq^{1-n}/k, q; q)_j}$$

$$= \frac{(aq, k^2/a^2 q; q)_n}{(k, k/a; q)_n}. \quad \text{(VII.54)}$$

$$\sum_{j=0}^{n} \frac{(a, q\sqrt{a}, -q\sqrt{a}, a\sqrt{\frac{q}{k}}, -a\sqrt{\frac{q}{k}}, \frac{a}{\sqrt{k}}, -\frac{aq}{\sqrt{k}}, \frac{k}{a}, kq^n, q^{-n}; q)_j q^j}{(\sqrt{a}, -\sqrt{a}, \sqrt{kq}, -\sqrt{kq}, q\sqrt{k}, -\sqrt{k}, \frac{a^2 q}{k}, \frac{aq^{1-n}}{k}, aq^{n+1}, q; q)_j}$$

$$= \frac{\left(aq, \sqrt{k}, \frac{k^2}{a^2}; q\right)_n}{\left(k, \frac{k}{a}, q\sqrt{k}; q\right)_n}. \quad \text{(VII.55)}$$

Two terminating $_8\phi_7$ summation formulae (page 127):

$$\sum_{j=0}^{n} \frac{1 - aq^{4j}}{1 - a} \frac{(a, b, bq, a/b^2, a^2 q^{2+2n}/b^2, q^{-2n}; q^2)_j}{(aq/b, aq^2/b, b^2 q^2, b^2 q^{-2n}/a, aq^{2+2n}, q^2; q^2)_j} \left(\frac{b^2 q}{a}\right)^j$$

$$= \frac{(-q, aq/b^2; q)_n (aq^2; q^2)_n}{(-bq, aq/b; q)_n (aq^2/b^2; q^2)_n}. \quad \text{(VII.56)}$$

$$\sum_{j=0}^{n} \frac{1 - aq^{4j}}{1 - a} \frac{(a, b, bq, aq^2/b^2, a^2 q^{2+2n}/b^2, q^{-2n}; q^2)_j}{(aq/b, aq^2/b, b^2, b^2 q^{-2n}/a, aq^{2+2n}, q^2; q^2)_j} \left(\frac{b^2}{aq}\right)^j$$

$$= \frac{(-q, aq^2/b^2; q)_n (aq^2; q^2)_n q^{-n}}{(-b, aq/b; q)_n (aq^2/b^2; q^2)_n}. \quad \text{(VII.57)}$$

A non-terminating $_8\phi_7$ summation formula (page 129):

$$\sum_{n=0}^{\infty} \frac{(1 - aq^{2n})(a, a/k; q)_n (k; q)_{2n}}{(1 - a)(kq, q; q)_n (a^2 q/k; q)_{2n}} \left(\frac{aq}{k}\right)^n = \frac{(aq, qa^2/k^2; q)_\infty}{(qa/k, qa^2/k; q)_\infty}. \quad \text{(VII.58)}$$

A terminating $_5\phi_4$ sum (page 129):

$$\sum_{r=0}^{n} \frac{(q\sqrt{a}, -q\sqrt{a}, a, b, q^{-n}; q)_r \, q^r}{(\sqrt{a}, -\sqrt{a}, \frac{qa}{b}, q^{2-n}b^2, q; q)_r} = \frac{\left(\frac{q\sqrt{a}}{\sqrt{q}b}, -\frac{q\sqrt{a}}{\sqrt{q}b}, \frac{a}{qb^2}, \frac{1}{qb}; q\right)_n}{\left(\frac{\sqrt{a}}{\sqrt{q}b}, -\frac{\sqrt{a}}{\sqrt{q}b}, \frac{1}{qb^2}, \frac{qa}{b}; q\right)_n}. \qquad \text{(VII.59)}$$

Selected Slater-type Identities

This section contains a number of identities of Rogers–Ramanujan–Slater type which are not proved in the text, but which are used in the proof of other identities. As elsewhere, the notation **S.n** refers to the identity labelled (n) in Slater's compendium of identities in [237].

$$\sum_{n=0}^{\infty} \frac{q^{n^2}}{(q;q)_{2n}} = \frac{(q^8, q^{12}, q^{20}; q^{20})_\infty (-q; q^2)_\infty}{(q^2; q^2)_\infty} \quad \text{(Rogers [219], \textbf{S.98})} \qquad \text{(18.165)}$$

$$\sum_{n=0}^{\infty} \frac{q^{n^2+2n}}{(q;q)_{2n+1}} = \frac{(q^4, q^{16}, q^{20}; q^{20})_\infty (-q; q^2)_\infty}{(q^2; q^2)_\infty} \quad \text{(Rogers [219], \textbf{S.96})}$$

$$\text{(18.166)}$$

$$\sum_{n=0}^{\infty} \frac{q^{n^2+n}}{(q;q)_{2n}} = \frac{(q, q^9, q^{10}; q^{10})_\infty (q^8, q^{12}; q^{20})_\infty}{(q; q)_\infty} \quad \text{(Rogers [217], \textbf{S.99})}$$

$$\text{(18.167)}$$

$$\sum_{n=0}^{\infty} \frac{q^{n^2+n}}{(q;q)_{2n+1}} = \frac{(q^3, q^7, q^{10}; q^{10})_\infty (q^4, q^{16}; q^{20})_\infty}{(q; q)_\infty} \quad \text{(Rogers [217], \textbf{S.94})}$$

$$\text{(18.168)}$$

The next two identities are, respectively, the analytic versions of the first- and second Göllnitz–Gordon partition identities.

$$\sum_{n=0}^{\infty} \frac{(-q; q^2)_n q^{n^2}}{(q^2; q^2)_n} = \frac{1}{(q, q^4, q^7; q^8)_\infty} \quad \text{(Slater [237] \textbf{S.36})} \qquad \text{(VII.150)}$$

$$\sum_{n=0}^{\infty} \frac{(-q; q^2)_n q^{n^2+2n}}{(q^2; q^2)_n} = \frac{1}{(q^3, q^4, q^5; q^8)_\infty} \quad \text{(Slater [237] \textbf{S.34})} \qquad \text{(VII.160)}$$

Bibliography

[1] Abel, N. (1826), Untersuchungen über die Reihe ..., *J. Reine Angew. Math.*, **1**, 311–339.

[2] Adiga, C.; Berndt, B. C.; Bhargava, S.; Watson, G. N. (1985), *Chapter 16 of Ramanujan's second notebook: theta-functions and q-series*. Mem. Amer. Math. Soc. **53**, no. 315, v+85 pp.

[3] Agarwal, R. P. (1993), Lambert series and Ramanujan. *Proc. Indian Acad. Sci. Math. Sci.* **103**, no. 3, 269–293.

[4] Agarwal, A. K.; Andrews, G. E.; Bressoud, D. M., The Bailey lattice. *J. Indian Math. Soc. (N.S.)* **51** (1987), 57–73.

[5] Alladi, K. (2010), A combinatorial study and comparison of partial theta identities of Andrews and Ramanujan. *Ramanujan J.* **23**, no. 1-3, 227–241.

[6] Alladi, K. (2013), Variants of classical q-hypergeometric identities and partition implications. *Ramanujan J.* **31**, no. 1-2, 213–238.

[7] Alladi, K.; Gordon B. (1994), Vanishing coefficients in the expansion of products of Rogers-Ramanujan type. Proc. Rademacher Centenary Conference, (G. E. Andrews and D. Bressoud, Eds.), *Contemp. Math.* **166**, 129–139.

[8] Andrews, G. E. (1965), A simple proof of Jacobi's triple product identity. *Proc. Amer. Math. Soc.* **16**, 333–334.

[9] Andrews, G. E. (1966), On basic hypergeometric series, mock theta functions, and partitions. I. *Quart. J. Math. Oxford Ser.* (2) **17**, 64–80.

[10] Andrews, G. E. (1966), q-identities of Auluck, Carlitz, and Rogers. *Duke Math. J.* **33**, 575–581.

[11] Andrews, G. E. (1966), On basic hypergeometric series, mock theta functions, and partitions. II. *Quart. J. Math. Oxford Ser.* (2) **17**, 132–143.

[12] Andrews, G. E. (1966), On the theorems of Watson and Dragonette for Ramanujan's mock theta functions. *Amer. J. Math.* **88**, 454–490.

[13] Andrews, G. E. (1967), Enumerative proofs of certain *q*-identities. *Glasgow Math. J.* **8**, 33–40.

[14] Andrews, G. E. (1968), On *q*-difference equations for certain well-poised basic hypergeometric series. *Quart. J. Math. Oxford Ser. (2)* **19**, 433–447.

[15] Andrews G. E. (1972), Two theorems of Gauss and allied identities proved arithmetically, *Pacific J. Math.* **41**, 563–578.

[16] Andrews, G. E. (1973), On the *q*-analog of Kummer's theorem and applications. *Duke Math. J.* **40**, 525–528.

[17] Andrews, G. E. (1974), Applications of basic hypergeometric functions. *SIAM Rev.* **16**, 441–484.

[18] Andrews, G. E. (1974), An analytic generalization of the Rogers-Ramanujan identities for odd moduli. *Proc. Nat. Acad. Sci. U.S.A.* **71**, 4082–4085.

[19] Andrews, G. E. (1976), On *q*-analogues of the Watson and Whipple summations. *SIAM J. Math. Anal.* **7**, no. 3, 332–336.

[20] Andrews, G. E. (1979), An introduction to Ramanujan's "lost" notebook. *Amer. Math. Monthly* **86**, no. 2, 89–108.

[21] Andrews, G. E. (1981), Ramanujan's "lost" notebook. I. Partial *θ*-functions. *Adv. in Math.* **41**, no. 2, 137–172.

[22] Andrews, G. E. (1984), Multiple series Rogers-Ramanujan type identities. *Pacific J. Math.* **114**, no. 2, 267–283.

[23] Andrews, G. E. (1984), Hecke modular forms and the Kac-Peterson identities. *Trans. Amer. Math. Soc.* **283**, no. 2, 451–458.

[24] Andrews, G. E. (1986), *q-series: Their Development and Application in Analysis, Number Theory, Combinatorics, Physics and Computer*

Algebra, C.B.M.S. Regional Conference Series in Math, No. **66**, American Math. Soc. Providence.

[25] Andrews, G. E. (1986), The fifth and seventh order mock theta functions. *Trans. Amer. Math. Soc.* **293**, no. 1, 113–134.

[26] Andrews, G. E. (1988), Ramanujan's fifth order mock theta functions as constant terms. Ramanujan revisited (Urbana-Champaign, Ill. 1987), 47–56, Academic Press, Boston, MA.

[27] Andrews, G. E. (1990), q-trinomial coefficients and Rogers-Ramanujan type identities. In Bruce Berndt *et al.*, editor, *Analytic Number Theory*, 1–11, Boston, Birkhauser.

[28] Andrews, G. E. (1991), Three aspects of partitions. *Séminaire Lotharingien de Combinatoire (Salzburg, 1990)*, 518, Publ. Inst. Rech. Math. Av., 462, Univ. Louis Pasteur, Strasbourg.

[29] Andrews, G. E. (1994), Schur's theorem, Capparelli's conjecture and q-trinomial coefficients. *The Rademacher legacy to mathematics (University Park, PA, 1992), 141–154, Contemp. Math.,* **166**, *Amer. Math. Soc., Providence, RI.*

[30] Andrews, G. E. (1998), *The theory of partitions.* Reprint of the 1976 original. Cambridge Mathematical Library. Cambridge University Press, Cambridge, 1998. xvi+255 pp.

[31] Andrews, G. E. (2001), Bailey's transform, lemma, chains and tree. *Special Functions 2000: Current Perspective and Future Directions*, NATO Sci. Ser. II Math. Phys. Chem., Tempe, AZ, vol. **30**, Kluwer Acad. Publ., Dordrecht, pp. 1–22.

[32] Andrews, G. E. (2012), q-Orthogonal polynomials, Rogers-Ramanujan identities, and mock theta functions *Proceedings of the Steklov Institute of Mathematics April 2012*, Volume **276**, Issue 1, pp. 21–32.

[33] Andrews, G. E.; Askey, R. (1978), A simple proof of Ramanujan's summation of the $_1\Psi_1$, *Aequationes Math.*, **18**, 333–337.

[34] Andrews, G. E., Askey, R. and Roy, R. (1999), *Special functions.* Encyclopedia of Mathematics and its Applications, 71. Cambridge University Press, Cambridge, xvi+664 pp.

[35] Andrews, G. E.; Baxter, R. J. (1987), Lattice gas generalization of the hard hexagon model. III. q-trinomial coefficients. *J. Statist. Phys.* **47**, no. 3–4, 297–330.

[36] Andrews, G. E.; Berkovich, A. (1998), A trinomial analogue of Bailey's lemma and $N = 2$ superconformal invariance. *Comm. Math. Phys.* **192**, no. 2, 245–260.

[37] Andrews, G. E.; Berkovich, A. (2002), The WP-Bailey tree and its implications. *J. London Math. Soc.* (2) **66**, no. 3, 529–549.

[38] Andrews, G. E.; Berndt, B. C. (2005), *Ramanujan's Lost Notebook. Part I.* Springer, New York, xiv+437 pp.

[39] Andrews, G. E.; Berndt, B. C. (2009), *Ramanujan's Lost Notebook, Part II.* Springer, New York, xii+418 pp.

[40] Andrews, G. E.; Berndt, B. C. (2012), *Ramanujan's Lost Notebook. Part III.* Springer, New York, xii+435 pp.

[41] Andrews, G. E.; Berndt, B. C.; Jacobsen, L.; Lamphere, R. L. (1992), *The continued fractions found in the unorganized portions of Ramanujan's notebooks.* Mem. Amer. Math. Soc. **99**, no. 477, vi+71 pp.

[42] Andrews, G. E.; Berndt, B. C.; Sohn, J.; Yee, A. J.; Zaharescu, A. (2003), On Ramanujan's continued fraction for $(q^2; q^3)_\infty / (q; q^3)_\infty$. *Trans. Amer. Math. Soc.* **355**, no. 6, 2397–2411.

[43] Andrews, G. E.; Bowman, D. (1999), The Bailey transform and D. B. Sears. *Quaest. Math.* **22**, no. 1, 19–26.

[44] Andrews, G. E.; Bressoud, D. M. (1979), Vanishing coefficients in infinite product expansions. *J. Austral. Math. Soc. Ser. A*, **27**, no. 2, 199-202.

[45] Andrews, G. E.; Eriksson, K. (2004), *Integer partitions.* Cambridge University Press, Cambridge, x+141 pp.

[46] Andrews, G. E.; Garvan, F. G. (1989), Ramanujan's "lost" notebook. VI. The mock theta conjectures. *Adv. in Math.* **73**, no. 2, 242–255.

[47] Andrews, G. E.; Hickerson, D. (1991), Ramanujan's "lost" notebook: the sixth order mock theta functions, *Adv. Math.* **89**, 60–105.

[48] Andrews, G. E.; Lewis, R.; Liu, Z.-G. (2001), An identity relating a theta function to a sum of Lambert series. *Bull. London Math. Soc.* **33**, no. 1, 25–31.

[49] Andrews, G. E.; Warnaar, S. O. (2007), The Bailey transform and false theta functions. *Ramanujan J.* **14**, no. 1, 173–188.

[50] Askey, R. (1980), Ramanujan's Extensions of the Gamma and Beta Functions. *Amer. Math. Monthly*, **87**, no. 5, 346–359.

[51] Askey, R. (1984), The very well poised $_6\psi_6$. II. *Proc. Amer. Math. Soc.* **90**, no. 4, 575–579.

[52] Askey, R.; Ismail, M. E. H. (1979), The very well poised $_6\psi_6$. *Proc. Amer. Math. Soc.* **77**, no. 2, 218–222.

[53] Askey, R.; Wilson, J. (1985), Some basic hypergeometric orthogonal polynomials that generalize Jacobi polynomials. *Mem. Amer. Math. Soc.* **54**, no. 319, iv+55 pp.

[54] Bailey, W. N. (1929), An identity involving Heine's basic hypergeometric series *J. London Math. Soc.*, **4**, pp. 254–257.

[55] Bailey, W. N. (1935), *Generalised Hypergeometric Series.* Cambridge, England: Cambridge University Press, reprinted Cambridge Tracts in Mathematics and Mathematical Physics, No. **32** Stechert-Hafner, Inc., New York 1964 v+108 pp.

[56] Bailey, W. N. (1936), Series of hypergeometric type which are infinite in both directions *Quart. J. Math. (Oxford)*, **7**, pp. 105–115.

[57] Bailey, W. N. (1941), A note on certain q-identities, *Quart. J. Math.*, Oxford Ser. **12**, 173–175.

[58] Bailey, W. N. (1947), Well-poised basic hypergeometric series, *Quart. J. Math. (Oxford)* **18**, 157–166.

[59] Bailey, W. N. (1947), Some identities in combinatory analysis, *Proc. London Math. Soc.* (2) **49**, 421–435.

[60] Bailey, W. N (1947), A transformation of nearly-poised basic hypergeometric series *J. London Math. Soc.*, **22**, 237–240.

[61] Bailey, W. N. (1948), Identities of the Rogers-Ramanujan type. *Proc. London Math. Soc.* (2) **50**, 1–10.

[62] Bailey, W. N. (1950), On the basic bilateral hypergeometric series $_2\psi_2$. *Quart. J. Math., Oxford Ser.* (2) **1**, 194–198.

[63] Bailey, W. N. (1950), On the analogue of Dixon's theorem for bilateral basic hypergeometric series. *Quart. J. Math., Oxford Ser.* (2) **1**, 318–320.

[64] Bailey, W. N. (1951), On the simplification of some identities of the Rogers-Ramanujan type. *Proc. London Math. Soc.* (3) **1**, 217–221.

[65] Bajpai, J.; Kimport, S.; Liang, J.; Ma, D.; Ricci, J. (2015), Bilateral series and Ramanujan's radial limits. *Proc. Amer. Math. Soc.* **143**, no. 2, 479–492.

[66] Bauer, G. (1872), Von einem Kettenbruch von Euler und einem Theorem von Wallis, *Abh. der Kgl. Bayr. Akad. der Wiss., München, Zweite Klasse*, **11**, 99–116.

[67] Berkovich, A.; McCoy, B. M.; Schilling, A. (1996), $N = 2$ supersymmetry and Bailey pairs. *Phys. A*, **228**, no. 1–4, 33–62.

[68] Berkovich, A.; Warnaar, S. O. (2005), Positivity preserving transformations for q-binomial coefficients. *Trans. Amer. Math. Soc.* **357**, no. 6, 2291–2351.

[69] Berndt, B. C. (1991), *Ramanujan's notebooks. Part III.* Springer-Verlag, New York, xiv+510 pp.

[70] Berndt, B. C. (2006), *Number theory in the spirit of Ramanujan.* Student Mathematical Library, **34**. American Mathematical Society, Providence, RI, xx+187 pp.

[71] Berndt, B. C. (2014), *Lectures Notes on Mock Theta Functions.*

http://www.math.uiuc.edu/~berndt/Berndt.pdf

[72] Berndt, B. C.; Chan, S. H. (2007), Sixth order mock theta functions. *Adv. Math.* **216**, no. 2, 771–786.

[73] Berndt, B. C.; Rankin, R. A. (1995), *Ramanujan: Letters and Commentary*, American Mathematical Society, Providence, RI, London Mathematical Society, London.

[74] Berndt, B. C.; Yee, A. J. (2003), On the generalized Rogers-Ramanujan continued fraction. *Rankin memorial issues. Ramanujan J.* **7**, no. 1-3, 321–331.

[75] Berndt, B. C.; Yee, A. J. (2003), Combinatorial proofs of identities in Ramanujan's lost notebook associated with the Rogers-Fine identity and false theta functions. *Ann. Comb.* **7**, no. 4, 409–423.

[76] Bernoulli D. (1775), *Disquisitiones ulteriores de indola fractionum continuarum, Novi comm.*, Acad. Sci. Imper. Petropol. **20**.

[77] Bhargava, S.; Adiga, C. (1984), On some continued fraction identities of Srinivasa Ramanujan. *Proc. Amer. Math. Soc.* **92**, no. 1, 13–18.

[78] Bhargava, S.; Adiga, C.; Somashekara, D. D. (1987), On some generalizations of Ramanujan's continued fraction identities, *Proc. Indian Acad. Sci. (Math. Sci.)* **97**, 31–43.

[79] Bhatnagar, G. (2014), How to prove Ramanujan's q-continued fractions. Ramanujan 125, *Contemp. Math.*, **627**, Amer. Math. Soc., Providence, RI, 49–68.

[80] Brillhart J. (2002), *Email to Bruce C. Berndt*, January 27, 2002.

[81] Borwein, J. M.; Garvan, F. G. (1997), Approximations to via the Dedekind eta function. Organic mathematics (Burnaby, BC, 1995), 89–115, *CMS Conf. Proc.*, **20**, Amer. Math. Soc., Providence, RI.

[82] Borwein, J. M. and Borwein, P. B. (1987), *Pi & the AGM: A Study in Analytic Number Theory and Computational Complexity*. New York: Wiley.

[83] Bressoud, D. M. (1980), Analytic and combinatorial generalizations of the Rogers-Ramanujan identities. *Mem. Amer. Math. Soc.* **24**, no. 227, 54 pp.

[84] Bressoud, D. M. (1981), Some identities for terminating q-series. *Math. Proc. Cambridge Philos. Soc.* **89**, no. 2, 211–223.

[85] Bressoud, D.; Ismail, M. E. H.; Stanton, D. (2000), Change of base in Bailey pairs. *Ramanujan J.* **4**, no. 4, 435–453.

[86] Bringmann, K.; Ono, K. (2006), The $f(q)$ mock theta function conjecture and partition ranks. *Invent. Math.* **165**, no. 2, 243–266.

[87] Bringmann, K.; Ono, K. (2010), Dyson's ranks and Maass forms. *Ann. of Math.* (2) **171**, no. 1, 419–449.

[88] Bringmann, K.; Ono, K. (2012), *Coefficients of harmonic Maass forms.* Partitions, *q*-series, and modular forms, 23–38, Dev. Math., **23**, Springer, New York.

[89] Cao, Z. (2011), Integer Matrix Exact Covering Systems and Product Identities for Theta Functions. *Int. Math. Res. Not.* no. **19**, 4471–4514.

[90] Carlitz, L. (1965), Note on some continued fractions of the Rogers-Ramanujan type. *Duke Math. J.* **32**, 713–720.

[91] Carlitz, L. (1974), Remark on a combinatorial identity. *J. Combinatorial Theory Ser. A* **17**, 256–257.

[92] Cauchy, A. L. (1843), Mémoire sur les fonctions dont plusieurs valeurs ... . *C.R. Acad. Sci. Paris, T. XVII*, pp. 523–531.

[93] Cauchy, A. L. (1843), Deuxieme Memoir sur les fonctions dont plusieurs valeurs. *C.R. Acad. Sci. Paris*, reprinted in Oeuvres de Cauchy, (1893), **8**, 50–55.

[94] Chan, H. H. (1995), On Ramanujan's cubic continued fraction. *Acta Arith.* **73**, no. 4, 343–355.

[95] Chan, S. H. (2004), Dissections of quotients of theta-functions. *Bull. Austral. Math. Soc.* **69**, no. 1, 19–24.

[96] Chan, S. H. (2005), Generalized Lambert series identities. *Proc. London Math. Soc.* (3) **91**, no. 3, 598–622.

[97] Choi, Y.-S. (1999), Tenth order mock theta functions in Ramanujan's lost notebook. *Invent. Math.* **136**, no. 3, 497–569.

[98] Choi, Y.-S. (2000), Tenth order mock theta functions in Ramanujan's lost notebook. II. *Adv. Math.* **156**, no. 2, 180–285.

[99] Choi, Y.-S. (2002), Tenth order mock theta functions in Ramanujan's lost notebook. IV. *Trans. Amer. Math. Soc.* **354**, no. 2, 705–733.

[100] Choi, Y.-S. (2007), Tenth order mock theta functions in Ramanujan's lost notebook. III. *Proc. Lond. Math. Soc. (3)* **94**, no. 1, 26–52.

[101] Choi, Y.-S. (2011), The basic bilateral hypergeometric series and the mock theta functions. *Ramanujan J.* **24**, no. 3, 345–386.

[102] Choi, Y.-S.; Kim, B. (2012), Partition identities from third and sixth order mock theta functions. *European J. Combin.* **33**, no. 8, 1739–1754.

[103] Chu, W. (2006), Bailey's very well-poised $_6\psi_6$-series identity, *J. Combin. Theory Ser. A*, **113**, 966–979.

[104] Chu, W. (2006), Bailey's very well-poised $_6\psi_6$-series identity. *J. Combin. Theory Ser. A.* **113**, no. 6, 966–979.

[105] Chu, W.; Wang, X. (2010), Proofs of Ramanujan's $_1\psi_1$-summation formula. *Ars Combin.* **97A**, 65–79.

[106] Clausen, T. (1828), Beitrag zur Theorie der Reihen... *Journal fr die reine und angewandte Mathematik* **3**, 92–95.

[107] Comtet, L. (1974), *Advanced combinatorics. The art of finite and infinite expansions. Revised and enlarged edition.* D. Reidel Publishing Co., Dordrecht, 1974. xi+343 pp.

[108] Cooper, S. (2006), The quintuple product identity. *Int. J. Number Theory* **2**, no. 1, 115–161.

[109] Daum, J. A. (1942), The basic analogue of Kummer's theorem. *Bulletin of the American Mathematical Society* **48**, no. 10, 711–713.

[110] Denis, R. Y. (1984), On generalization of certain q-series results of Ramanujan. *Math. Student* **52**, no. 1–4, 47–58.

[111] Denis, R. Y. (1985), On certain q-series and continued fraction identities. *Math. Student* **53**, no. 1–4, 243–248.

[112] Dragonette, L. A. (1952), Some asymptotic formulae for the mock theta series of Ramanujan. *Trans. Amer. Math. Soc.* **72**, 474–500.

[113] Eisenstein, G. (1844), Transformations remarquables de quelques séries *J. Reine Angew. Math.*, **27**, 193–197.

[114] Euler, L. (1748), *Introductio in Analysin Infinitorum*, Marcum - Michaelem Bousquet, Lausannae.

[115] Fine, N. J. (1988), *Basic hypergeometric series and applications.* Mathematical Surveys and Monographs, **27**. American Mathematical Society, Providence, RI, xvi+124 pp.

[116] Folsom, A. (2008), A short proof of the mock theta conjectures using Maass forms. *Proc. Amer. Math. Soc.* **136**, no. 12, 4143–4149.

[117] Folsom, A.; Ono, K.; Rhoades, R. C. (2013), Mock theta functions and quantum modular forms. *Forum Math. Pi* **1**, e2, 27 pp.

[118] Frank, E. (1946), Corresponding type continued fractions. *Amer. J. Math.* **68**, 89–108.

[119] Franklin, F. (1881), Sure le développement du produit infini $(1-x)(1-x^2)(1-x^3)\dots$, *C. R. Acad. Paris Ser. A* **92**, 448–450.

[120] Gasper, G. (1989), Summation, transformation, and expansion formulas for bibasic series. *Trans. Amer. Math. Soc.* **312**, no. 1, 257–277.

[121] Gasper, G. (1997), Elementary derivations of summation and transformation formulas for q-series, In: *Special Functions, q-series and Related Topics (Toronto, ON, 1995)*, 55–70, Fields Inst. Commun., **14**, Amer. Math. Soc., Providence, RI.

[122] Gasper, G.; Rahman, M. (1986), Positivity of the Poisson kernel for the continuous q-Jacobi polynomials and some quadratic transformation formulas for basic hypergeometric series. *SIAM J. Math. Anal.* **17**, no. 4, 970–999.

[123] Gasper, G.; Rahman, M. (2004), *Basic hypergeometric series.* With a foreword by Richard Askey. Second edition. Encyclopedia of Mathematics and its Applications, 96. Cambridge University Press, Cambridge, xxvi+428 pp.

[124] Gauss, C. F. (1866), Zur Theorie der neuen Transscendenten. II, *Werke*, vol. **3**, Gttingen, pp. 436–445.

[125] Gauss, C. F. (1876), Hundert Theoreme uber die neuen Transscendenten, *Werke*, vol. **3**, Gttingen, 461–469.

[126] Gauss, C. F. (2005), *Mathematisches Tagebuch, 1796–1814.* Ostwalds Klassiker der Exakten Wissenschaften, **256**. Verlag Harri Deutsch, Frankfurt am Main, 235 pp.

[127] Gordon, B. (1961), Some identities in combinatorial analysis. *Quart. J. Math. Oxford Ser.* (2) **12**, 285–290.

[128] Gordon, B. (1961), A combinatorial generalization of the Rogers-Ramanujan identities. *Amer. J. Math.* **83**, 393–399.

[129] Gordon, B. (1965), Some continued fractions of the Rogers-Ramanujan type. *Duke Math. J.* **32**, 741–748.

[130] Gordon, B.; McIntosh, R. J. (2000), Some eighth order mock theta functions. *J. London Math. Soc.* (2) **62**, no. 2, 321–335.

[131] Gordon, B.; McIntosh, R. J. (2003), Modular transformations of Ramanujan's fifth and seventh order mock theta functions. Rankin memorial issues. *Ramanujan J.* **7**, no. 1–3, 193–222.

[132] Gordon, B.; McIntosh, R. J. (2012), *A survey of classical mock theta functions.* Partitions, q-series, and modular forms, 95–144, Dev. Math., **23**, Springer, New York.

[133] Gould, H. W. (1972), A new symmetrical combinatorial identity. *J. Combinatorial Theory Ser. A* **13**, 278–286.

[134] Griffin, M.; Ono, K.; Rolen, L. (2013), Ramanujan's mock theta functions. *Proc. Natl. Acad. Sci. USA* **110**, no. 15, 5765–5768.

[135] Griffin, M. J.; Ono, K.; Warnaar, S. O. (2016), A framework of Rogers-Ramanujan identities and their arithmetic properties. *Duke Math. J.* **165**, no. 8, 1475–1527.

[136] Guo, V. J. W.; Zeng, J. (2007), Short proofs of summation and transformation formulas for basic hypergeometric series. *J. Math. Anal. Appl.* **327**, no. 1, 310–325.

[137] Hahn, W. (1949), Beiträge zur Theorie der Heineschen Reihen. *Math. Nachr.* **2**, 340–379.

[138] Hardy G. H. (1940), *Ramanujan: Twelve lectures on subjects suggested by his life and work,* Cambridge University Press, Cambridge, 1940; AMS Chelsea, New York (1999).

[139] Hardy, G. H.; Wright, E. M. (1979), *An introduction to the theory of numbers.* Fifth edition. The Clarendon Press, Oxford University Press, New York, 1979. xvi+426 pp.

[140] Heine, E. (1847), Untersuchungen über die Reihe ... , *J. reine angew. Math.* **34**, 285–328.

[141] Hickerson, D. (1988), A proof of the mock theta conjectures. *Invent. Math.* **94**, no. 3, 639–660.

[142] Hickerson, D. (1988), On the seventh order mock theta functions. *Invent. Math.* **94**, no. 3, 661–677.

[143] Hickerson, D. R.; Mortenson, E. T. (2014), Hecke-type double sums, Appell-Lerch sums, and mock theta functions, I. *Proc. Lond. Math. Soc.* (3) **109**, no. 2, 382–422.

[144] Hirschhorn, M. D. (1974), A continued fraction. *Duke Math. J.* **41**, 27–33.

[145] Hirschhorn, M. D. (1980), Developments in Theory of Partitions. *PhD Thesis*, University of New South Wales.

[146] Hirschhorn, M. D. (1980), A continued fraction of Ramanujan. *J. Austral. Math. Soc. Ser. A* **29**, no. 1, 80–86.

[147] Hirschhorn, M. D. (1999), Jacobi's two-square theorem and related identities. *Ramanujan J.* **3**, no. 2, 153–158.

[148] Ismail, M. E. H. (1977), A simple proof of Ramanujan's $_1\psi_1$ sum. *Proc. Amer. Math. Soc.* **63**, no. 1, 185–186.

[149] Ismail, M. E. H.; Stanton, D. (2006), Ramanujan continued fractions via orthogonal polynomials. *Adv. Math.* **203**, no. 1, 170–193.

[150] Jackson, F. H. (1910), Transformations of q-series, *Messenger of Math.*, **39**, 145–153.

[151] Jackson, F. H. (1921), Summation of q-hypergeometric series *Messenger Math.*, **57**, 101–112.

[152] Jackson, M. (1950), On Lerch's transcendant and the basic bilateral hypergeometric series $_2\Psi_2$. *J. London Math. Soc.* **25**, 189–196.

[153] Jacobi, C. G. J. (1829), *Fundamenta Nova Theoriae Functionum Ellipticarumm*, Sumtibus fratrum Borntraeger, Regiomonti, reprinted in Jacobi's Cesammelte Werke, vol. 1, (Reimer, Berlin, 1881–1891), pp. 49–239; reprinted by Chelsea (New York, 1969).

[154] Jacobi, C. G. J. (1969), *Gesammelte Werke. Bnde I*, Herausgegeben auf Veranlassung der Königlich Preussischen Akademie der Wissenschaften. Zweite Ausgabe, Chelsea, New York.

[155] Jacobsen, L. (1986), General convergence of continued fractions, *Trans. Amer. Math. Soc.* **294** (2), 477–485.

[156] Jacobsen, L. (1987), Convergence of limit k-periodic continued fractions in the hyperbolic or loxodromic case. *Skr. K. Nor. Vidensk. Selsk.* no. **5**, 23 pp.

[157] Jacobsen, L. (1989), Domains of validity for some of Ramanujan's continued fraction formulas. *J. Math. Anal. Appl.* **143**, no. 2, 412–437.

[158] Jacobsen, L. (1990), On the Bauer-Muir transformation for continued fractions and its applications. *J. Math. Anal. Appl.* **152** no. 2, 496–514.

[159] Jain, V. K. (1981), Some transformations of basic hypergeometric functions. II. *SIAM J. Math. Anal.* **12**, no. 6, 957–961.

[160] Jain, V. K. (1982), Certain transformations of basic hypergeometric series and their applications. *Pacific J. Math.* **101**, no. 2, 333–349.

[161] Jain, V. K.; Verma, A. (1980), Transformations between basic hypergeometric series on different bases and identities of Rogers-Ramanujan type. *J. Math. Anal. Appl.* **76**, no. 1, 230–269.

[162] Jones, W. B.; Thron, W.J. (1980), *Continued Fractions Analytic Theory and Applications*, Addison-Wesley, London-Amsterdam-Ontario-Sydney-Tokyo.

[163] Jouhet, F. (2010), Shifted versions of the Bailey and well-poised Bailey lemmas. *Ramanujan J.* **23**, no. 1–3, 315–333.

[164] Jouhet, F.; Schlosser, M. (2005), Another proof of Bailey's $_6\psi_6$ summation. *Aequationes Math.* **70**, no. 1-2, 43–50.

[165] Khovanskii A. N. (1963), *The application of continued fractions and their generalizations to problems in approximation theory*, Translated by Peter Wynn, P. Noordhoff N. V., Groningen, xii + 212 pp.

[166] Komatsu, T. (2003), On Tasoev's continued fractions. *Math. Proc. Cambridge Philos. Soc.* **134**, no. 1, 1–12.

[167] Komatsu, T. (2003), On Hurwitzian and Tasoev's continued fractions. *Acta Arith.* **107**, no. 2, 161–177.

[168] Komatsu, T. (2004), Tasoev's continued fractions and Rogers - Ramanujan continued fractions. *J. Number Theory* **109**, no. 1, 27–40.

[169] Komatsu, T. (2005), Hurwitz and Tasoev continued fractions. *Monatsh. Math.* **145**, no. 1, 47–60.

[170] Komatsu, T. (2006), Hurwitz and Tasoev continued fractions with long period. *Math. Pannon.* **17**, no. 1, 91–110.

[171] Komatsu, T. (2008), Tasoev continued fractions with long period. *Far East J. Math. Sci. (FJMS)* **28**, no. 1, 89–121.

[172] Komatsu, T. (2008), More on Hurwitz and Tasoev continued fractions. *Sarajevo J. Math.* **4(17)**, no. 2, 155–180.

[173] Komatsu, T. (2012), Some exact algebraic expressions for the tails of Tasoev continued fractions. *J. Aust. Math. Soc.* **92**, no. 2, 179–193.

[174] Koornwinder, T. H. (1990), Jacobi functions as limit cases of q-ultraspherical polynomials. *J. Math. Anal. Appl.* **148**, no. 1, 44–54.

[175] Lam, H. Y. (2007), The number of representations by sums of squares and triangular numbers. *Integers* **7**, A28, 14 pp.

[176] Lebesgue, V. A. (1840), Summation de quelques series, *J. de math, pures appliqus (Sér. 1)*,**5**, 42–71.

[177] Lee, J.; Sohn, J. (2005), Some continued fractions in Ramanujan's lost notebook. *Monatsh. Math.* **146**, no. 1, 37–48.

[178] Lee, J; Mc Laughlin, J.; Sohn J. (2016), Applications of the Heine and Bauer-Muir transformations to Rogers-Ramanujan type continued fractions. *submitted.*

[179] Lehmer, D. H. (1973), Continued fractions containing arithmetic progressions. *Scripta Math.* **29**, 17–24.

[180] Littlewood, D. E. (1961), On certain symmetric functions. *Proc. London Math. Soc. (3)* **11**, 485–498.

[181] Liu, Q.; Ma, X. (2009), On a characteristic equation of well-poised Bailey chains. *Ramanujan J.* **18**, no. 3, 351–370.

[182] Lorentzen, L. (1994), Divergence of continued fractions related to hypergeometric series. *Math. Comp.* **62**, no. 206, 671–686.

[183] Lorentzen, L.; Waadeland, H. (1992), *Continued fractions with applications*. Studies in Computational Mathematics, **3**. North-Holland Publishing Co., Amsterdam, xvi+606 pp.

[184] Lorentzen, L.; Waadeland, H. (2008), Continued fractions. Vol. 1. Convergence theory. Second edition. Atlantis Studies in Mathematics for Engineering and Science, **1**. Atlantis Press, Paris; World Scientific Publishing Co. Pte. Ltd., Hackensack, NJ, xii+308 pp.

[185] Lovejoy, J. (2004), A Bailey lattice. *Proc. Amer. Math. Soc.* **132**, no. 5, 1507–1516.

[186] Lovejoy, J. (2010), On identities involving the sixth order mock theta functions. *Proc. Amer. Math. Soc.* **138**, no. 7, 2547–2552.

[187] Macdonald, I. G. (1995), *Symmetric functions and Hall polynomials*. Second edition. With contributions by A. Zelevinsky. Oxford Mathematical Monographs. Oxford Science Publications. The Clarendon Press, Oxford University Press, New York. x+475 pp.

[188] MacMahon, P. A. (1918), *Combinatory Analysis*, volume 2. Cambridge University Press, London.

[189] McIntosh, R. J. (2007), Second order mock theta functions. *Canad. Math. Bull.* **50**, no. 2, 284–290.

[190] McIntosh, R. J. (2012), The H and K family of mock theta functions. *Canad. J. Math.* **64**, no. 4, 935–960.

[191] McIntosh, R. J. (preprint), *Modular transformations of Ramanujan's sixth order mock theta functions*, preprint.

[192] Mc Laughlin, J. (2008), Some new families of Tasoevian and Hurwitzian continued fractions. *Acta Arith.* **135**, no. 3, 247–268.

[193] Mc Laughlin, J. (2010), Some new transformations for Bailey pairs and WP-Bailey pairs. *Cent. Eur. J. Math.* **8**, no. 3, 474–487.

[194] Mc Laughlin, J. (2013), A Generalization of Schröter's Formula - submitted.

[195] Mc Laughlin, J. (2013), Further Results on Vanishing Coefficients in Infinite Product Expansions - to appear.

[196] McLaughlin, J.; Sills, A. V.; Zimmer, P. (2008), Rogers-Ramanujan-Slater Type Identities. *Electron. J. Combin. Dynamic Survey* #**DS15**, 59 pp.
www.combinatorics.org/ojs/index.php/eljc/article/view/DS15/pdf.

[197] McLaughlin, J.; Sills, A. V.; Zimmer, P. (2010), Some implications of Chu's $_{10}\psi_{10}$ extension of Bailey's $_{10}\psi_{10}$ summation formula. *Online J. Anal. Comb. No.* **5**, 24 pp.

[198] McLaughlin, J.; Wyshinski, N. J. (2005), Real numbers with polynomial continued fraction expansions. Acta Arith. **116**, no. 1, 63–79.

[199] McLaughlin, J.; Wyshinski, N. J. (2005), Ramanujan and the regular continued fraction expansion of real numbers. *Math. Proc. Cambridge Philos. Soc.* **138**, no. 3, 367–381.

[200] Mc Laughlin, J.; Zimmer, P. (2008), Some identities between basic hypergeometric series deriving from a new Bailey-type transformation. *J. Math. Anal. Appl.* **345**, no. 2, 670–677.

[201] Mc Laughlin, J.; Zimmer, P. (2010), General WP-Bailey Chains, *Ramanujan J.* **22**, no. 1, 11–31.

[202] Mc Laughlin, J.; Zimmer, P. (2012), A reciprocity relation for WP-Bailey pairs. *Ramanujan J.* **28**, no. 2, 155–173.

[203] Menon, P. K. (1965), On Ramanujan's continued fraction and related identities, *J. London Math. Soc.* **40**, 49–54.

[204] Milne, S. C. (1982), A generalization of Andrews' reduction formula for the Rogers-Selberg functions. *Amer. J. Math.* **104**, no. 3, 635–643.

[205] Muir, T. (1877), A Theorem in Continuants, *Phil. Mag.*, (5) **3**, 137–138.

[206] Ono, K. (2008), *Mock theta functions, ranks, and Maass forms.* Surveys in number theory, 119–141, Dev. Math., **17**, Springer, New York.

[207] Ono, K. (2009), *Unearthing the visions of a master: harmonic Maass forms and number theory.* Current developments in mathematics, 2008, 347–454, Int. Press, Somerville, MA.

[208] Pak, I. (2006), Partition bijections, a survey. *Ramanujan J.* **12**, no. 1, 5–75.

[209] Paule, P. (1985), On identities of the Rogers-Ramanujan type, *J. Math. Anal. Appl.* **107**, no. 1, 255–284.

[210] Paule, P. (1987), A note on Bailey's lemma. *J. Combin. Theory Ser. A*, **44**, no. 1, 164–167.

[211] Pringsheim, A. (1899), *Über die Konvergenz unendlicher Kettenbrüche.* Bayer. Akad. Wiss., Math.- Natur. Kl. **28**, pp. 295–324.

[212] Ramanujan, S. (1957), *Notebooks* (2 volumes), Tata Institute of Fundamental Research, Bombay.

[213] Ramanujan, S. (1988), *The Lost Notebook and Other Unpublished Papers*, Narosa, New Delhi.

[214] Ramanujan, S. (2000), *Collected papers of Srinivasa Ramanujan.* Edited by G. H. Hardy, P. V. Seshu Aiyar and B. M. Wilson. Third printing of the 1927 original. With a new preface and commentary by Bruce C. Berndt. AMS Chelsea Publishing, Providence, RI. xxxviii+426 pp.

[215] Rhoades, R. C. (2013), On Ramanujan's definition of mock theta function. *Proc. Natl. Acad. Sci. USA* **110**, no. 19, 7592–7594.

[216] Richmond, B.; Szekeres, G. (1978), The Taylor coefficients of certain infinite products. *Acta Sci. Math. (Szeged)* **40**, no. 3–4, 347–369.

[217] Rogers, L. J. (1894), Second memoir on the expansion of certain infinite products. *Proc. London Math. Soc.* **25**, 318–343.

[218] Rogers, L. J. (1895), Third memoir on the expansion of certain infinite products, *Proc. London Math. Soc.*, **26**, 15–32.

[219] Rogers, L. J. (1917), On two theorems of combinatory analysis and some allied identities, *Proc. London Math. Soc (2).* **16**, 315–336.

[220] Rothe, H. A. (1811), *Systematisches Lehrbuch der Arithmetik*, Barth, Leipzig.

[221] Rowell, M. J. (2008), A new general conjugate Bailey pair. *Pacific J. Math.* **238**, no. 2, 367–385.

[222] Schilling, A.; Warnaar, S. O. (1997), A higher-level Bailey lemma. Proceedings of the Conference on Exactly Soluble Models in Statistical Mechanics: Historical Perspectives and Current Status (Boston, MA, 1996). *Internat. J. Modern Phys. B* **11**, no. 1–2, 189–195.

[223] Schilling, A.; Warnaar, S. O. (1998), A higher level Bailey lemma: proof and application. *Ramanujan J.* **2**, no. 3, 327–349.

[224] Schilling, A.; Warnaar, S. O. (2002), Conjugate Bailey pairs: from configuration sums and fractional-level string functions to Bailey's lemma. Recent developments in infinite-dimensional Lie algebras and conformal field theory (Charlottesville, VA, 2000), 227–255, *Contemp. Math.*, **297**, Amer. Math. Soc., Providence, RI.

[225] Schlosser, M. (2002), A simple proof of Bailey's very-well-poised $_6\psi_6$ summation. *Proc. Amer. Math. Soc. 130*, no. 4, 1113–1123.

[226] Schur, I. (1917), Ein Beitrag zur additeven Zahlentheorie und zur Theorie der Kettenbrüche. *Sitzungsberichte der Berliner Akademie*, 302–321.

[227] Sears, D. B. (1951), Transformations of basic hypergeometric functions of special type, *Proc. London Math. Soc.* (2) **52**, 467–483.

[228] Sears, D. B. (1951), On the transformation theory of basic hypergeometric functions. *Proc. London Math. Soc.* (2) **53**, 158–180.

[229] Selberg, A. (1936), Über Einige Arithmetische Identitäten, *Av- handlinger Norske Akad*, Vol. **8**, 1–23.

[230] Selberg, A. (1938), Über die Mock-Thetafunktionen siebenter Ordnung. *Arch. Math. og Naturvidenskab*, **41**, 3–15.

[231] Shen, L.-C. (1994), On the additive formulae of the theta functions and a collection of Lambert series pertaining to the modular equations of degree 5. *Trans. Amer. Math. Soc.* **345**, no. 1, 323–345.

[232] Shen, L.-C. (1994), On the modular equations of degree 3. *Proc. Amer. Math. Soc.* **122**, no. 4, 1101–1114.

[233] Sills, A. V. (2003), Finite Rogers-Ramanujan type identities. *Electron. J. Combin.* **10**, Research Paper 13, 122 pp.

[234] Singh, U. B. (1994), A note on a transformation of Bailey. *Quart. J. Math. Oxford Ser.* (2) **45**, no. 177, 111–116.

[235] Singh, V. N. (1959), The basic analogues of identities of the Cayley-Orr type. *J. London Math. Soc.* **34**, 15–22.

[236] Slater, L. J. (1951), A new proof of Rogers's transformations of infinite series, *Proc. London Math. Soc.* (2) **53**, 460–475.

[237] Slater, L. J. (1952), Further identities of the Rogers-Ramanujan type, *Proc. London Math. Soc.* (2) **54**, 147–167.

[238] Slater, L. J.; Lakin, A. (1956), Two proofs of the $_6\psi_6$ summation theorem. *Proc. Edinburgh Math. Soc.* (2) **9**, 116–121.

[239] Śleszyński, J. V. (1889), Zur Frage von der Konvergenz der Kettenbrüche *Mat. Sb.,* **14**, pp. 337–343.

[240] Sparks, R. C. (2014), A Collection of Basic Hypergeometric Identities. - Master's thesis.

[241] Srivastava, B. (2005), Hecke modular form expansions for eighth order mock theta functions. *Tokyo J. Math.* **28**, no. 2, 563–577.

[242] Srivastava, B. (2005), A comprehensive study of second order mock theta functions. *Bull. Korean Math. Soc.* **42**, no. 4, 889–900.

[243] Srivastava, B. (2007), Partial second order mock theta functions, their expansions and Pad approximants. *J. Korean Math. Soc.* **44**, no. 4, 767–777.

[244] Stanley, R. P. (1972), *Ordered structures and partitions.* Memoirs of the American Mathematical Society, No. **119**. American Mathematical Society, Providence, R.I., iii+104 pp.

[245] Stanton, D. (2001), The Bailey-Rogers-Ramanujan group. *q-series with applications to combinatorics, number theory, and physics (Urbana, IL, 2000),* 55–70, Contemp. Math., **291**, Amer. Math. Soc., Providence, RI,

[246] Starcher, G. W. (1930), On identities arising from solutions of q-difference equations and some interpretations in number theory, *Amer. J. Math. 53*, 801–816.

[247] Tannery, J. (1886), Introduction a la Theorie des Fonctions d'une Variable, Sec. 183.

[248] Tasoev, B. G. (1984), Certain problems in the theory of continued fractions. (Russian) *Trudy Tbiliss. Univ. Mat. Mekh. Astronom.* No. **16-17**, 53–83.

[249] Tasoev, B. G. (2000), On rational approximations of some numbers. (Russian) *Mat. Zametki*, **67**, no. 6, 931–937; translation in Math. Notes **67**, no. 5-6, 786–791.

[250] Thron, W. J. (1990), Some results on separate convergence of continued fractions. Computational methods and function theory (Valparaso, 1989), 191–200, *Lecture Notes in Math.*, **1435**, Springer, Berlin.

[251] van der Poorten, A. J. (1994), Explicit Formulas for Units in Certain Quadratic Number Fields. *Algorithmic number theory (Ithaca, NY, 1994)*, 194–208, Lecture Notes in Comput. Sci., **877**, Springer, Berlin.

[252] Wall, H. S. (1948), *Analytic Theory of Continued Fractions*, D. Van Nostrand Company, Inc., New York, N. Y., xiii+433 pp.

[253] Warnaar, S. O. (1999), q-trinomial identities. *J. Math. Phys.* **40**, no. 5, 2514–2530.

[254] Warnaar, S. O. (2001), 50 years of Bailey's lemma, *Algebraic combinatorics and applications* (Gössweinstein, 1999), Berlin, New York: Springer-Verlag, pp. 333–347.

[255] Warnaar, S. O. (2003), Partial theta functions. I. Beyond the lost notebook. *Proc. London Math. Soc.* (3) **87**, no. 2, 363–395.

[256] Warnaar, S. O. (2003), Extensions of the well-poised and elliptic well-poised Bailey lemma. *Indag. Math. (N.S.)* **14**, no. 3-4, 571–588.

[257] Watson, G. N. (1929), A New Proof of the Rogers-Ramanujan Identities. *J. London Math. Soc.* **4**, 4–9.

[258] Watson, G. N. (1929), Theorems stated by Ramanujan. VII: Theorems on continued fractions., *J. London Math. Soc.*, **4** (1), 39–48.

[259] Watson, G. N. (1936), The final problem: An account of the mock theta functions, *J. London Math. Soc.* **11**, 55–80.

[260] Watson, G. N. (1937), The mock theta functions (2), *Proc. London Math. Soc.* **42**, 274–304.

[261] Werley, E. M. (2013), Finite Versions of the Rogers-Ramanujan-Slater Type Identities - submitted.

[262] Worpitzky, J. (1865), Untersuchungen über die Entwickelung der monodromen und monogenen Functionen durch Kettenbrüche. In: Friedrichs-Gymnasium und Realschule: Jahresbericht, 3–39.

[263] Zwegers, S. (2001), Mock θ-functions and real analytic modular forms, q-series with applications to combinatorics, number theory, and physics. *Contemp Math.* **291**, 268–277.

[264] Zwegers, S. (2002), *Mock theta functions.* PhD thesis (Univ of Utrecht, Utrecht, The Netherlands).

Author Index

Subject Index

386

Printed in the United States
By Bookmasters